WORSE THAN A MONOLITH

PRINCETON STUDIES IN INTERNATIONAL HISTORY AND POLITICS

Series Editors G. John Ikenberry and Marc Trachtenberg

Recent Titles

WORSE THAN A MONOLITH

Alliance Politics and Problems of Coercive Diplomacy in Asia

Thomas J. Christensen

PRINCETON UNIVERSITY PRESS

PRINCETON AND OXFORD

Published by Princeton University Press,
41 William Street, Princeton, New Jersey 08540
In the United Kingdom: Princeton University Press,
6 Oxford Street, Woodstock, Oxfordshire OX20 1TW

press.princeton.edu

Library of Congress Cataloging-in-Publication Data

Christensen, Thomas J., 1962-
Worse than a monolith : alliance politics and problems of coercive diplomacy in Asia /
Thomas J. Christensen.
p. cm. — (Princeton studies in international history and politics)
Includes bibliographical references and index.
ISBN 978-0-691-14260-9 (hardcover : alk. paper) — ISBN 978-0-691-14261-6 (pbk. : alk. paper)
1. Asia—Foreign relations—1945- 2. Alliances—History—20th century. I. Title.
DS35.2.C47 2011
327.5—dc22

British Library Cataloging-in-Publication Data is available

This book has been composed in Minion
Printed on acid-free paper. ∞
Printed in the United States of America

10 9 8 7 6 5 4 3 2

For Barbara, Theresa, and Will

Contents

Acknowledgments

This book has been several years in the making. The research and writing process was interrupted by many other research projects, two family moves, and a public service leave during which I served as Deputy Assistant Secretary of State for East Asian and Pacific Affairs. The views expressed in the book are purely my own and do not represent the views of the U.S. government or the U.S Department of State. Certain sections of the book are culled from articles that appeared in the journals *Asian Security* (vol. 1, no. 1, January 2005) and *International Security* (vol. 23, no. 4, spring 1999).

For comments and research assistance, I am grateful to David Bachman, Gary J. Bass, Yan Bennett, Chen Jian, Chen Zhiming, Ja Ian Chong, Milton Esman, Taylor Fravel, Aaron Friedberg, Lily Fu, Michael Glosny, James Goldgeier, Avery Goldstein, Steven Goldstein, Todd Hall, Hope Harrison, Yinan He, Serene Hung, Samuel P. Huntington, Robert Jervis, Alastair Iain Johnston, Peter Katzenstein, Carl Kaysen, Andrew Kennedy, Jonathan D. Kirshner, Li Danhui, Li Hong, Adam Liff, Jennifer Lind, Liu Dongsheng, Oriana Mastro, Niu Jun, Barry Posen, Lucian Pye, Alan Romberg, Stephen Peter Rosen, Robert Ross, Richard Samuels, Randall Schweller, Adam Segal, Shen Zhihua, Jack L. Snyder, Christopher Twomey, Stephen Van Evera, Lynn White, Allen Whiting, Xu Xin, Dali Yang, Suisheng Zhao, two anonymous reviewers, the department of Politics at Princeton University, the faculty and fellows at the Security Studies Program at MIT, the helpful staff at the Ministry of Foreign Affairs Archives in Beijing, and my many students in China's Foreign Relations and International Relations of East Asia at Cornell, Princeton, and MIT. Chuck Myers at Princeton University Press has been an enormous source of ideas, encouragement, and editorial suggestions. It is easy to see why the press is a leader in the field. Linda Truilo and Terri O'Prey did a very fine job editing the final manuscript for the Press. I also thank the independent editor Jennifer Camille Smith for offering expert advice even when she was very busy and for motivating me to move this from draft manuscript to final product. I am grateful to the Smith Richardson Foundation for financial support for my archival research in China in 2004. I apologize to all of those many people who helped me over the years but whom I neglected to mention here.

The book is dedicated to my family: my wife, Barbara Edwards, my daughter Theresa, and my son, William. They have been a fountain of joy, support, and inspiration, especially while I was away from home in Washington wrestling with some of the policy issues discussed in the latter part of this book.

WORSE THAN A MONOLITH

Chapter 1

INTRODUCTION

It should seem obvious that the more united and organized one's enemies, the worse one's own lot. This book makes the counterintuitive assertion that this need not always be the case. Poor coordination, internal mistrust, and intramural rivalry in enemy alliances can be dangerous for one's own side because such internal divisions make engaging in successful coercive diplomacy with those enemies more difficult. During all-out war—pure competitions of brute force—internal divisions and lack of coordination within the enemy camp are clearly to one's advantage. But such wars are the exception, not the rule, in international security politics. More commonplace is coercive diplomacy—the use of threats and assurances in combination to influence the behavior of real or potential adversaries. In such instances it is often more difficult to achieve one's goals at acceptably low costs and to limit the duration and scope of existing conflicts when an adversarial alliance is ill formed, in flux, or internally divided than when it is well organized and hierarchically structured.

This book focuses on the alliance dynamics of Cold War East Asia from 1949–69 and concludes with a chapter on post–Cold War East Asia. The argument of the book is that disunity, lack of coordination, and intra-alliance rivalry increased both the chance that regional conflicts would occur and the likelihood that existing conflicts would persist and escalate. In their formative years, both the U.S.-led alliance system and the Asian communist alliance sent dangerously confusing signals regarding the cohesion, resolve, and intent of their respective blocs. Those signals undercut coercive diplomacy in Asia and created conditions for both crisis and war. From 1958 to 1969 a different phenomenon destabilized relations across the Cold War divide: the ideological rivalry between the Soviet Union and the PRC ("the Sino-Soviet split") actually harmed U.S. national security interests in Indochina and beyond by catalyzing the two competing communist giants to increase support for revolutionaries in the developing world and to scuttle peace talks to end existing conflicts. This condition changed only in 1969, when Moscow and Beijing turned their guns directly on each other.

The book also provides a brief study of the legacies of U.S. Cold War alliances for contemporary Sino-American relations. Although China and the United States today are very far from being enemies as they were in the 1950s and 1960s, they do engage in mutual coercive diplomacy over issues like relations across the Taiwan Strait. Certain potentially destabilizing dynamics related to the early formation of the U.S. alliance system in Asia have reemerged, albeit in a more manageable form, as Washington and its regional partners have adjusted to the changes in the global and regional security structure following the collapse of the Soviet Union.

WHY ENEMY DISUNITY AND RIVALRY IS DANGEROUS

This book addresses two forms of dangerous dynamics among enemy alliances: poor coordination and, in the case of revisionist alliances, the catalyzing effect of ideology and the pursuit of prestige on aggression toward enemies.

The first set of theoretical arguments in the book focuses on how weakly formed and poorly organized alliances send signals that can undercut the key components of successful coercive diplomacy: the use of clear and credible threats and assurances in combination to dissuade target countries from undesirable behavior. The problems of coordination and weak signaling explained here can apply to any alliance, whether or not the participants have revisionist goals.

A second set of theoretical arguments is specifically relevant to the study of alliances that were formed with transnational revisionist goals in mind. When such an alliance is first forming and when it is fraught with internal rivalries for leadership, the shared revisionist ideology of the alliance members creates dynamics that make the alliance as a whole aggressive and hard to contain through the use of coercive diplomacy—much more so than either status quo alliances or revisionist alliances that are more firmly established and enjoy clear leadership in a hierarchical structure. In the alliance's formative phase, individual candidates for membership may take unusually aggressive acts toward international enemies to demonstrate that they are bona fide internationalists, not simply nationalists who happen to share domestic political preferences with their potential allies. When there is competition for leadership within such an alliance between two or more members, that competition will often take the form of outbidding rivals by demonstrating support for revolution or revisionism in third areas, thus catalyzing the overall movement's aggression toward the outside world and making containment through coercive diplomacy more costly and more difficult for status quo enemies.

It is important here to make a distinction between coercive diplomacy and the brute force fighting of total wars, in which the central aim is the annihilation of one's enemies, such as World War II and the war against the Al Qaeda network. Such wars are very important but also are, fortunately, quite rare. More commonly, actual or potential enemies are involved in coercive diplomacy (either deterrence to prevent a change in the status quo, or compellence to cause such a change). This is true in peacetime and during limited wars, which constitute the great majority of armed conflicts and during which opposing sides bargain over the terms of peace through a combination of physical force and diplomatic negotiation. In brute force wars or total wars for survival, disunity among one's adversaries is a clear benefit because it renders the enemy alliance physically weaker. But in the world of coercive diplomacy, threats and assurances must be balanced through a process of clear and credible signaling, and enforceable bargains must be struck short of total defeat or victory for either side. Without credible threats, coercion is obviously ineffective. But what is less well understood is that coercion is also unlikely to be effective without simultaneously transmitted credible assurances that the threat is fully conditional upon the target's behavior and that the target's key security interests will not be harmed if it complies with the demands of those leveling the threats. Without receiving both threats and assurances in concert, the target of a coercive threat has little incentive to comply with the demands being made.[1] Since there is often a tension between these two central aspects of coercive diplomacy, blending threats and assurances effectively is not an easy task in coercive bargaining. Even in the simplest bilateral or "dyadic" relationships, such an effective blend is hard to achieve; but coercive bargaining among adversarial alliances is much more complex still, and the divisions and political jockeying within one or both of the opposing alliances can make such bargaining very difficult indeed. Enemy disunity has two potentially negative implications for one's own security: wars will be more likely to happen because diplomatic solutions to differences short of war will be more difficult to achieve; and limited wars are more likely to endure and even escalate because of the added complexity of intra-war coercive diplomacy.[2]

[1] For the original theoretical work that specifies the need for both credibility of threat and credibility of reassurance, see Thomas C. Schelling, *Arms and Influence* (New Haven: Yale University Press, 1966). Also see James Davis, *Threats and Promises: The Pursuit of International Influence* (Baltimore: Johns Hopkins University Press, 2000). For more recent related work in the rational choice tradition on the importance of two factors—transparency (complete information) and enforceable commitments—in preventing conflict, see James Fearon, "Rationalist Explanations for War," *International Organization* 49, no. 3 (summer 1995): 379–414; Erik Gartzke, "War Is in the Error Term," *International Organization* 53, no. 3 (summer 1999): 567–87; and Robert Powell, "Bargaining Theory and International Conflict," *Annual Review of Political Science* 5 (June 2002): 1–30. In a sense, the alliance politics discussed in this book make alliances both less transparent and less capable of making clear, credible, and enforceable commitments to enemies. Jonathan Kirshner offers an approach that is critical of the recent rational choice arguments that emphasize transparency and credibility of commitment, instead emphasizing the danger of miscalculation in the highly complex world of international security politics, even if actors were somehow to enjoy perfect information and enforceable commitments. In Kirshner's thesis, the alliance politics discussed here would simply add greatly to the complexity of international relations and would, therefore, increase the likelihood of dangerous miscalculations by members of an alliance or their enemies, regardless of the robustness of available information. See Jonathan D. Kirshner, "Rational Explanation for War?" *Security Studies* 10, no. 1 (autumn 2000): 143–50.

[2] Glenn H. Snyder, "Deterrence and Defense," pp. 25–44 in Robert J. Art and Kenneth N. Waltz, eds., *The Use of Force: Military Power and International Politics*, 3rd ed. (New York: University Press of America, 1988). Schelling, *Arms and Influence*, 2–3.

Alliance Cohesion, Clarity, and Coercive Diplomacy

The first problem discussed in this book is how a lack of coordination and clarity of commitment in alliances renders groups as a whole less transparent to enemies and, therefore, makes it harder for opposing alliances to engage each other in effective, mutual coercive diplomacy. Such divisions and uncertainties, common in the formative stages of an alliance or after major changes in the international system, can create in enemy capitals dangerous misperceptions regarding the capabilities, resolve, or intentions of the alliance, with negative implications for crisis management. Such misperceptions can lead to overestimations or underestimations of the challenge posed by the enemy alliance. The problem is only exacerbated if, as was the case in the early Cold War in East Asia, two opposing alliances (or alignments) are both in a formative stage and, therefore, suffer from poorly coordinated policies and send confusing signals to each other about power and purpose.

When coordination is poor, the most determined and aggressive actors within an alliance are most capable of dragging their partners into conflicts. Poor coordination also increases the likelihood that the alliance will send unclear and misleading signals to adversaries. The inherent complexity of alliances and alignments can be exacerbated by poor coordination among the allies, thus making it particularly difficult for an alliance and its adversary to find an effective balance between credible threats and credible assurances. Especially when poorly organized, alliances can send messages that undercut either credible threats or credible assurances, thus making stable coercive diplomacy with enemies (short of war) harder to maintain than it would be if the alliances were better coordinated and exhibited clearer leadership. What is worse still, in cases where an alliance seems currently weak but potentially strong and aggressive in the future, both credible threats and credible assurances can be undercut simultaneously, significantly increasing both the likelihood of new conflicts and the escalation of existing conflicts.

Poor coordination tends to be prevalent in the formative stages of alliances, when security alliances and alignments are often still informal, mutual suspicions among security partners about each other's near-term and long-term goals and reliability are strongest, and burden-sharing arrangements within alliances have not been clearly delineated. Coordination problems can also arise when the alliance's original mission has disappeared and the alliance must adjust to fundamentally new conditions.

Revisionist Alliance Dynamics and Coercive Diplomacy

The second form of internal alliance dynamics studied in this book is mutual mistrust and intramural competition for leadership in revisionist alliances. Alliances with revisionist, internationalist goals can be even more aggressive and harder to pacify through coercive diplomacy when they are poorly formed or rife with internal rivalries than they would be if they were under one actor's clear leadership. There are at least three reasons that this is the case.

First, actors in an alliance, even a revisionist one, often have differing interests. Some will have greater incentives to pursue revisionism and spread revolution violently than other actual or potential partners in the movement. Even if a status quo state and its allies can successfully deter the strongest, leading member of the revisionist alliance through threats and assurances (and this is not always the case), in the absence of a clearly established hierarchy and close coordination within the revisionist alliance, it will be difficult to deter all members of the revisionist alliance simultaneously. Some of the revisionist allies might feel insufficiently threatened and others insufficiently reassured to keep the peace with real or potential adversaries outside the camp. Not only will different revisionist actors interpret enemy threats and assurances differently, but they will often have quite different preference orderings based either on their particular national interests or geographic locations or on their desire to secure or improve their reputations as internationalist actors within the revisionist alliance. For example, revolutionary political movements involved in civil wars in divided countries will have a greater stake in unification of their nations under their rule than will their foreign ideological allies. Among those foreign allies, states geographically adjacent to the sites where local revolutionaries operate might be much more aggressive in support of those local revolutionaries than would geographically more distant states. So, when alliances exhibit internal disunity, the most aggressive actors may be more difficult for their partners to restrain, and aggressive actors may find it easier to drag their more conservative partners into conflicts.

Second, in the formative phases of the alliance, the leaders of a revisionist state will often feel the need to prove to their prospective foreign revisionist allies that they are full-fledged members of the internationalist movement, not merely parochial leaders of "national liberation" efforts. New members of the revisionist alliance might do this by taking aggressive actions toward shared enemies and in support of foreign allies. This revolutionary activity may surpass in intensity the expectations of a state merely acting in its own national interest and may exceed the level of risk considered prudent by other members of the alliance.

Third, once the alliance is formed, member states' concerns about their revolutionary prestige and the competition for leadership of the international movement can catalyze the alliance's revolutionary activity, rendering the alliance as a whole even more aggressive and harder to constrain peacefully through coercive diplomacy than would be a more hierarchically ordered alliance. Competition will often take the form of one or more states attempting to appear the most resolute in confronting shared enemies and the most active in supporting other revisionists in the international system.[3] Internal rivalries within alliances tend to form and become dangerous when the ideology binding an alliance together is transnational and revisionist in nature. Their ideologies—e.g., Marxism-Leninism, pan-Arabism, and militant Islamic fundamentalism—make such alliances prone to intramural competition for leadership. For states attempting to contain revisionist alliances through coercive diplomacy, such rivalries pose real problems. The intramural competition within the revisionist alliance revolves around which ally can prove itself most revolutionary and most resolute in overthrowing the status quo via belligerence toward common enemies. The competition among revisionist actors to appear the most uncompromising toward the enemy will lead to a ratcheting up of revolutionary fervor within the international movement, as more radical members of the movement catalyze the generally more moderate members of the movement into more aggressive activity than we would otherwise expect.

All of these factors can lead to a situation in which tails wag dogs and competitions to induce fervor undercut proposals for compromise raised by the most cautious and moderate capitals within the alliance. For enemies of the revisionist alliance, these problems will persist until the intramural competition escalates into total alliance breakdown and, perhaps even military conflict among the former allies.

There are some reasons to expect that, all things being equal, the strongest and most influential actor in any international revisionist movement will be more willing to moderate its behavior and therefore will be easier than its weaker and ambitious allies for non-allied states to engage in coercive diplomacy. First, such established leaders might simply have more to lose from the escalation of conflict with the enemy than less-established local revolutionary allies, some of whom might perceive themselves locked into struggles to the death with local foes and unable to compromise at least until they win their local battles. Even if the leading state is still highly unsatisfied with the status quo, it might still have a more globally oriented, longer-term, and therefore more cautious strategy than its less-secure, less-experienced, and more locally focused allies. Finally, other states within the international movement that have stronger incentives to improve their prestige and rankings within the movement might also be much more supportive of the most radical local revolutionaries than would the more established leaders in the movement, who should be more satisfied with their position within the movement's existing hierarchy.

---

[3] The ratcheting effect in international revisionist alliances is similar to the competitive bidding for nationalist credentials in immature democracies, as analyzed in Jack Snyder's path-breaking work on democratization and war. Jack Snyder, *From Voting to Violence: Democratization and Nationalist Conflict* (New York: Norton, 2000), 68–69.

When challenged by upstart rivals in their own movements, leading states in revisionist alliances might have to jettison their caution as they have to worry not only about their reputation for resolve with the enemy alliance, but also their prestige within their own revolutionary movement. They may, therefore, be dragged more deeply into conflicts by their more activist allies than we would otherwise expect. In this sense, from the perspective of achieving peace through coercive diplomacy, all things being equal, rivalries and differences within revisionist movements make them worse than a monolith.

It is, of course, possible that the most powerful leader of the movement could be even more aggressive than its weaker allies. Under such circumstances, of course, coercive diplomacy as a tool to contain such an alliance might be very difficult, if not impossible, for the status quo alliance, regardless of the cohesion of the revisionist alliance. Weaker but more conservative allies of a strong but highly radical leading actor would likely find it hard to restrain their more powerful ally. Large-scale war might simply naturally ensue between the two camps, and we would then leave the world of coercive diplomacy and enter the world of simple brute force, in which splits in the revisionist camp are more clearly to the advantage of the camp's adversaries.

The very nature of some revisionist movements may render moot the problems of coercive diplomacy analyzed in this book. In some senses, the struggle against the Al Qaeda network might be seen as one against a revisionist alliance. But the approach here does not apply to the fight against Al Qaeda because that struggle is arguably much more one of brute force than of coercive diplomacy. Coercive diplomacy is a form of bargaining, even when it occurs among bitter enemies, as it did during the Cold War. A prerequisite of such a bargaining environment is there being at least some potential common ground between the interests of the enemy camps to allow for negotiation. So, despite severe tensions and persistent security competition, each Cold War superpower learned to live with the existence of the other as long as its own survival was guaranteed. This acceptance of an unhappy but tolerable status quo allowed for mutual nuclear coercion, commonly described as "Mutually Assured Destruction," a condition that arguably deterred not only nuclear war between the United States and the Soviet Union, but also large-scale, direct conventional engagements.[4] It is difficult to imagine Al Qaeda and its many enemies—moderate states in the Islamic world, the United States, Europe, and Israel—finding enough common ground to settle into such a pattern of coercive diplomatic bargaining. Instead, the struggle seems zero-sum, without a readily imaginable bargaining space. Al Qaeda will be satisfied with nothing less than overthrowing its enemies, and its enemies

[4] For a classic work analyzing this concept, see Robert Jervis, *The Meaning of the Nuclear Revolution: Statecraft and the Prospect of Armageddon* (Ithaca: Cornell University Press, 1989).

are primarily interested not simply in containing the Al Qaeda network, but weakening it, keeping it off balance and less capable of leveling devastating attacks, and then, ultimately, destroying it.

THE BOOK'S CLAIMS AND THE EXISTING THEORETICAL LITERATURE

The book offers two sets of theoretical approaches toward alliances and coercive diplomacy. Each builds upon, revises, or melds existing theories in the literature.

Alliance Cohesion and Coercive Diplomacy

The first theoretical strand in this exploration of alliance cohesion and coercive diplomacy integrates theoretical concepts about dilemmas of alliance maintenance with arguments about clarity of signaling in deterrence theory. Much of the literature to date has emphasized how alliances are fraught with rather paradoxical ailments: fears of abandonment in time of need balanced against fears of entrapment in conflicts unnecessarily provoked by one's own allies.[5] But these arguments about alliance maintenance have only rarely been tied into theories about how alliances interact with outsiders in relationships involving coercive diplomacy. One notable exception, on which I attempt to build here both theoretically and empirically, is Glenn H. Snyder's path-breaking concept of "the composite security dilemma." In a book that is largely about internal alliance dynamics, Snyder also argues more briefly that individual allies need to manage simultaneously both their relations with allies and their coercive diplomacy with enemies at the same time (hence the adjective "composite").[6] Changes designed to shore up alliance cohesion or redistribute burden-sharing within an alliance can have unintended deleterious effects on the alliance's coercive diplomacy toward outsiders.

---

[5] For the original formulation, see Glenn H. Snyder, "The Security Dilemma in Alliance Politics," *World Politics* 36 (July 1984): 461–95. This theme is developed further in Glenn H. Snyder, *Alliance Politics* (Ithaca: Cornell University Press).

[6] Snyder, *Alliance Politics*, 192–98, 330–55. Snyder's book does not focus primarily on this issue. Some of my arguments about alliance divisions and coercive diplomacy will be very much in the spirit of those sections of his work, but will focus on alliance politics in a very different international context. Snyder focuses on cases of international multipolarity before the Cold War, not the Cold War or post–Cold War periods. Moreover, he focuses on formal alliances in the book while I will also discuss less formal alignments. Another related contribution to the field is Timothy Crawford's excellent book on what he calls "pivotal deterrence" which addresses how third parties attempt to deter both sides in a conflict, often through an ambiguous mix of threats and assurances that keep either party from taking the first step toward disaster. Timothy W. Crawford, *Pivotal Deterrence: Third-Party Statecraft and the Pursuit of Peace* (Ithaca: Cornell University Press. 2003).

Other works explore how free-riding or buck-passing within alliances can encourage piecemeal aggression by adversaries or how allies can sometimes drag each other into war as if they were tethered in a "chain gang" in a way that makes escalation of conflicts harder to avoid.[7] The relationship between alliances and coercive diplomacy has also been addressed in the literature on Cold War nuclear deterrence. That literature notes that, because of the underlying condition of mutually assured destruction during much of the Cold War, it was easier for each superpower to deter an attack by the other against its homeland than to deter aggression against its allies (this was labeled the problem of "extended deterrence").[8]

Little work has been done to date, however, on the relationship between internal alliance dynamics and the coercive diplomacy of alliances toward adversaries. There is a related hole in the literature on coercive diplomacy itself. Deterrence theories have generally focused on simpler dyadic relationships, eschewing the complexity caused by groups of allies facing off in crisis or limited war.[9] How members of alliances and less-formal security alignments deal with the internal problems of abandonment and entrapment affects the ability of the alliance or alignment to coerce common enemies effectively. Excessive fears of abandonment can lead to tighter and more aggressive postures that may unintentionally signal hostility toward adversaries and may, thereby, undercut

[7] On the problems of buck-passing and chain-ganging in multipolar alliances, see Kenneth N. Waltz, *Theory of International Politics* (Reading, Mass.: Addison-Wesley, 1979); Barry Posen, *The Sources of Military Doctrine: France, Britain, and Germany Between the Wars* (Ithaca: Cornell University Press, 1984); and Thomas J. Christensen and Jack Snyder, "Chain Gangs and Passed Bucks: Predicting Alliance Patterns in Multipolarity," *International Organization* 44 (1990): 137–68.

[8] On the problem of "extended deterrence," or deterrence umbrellas extended to alliances in worlds of mutually assured destruction, see Robert Jervis, *The Illogic of American Nuclear Strategy* (Ithaca: Cornell University Press, 1984), 29–34. The notion of an extended deterrence problem flows from what Glenn Snyder labeled the "stability-instability paradox." The threat of mutually assured nuclear destruction against each other's homelands among the two superpowers opened up the possibility of conflicts involving third parties or at lower levels of violence. See Glenn H. Snyder, "The Balance of Power and the Balance of Terror," pp. 184–201 in Paul Seabury, ed., *The Balance of Power* (San Francisco: Chandler, 1965).

[9] A review of Schelling's *Arms and Influence* revealed only two references to alliances in the index. Alliance politics, especially the external effects of internal alliance dynamics, also receive relatively sparse coverage in Jervis's classic *Perception and Misperception in International Politics* (Princeton: Princeton University Press, 1967). The same is true for Stephen Van Evera's much-noted *Causes of War: Power and the Roots of Conflict* (Ithaca: Cornell University Press, 1999). I raise this not to criticize these excellent books, but to point out that this topic has not yet received the attention it likely deserves in the existing literature. The literature emphasizing the destabilizing nature of multipolarity in international politics, however, employs the complexity of alliance politics among multiple great powers as a transmission belt between the independent variable—system polarity—and the dependent variable—system-wide stability or instability. See, for example, Waltz, *Theory of International Politics*.

reassurances. This dynamic may provoke adversaries into thinking in terms of closing windows of opportunity or opening windows of vulnerability if they wait until the alliance is fully solidified. Fears of abandonment can also lead more conservative members of an alliance to reduce restraints on aggressive allies, thereby allowing them to more readily drag the alliance as a whole into war with enemies. On the other side of the coin, excessive fears about entrapment and excessive conditionality on commitments among allies can make the alliance appear to real or potential enemies to be insufficiently committed to defense of certain interests, thereby rendering less credible threats of military response to provocations and probes by the enemy.[10]

All things being equal, alliance leaders like the United States and the Soviet Union during the Cold War would have liked their weaker allies to shoulder a good bit of the alliance burden. On the other hand, they would have liked the allies to respect and follow the alliance leader's general line on policies toward adversaries and neutrals. One major problem is that to the degree any ally contributes to its own defense and to the operations of the alliance as a whole, it will likely also want a degree of political voice in the alliance or, perhaps, even a degree of political independence from the alliance. An ally's increased independence reduces the ability of its partners to restrain it and thereby increases the risk of entrapment for those partners if the maverick ally behaves in belligerent ways. At the same time, increased independence of one's ally increases the risk of abandonment, if the maverick ally seeks a separate peace or conciliation with certain members of an enemy alliance.[11]

---

[10] Using somewhat different terms, Glenn Snyder makes similar arguments about his multipolar alliances in a conventional world. See *Alliance Politics*, especially chapters 1, 6, and 9. On the dangers that tightening of alliances play in exacerbating existing tensions and increasing the likelihood of war, see John Vasquez, *The War Puzzle* (Cambridge: Cambridge University Press, 1993), 166.

[11] In *Alliance Politics*, Glenn Snyder tends to emphasize the danger of entrapment flowing from one ally's lack of restraint over another. See, for example, page 321. Snyder's point is well taken, but there is no reason why the increased independence of one's ally might not also lead to fears of abandonment as well, particularly if the national interests of allies or the preferred strategies of allies do not fully overlap. A more independent ally, more capable of defending itself, might seek to make accommodations with select members of an enemy coalition against the wishes of its own alliance leader or may decide that it is able to go it alone, choosing neutrality between the camps, rather than alliance. The dilemma of burden-sharing versus political cohesion sets up a strategic bargaining situation whereby there can be trade-offs between what an ally agrees to contribute to an alliance and how much it is willing to toe the alliance leader's line in its external relations. The details of such bargaining have implications for the credibility of both threats and reassurances. Domestic politics, particularly in democracies, can play an important role in such bargaining. The political science literature on alliances sometimes states too starkly the domestic trade-offs between "arms and alliances" in any country's national security portfolio, but it is certainly true that domestic political concerns affect choices about the proper mix in national grand strategy. For an excellent book on the domestic politics of grand strategy in the Cold War, see Aaron Friedberg's *In the Shadow of the Garrison State* (Princeton: Princeton University Press, 2000). For the argument that there is a trade-off between arms and alliances, see James D. Morrow, "Arms versus Allies: Trade-offs in the

Most of the important alliance dynamics pose true dilemmas, with factors pulling in opposite directions. Depending on the political context, any single adjustment in an alliance can undercut either deterrent threats or assurances. Keeping smaller allies weak and dependent does more than increase the burden of the alliance on the alliance leader. This strategy can undercut deterrence for the alliance as a whole if it makes the alliance overall seem weak or irresolute, particularly in the regions in which the weaker allies reside. But efforts to get allies to do more can also trigger unintended negative reactions among allies and adversaries alike. A more robust military posture by a local ally as part of burden-sharing can send provocative signals to regional adversaries about the long-term dangers posed by the ally in question and the alliance as a whole. Such signals might undercut allied coercive diplomacy if an adversary's perception of an increase in the ally's power over time undercuts that alliance's efforts at reassurance. At other times, the dilemma might cut in the other direction. If the leader appears increasingly reliant on the newly mobilized allies, this apparent need for military help can actually undercut deterrence by making the leader seem too weak or irresolute to get involved directly in a local conflict. To the degree that it looks like the leader needs its weaker allies to fight, its own ability to mobilize effectively for war might appear to be reduced, thus undercutting deterrence for the overall alliance.

As we can see from the foregoing discussion, alliances are always complex management challenges for actors engaging in coercive diplomacy (and alliances almost always are so engaged). They can send unintended signals that lead others to exaggerate their aggressiveness or underestimate their resolve or power in dangerous ways. The argument here is that such dangerous signals are most likely to be sent in periods of change or uncertainty in which alliances are not yet fully and clearly formed or in which the continued leadership capabilities of the most powerful ally are somehow called into question.

The Special Problems for Coercive Diplomacy Posed by Revisionist Alliances

A second theoretical line in this book melds two existing theoretical approaches about alliances to create a new explanation about the behavior of coalitions that are revisionist and ideologically driven. In the current literature, ideology is treated as an independent variable in two ways: ideology is either a factor that contributes to cohesion or breakdown of security alliances or a factor that makes

---

Search for Security," *International Organization* 47, no. 2 (spring 1993): 207–33; for my reservations with this approach, see Thomas J. Christensen, "Perceptions and Alliances in Europe, 1865–1940," *International Organization* 51, no. 1 (winter 1997): 65–97.

alliances relatively aggressive or defensive toward outsiders, depending on the nature of the shared ideology of the allies.

Scholars from various theoretical perspectives have argued that ideological factors may serve either as glue or wedges within alliances, thus making the alliance stronger or weaker overall than a realpolitik analysis would otherwise expect. In other words, ideological factors can help us understand how easily alliances will form and, if they do, whether they will hold together or not. Even realists, who generally downplay the importance of ideology, often accept its importance in weakening or strengthening alliances.[12] In his important realist work, *Origins of Alliances*, Stephen Walt has noted that certain ideologies, such as communism, encourage intra-alliance disputes more readily than do others, such as liberal democracy. Walt argues that this is the case because some ideologies are hierarchical and the competition for leadership in the group increases intra-alliance strife. Communist states during the Cold War, for example, had an incentive to compete within the international ideological hierarchy, while alliances of liberal democracies tend not to have such internal rivalries as their political philosophy lends itself neither to hierarchy nor rivalry.[13] Walt is interested primarily in how internal divisions or cohesion affect the overall power of an alliance. He does not explore how the process of forming a hierarchical ideological alliance or the internal competition for leadership within such an alliance might affect the allies' security policies toward opponents.

In an innovative two-level argument, Randall Schweller addresses why alliances with revisionist ideology are more aggressive than other types of security partnerships. He argues that, even in cases of intense mutual distrust, if a group of allies all have revisionist political aims, weaker revisionist allies may "bandwagon for profit" with stronger aggressive states. Schweller points out that although they have autonomy of action, weaker revisionist states, such as fascist Italy in World War II, will cooperate early and actively with stronger ones, such as Hitler's Germany, so as to share in the spoils of overturning the international status quo. He contends that, because they more readily avoid the problem of

[12] For a realist account, see, for example, Hans J. Morgenthau, "Alliances in Theory and Practice," in Arnold Wolfers, ed., *Alliance Policy in the Cold War* (Baltimore, MD: Johns Hopkins University Press, 1959). Others treat ideology as an important prerequisite for alliance cohesion. See, for example, George Liska, *Nations in Alliance: The Limits of Interdependence* (Baltimore: Johns Hopkins University Press, 1968), ch. 2; and Paul Nitze, "Coalition Policy and the Concept of World Order," in Arnold Wolfers, ed., *Alliance Policy and the Cold War* (Baltimore: Johns Hopkins University Press, 1969).

[13] Stephen M. Walt, *The Origins of Alliances* (Ithaca: Cornell University Press, 1987), 35–37; for a similar argument, see Ole Holsti, P. Terrence Hopmann, and John D. Sullivan, *Unity and Disintegration in International Alliances: Comparative Studies* (New York: John Wiley and Sons, 1973), 15–16, 30.

buck-passing, which sometimes renders status quo alliances too lethargic to counter threats effectively, revisionist alliances are tighter and more aggressive than the defensive, status quo–oriented alliances commonly discussed in the international relations literature.[14]

Schweller's excellent article explains why some revisionist actors bandwagon with stronger neighbors rather than balance against them, but it does not address the catalyzing effects of intramural rivalry for leadership within revisionist, ideologically driven alliances. In this book I adopt a different approach that connects Walt's concerns about intramural competition with Schweller's concerns about the aggressiveness of revisionist alliances. In my discussions of intramural competition, I take the revisionist and revolutionary nature of the ideology as given. I then treat the relative cohesion of the movement at any given time as an independent variable. The relative degree of aggressiveness toward outsiders and the related difficulty of containment of the movement by means of coercive diplomacy is the dependent variable. As one would expect, movements will be easier to contain after they have fully devolved into direct confrontation and, in some severe cases, even into shooting wars among their members. But I argue that under conditions short of that open intramural conflict, internal fragmentations and lack of clear leadership render revisionist alliances more difficult and more costly for status quo alliances to contain through coercive diplomacy (the combined use of credible threats and credible assurances regarding the conditionality of those threats). In a counterintuitive sense, then, containment of revisionist coalitions through coercive diplomacy will be easier for opponents when the revisionist alliance's membership and leadership are more clear and less contested, and its internal cohesion is high, than when membership has not been fully established or when rival revolutionaries are vying for power within a competitive alliance.

"Veto Players" and "Catalytic Players": Revisionist Alliances and Coercive Diplomacy

There is a rich literature on how alliances can create instability in international politics either by encouraging excessive reactions to relatively minor systemic disturbances (such as the problems in the Balkans in 1914) or dangerously lethargic reactions to more pressing security threats (such as Hitler's rampage in Eastern Europe in 1938–39). But there is also important theoretical work that emphasizes the stabilizing role of alliances as restraints on the most aggres-

[14] Randall L. Schweller, "Bandwagoning for Profit: Bringing the Revisionist State Back In," *International Security* 19, no. 1 (1994): 72–107.

sive actors in the system. Alliances can serve as a "drunk tank" for the most belligerent members.[15] In fact, as Patricia Weitsman argues, actual or potential adversaries can build trust, transparency, and stability in their relations by forming alliances with each other. Such "tethering" can reduce security dilemmas between them.[16]

This literature on alliances as restraining factors ties in well to the work in comparative politics on the conservative and stable nature of pluralistic systems of domestic governance. In a fascinating book, George Tsebelis formalizes some age-old wisdom about the stabilizing role of diffusion of power and interest groups in well-institutionalized settings. The greater the number of veto players in any political system and the greater difference among them in terms of ideology and interests, the less likely there will be significant changes in the fundamental policies or constitution of polities.[17]

Tsebelis and authors who have built upon his work make clear one important point: stability should not be equated with "good." Increasing the number of veto players can make it much more difficult to reach new international and domestic agreements that increase economic cooperation abroad or reduce the chance for civil conflict at home, especially in ethnically divided states.[18] In existing international conflicts, what is a factor for stability within well-functioning democracies can be a hindrance to international conflict resolution, especially since in those situations war is the status quo and "policy stability" means the continuation of war. In other words, ending war often requires novelty and innovation

---

[15] On the potentially moderating influence of alliances on their more belligerent members, who can be placed in a "drunk tank" by more moderate allies, see Paul W. Schroeder, "Alliances, 1814–1945: Weapons of Power and Tools of Management," pp. 227–62 in Klaus Knorr, ed., *Historical Dimensions of National Security Problems* (Lawrence: University of Kansas Press, 1976); Stephen Van Evera, "Primed for Peace: Europe after the Cold War," *International Security* 15, no. 3 (winter 1990/1991): 7–57, esp. 39; Jeremy Pressman, *Warring Friends: Alliance Restraint in International Politics* (Ithaca: Cornell University Press, 2008); and Christopher Gelpi, "Alliances as Instruments of Intra-Allied Control," pp. 107–39 in Helga Haftendorn, Robert O. Keohane, and Celeste A. Wallender, eds., *Imperfect Unions: Security Institutions over Time and Space* (Oxford: Oxford University Press, 1999).

[16] For the concept of "tethering" in alliance politics, see Patricia A. Weitsman, *Dangerous Alliances: Proponents of Peace, Weapons of War* (Stanford: Stanford University Press, 2004).

[17] George Tsebelis, *Veto Players: How Political Institutions Work* (Princeton: Princeton University Press, 2002).

[18] For interesting applications of Tsebelis's insights, see Edward Mansfield, Helen Milner, and Jon C. Pevehouse, "Vetoing Cooperation: The Impact of Veto Players on International Trade Agreements," *British Journal of Political Science* 36, no. 4 (December 2006): 403–32; and Kathleen Gallagher Cunningham, "Divided and Conquered: Why States and Self-Determination Groups Fail in Bargaining over Autonomy," Ph.D. diss., University of California, San Diego, 2007; for a related argument about why and how splinter groups within insurgencies can undercut the negotiating efforts of more moderate leaders, see Andrew Kydd and Barbra F. Walter, "Sabotaging the Peace: The Politics of Extremist Violence," *International Organization* 56, no. 2 (spring 2002): 263–96.

that might be vetoed by members of an alliance already involved in conflict. The problem of disorganization and diffusion is compounded in peace negotiations if one member of an alliance can effectively veto any proposed peace agreement. Holding such veto power, however, might require the ally being able to fight without the political or material support of others. Most small local powers, however, would be unable to take on a strong enemy alliance without some external support, so in such an instance they would need to convince a sufficient number of their less-aggressive allies to provide sufficient support for continued belligerence even in the face of the high costs of war.

The good news is that a larger number of veto players could serve also to make wars harder to start, since the move from peace to war is often a controversial and costly change from the preexisting status quo. Alliances with avowedly revisionist ideologies, however, pose a particularly knotty problem not found in other alliances. First, the "normal" baseline behavior of a member of an internationalist revisionist alliance is to support violence against the allegedly illegitimate status quo. So, in revolutionary alliances, restraint, not aggression, is what needs to be justified at home and abroad. Put another way, in an intellectual and spiritual sense normal, or "status quo," behavior in a revisionist international alliance is to actively spread an ideology and overturn the international political or geographic status quo. Moreover, revisionism is often a conspiratorial and dangerous business and members of an alliance must prove their mettle before being trusted as full-fledged members of the movement. For the purpose of gaining such status, freshman members of the movement might make sacrifices and take risks that might exceed immediate parochial national self-interests or even the overall interests of the alliance as a whole.

Finally, and from a theoretical point of view, most interestingly, competition for leadership within a divided international revisionist alliance can push the entire movement in a more aggressive direction than one would expect, even from a revisionist alliance with more cohesion. The literature to date on veto players seems to assume that all parties to a potential deal enter into the negotiations with interests that are set in advance of their interaction. They will then accept a new status quo only if the agreement in question already overlaps with their predetermined interests, or if some kind of material or political side payment can be made by other players to get them on board. In my opinion, while generally useful, this approach misses some key aspects of what is explored in this book. One problem is that in many cases, revolutionary goals are neither material in nature nor clearly divisible (thus calling into question the utility of the economist's concept of "side payments" so often used by political scientists).[19] More important

---

[19] See, for example, Cunningham, "Divided and Conquered," especially pages 26–39. As with so much of the rational choice literature in international relations, which often mimics the economics

still, the potentially competitive nature of revolutionary alliances of revisionists is missed by a model that assumes fixed bargaining preferences between a set of revisionist actors and outside parties. In such a competition each member ("player") in the revisionist alliance may value something outside the details of the negotiation in question, namely the creation, preservation, or bolstering of the member's reputation as a loyal and pure supporter of revolution against the status quo. One goal may be to prove that one is a more vigorous supporter of revisionism than other members of the alliance, who initially might not be as aggressive in supporting violence and might be more willing to compromise with the enemy camp. Concerned for their own positions within the alliance, the initially more moderate "players" might be catalyzed by the process of such a competition and thereby become more aggressive toward the enemy camp than they were initially.

Competition for leadership among belligerent revisionists may be aimed at winning over political support for one's own party or nation from third-nation revolutionary leaders or from subnational actors or forces within one's own population who might favor aggressive revisionism or, at least, oppose accommodation. In other words, the existence of multiple relatively powerful and relatively autonomous players in a revolutionary alliance setting may allow individual actors to be "catalytic players" rather than "veto players." While veto players prevent change, catalytic players cause it because their pursuit of revolutionary policies can place more accommodating or moderate members of the alliance in an uncomfortable position that, if maintained, might cost them in the intramural competition for the hearts and minds of salient foreign and domestic audiences. This international ratcheting process is quite similar to what Jack Snyder explores in his analysis of belligerent forms of nationalism in weakly institutionalized young democracies. In those cases, it is not always the initial interest of subnational actors in international conflict that produces obstreperous policy postures and increases the likelihood of war, but the competitive process of winning the hearts, minds, and votes of an ideologically mobilized public through nationalistic posturing that creates a foreign policy resultant of external belligerence.[20]

---

literature, Cunningham implicitly assumes both fixed preferences and the possibility of side payments, often in the form of material compensation, for changed positions of the actors in question (in her case governments and autonomy movements within states). It is this author's opinion that the notion of divisible goods and the possibility of side payments to prevent security conflict is a major limiting factor in rational actor approaches to war and peace in the field. For the best explication of this position, see Kirshner, "A Rational Theory of War?"

[20] *From Voting to Violence*, esp. pp. 68–69; and Edward Mansfield and Jack Snyder, "Democratization and the Danger of War," *International Security* 20, no. 1 (summer 1995): 5–38.

OUTLINE FOR THE BOOK

In the chapters that follow I use a historical review of alliances and coercive diplomacy in East Asia to demonstrate how poor coordination and unclear signals by both communist and anticommunist coalitions negatively affected stability in this region during the early Cold War in the Korean War and the 1954–55 Taiwan Strait crisis. I will discuss how the communist alliance was the easiest and cheapest for anticommunist allies to contain in Indochina after the communist alliance had solidified and Beijing and Moscow were so closely coordinating their policy (in the period 1954–57). When the Sino-Soviet relationship started to break down and the rivalry for leadership in the international communist movement evolved in the late 1950s and early 1960s, the communist alliance became much more dangerous from an American perspective. As we will see, that discord and rivalry catalyzed Beijing's and, eventually, Moscow's support of Third World revolutionaries in Vietnam, Laos, and even faraway Cuba. For the post–Cold War world I will discuss how uncertainties about the durability, structure, and purpose of the U.S. defense posture in East Asia following the collapse of the Soviet Union affected Beijing security analysts' views of the future prospects for conflict across the Taiwan Strait, and how these uncertainties also affected competition among the great powers in the region more generally.

In the cases explored here, I argue that the period in which the communists in East Asia were easiest to contain was from 1954 to 1957, when Mao Zedong had already proven quite handily that he was no Tito and that he was very much a member of the international communist movement—at a time when Soviet leadership within that movement was widely accepted, and Sino-Soviet relations were still quite warm. The process of getting to that point was fraught with dangers for the anticommunist alliance in both Korea and Vietnam. Moreover, when the intra-alliance cohesion broke down in the late 1950s, the East Asian communist movement became very hard to contain through coercive diplomacy. The ensuing internal competition between the Soviets and the Chinese communists catalyzed the communist alliance's aggressive actions in ways that were detrimental to the security interests of the anticommunist alliance. For this reason, the communist alliance system was even more difficult to counter than Schweller's more generic pack of revisionists, whom he compares to a pack of "jackals" trying to maximize their individual share of a kill. Not only did the communist bloc comprise a group of revisionists, but this was a group of revisionists with an international ideology that was hierarchical and competitive in nature. Mao supported East Asian revolutionaries not only to improve his own nation's international security position, but also because it assisted his personal prestige and his party's revolutionary reputation in the international communist movement. The Soviets in turn were more supportive of their far-flung allies than they otherwise would have preferred to be, in part because Moscow thought it needed to compete with Beijing for leadership in the international movement. Intra-alliance competition within the communist bloc fed rather than starved the expansionism of the movement, at least until 1969, when the intramural competition led to overt fighting among the former allies. It was only then that the United States could truly enjoy the benefits of the Sino-Soviet split.

Chapters 2 and 3: Quasi-Alliances and Real Threats

Chapters 2 and 3 address how lack of coordination in both communist and anti-communist camps in their formative stages led to dangerous misperceptions on both sides of the Cold War divide, particularly in Korea. These problems helped create the conditions for the North Korean invasion of South Korea and helped set the stage for escalation of the Korean War in autumn 1950. The process of alliance formation among Asian communists arguably also allowed Ho Chi Minh to acquire more support for his revolutionary efforts in Vietnam in the early 1950s than he would likely have received if the Soviets had been managing the Southeast Asia portfolio more actively from Moscow.

In 1950 Washington had defense relationships with the Republic of Korea (South Korea) and the Republic of China (Taiwan), offering weapons, aid, and military training, but the United States eschewed direct military commitments to either government. Truman administration officials feared entrapment by the United States' smaller security partners in wars with secondary enemies in areas of limited U.S. power. This concern was one factor preventing the administration from offering more military assistance and a tighter defense commitment to either Chiang Kai-shek (in Taiwan) or Syngman Rhee (in South Korea). After failing to intervene directly on the Chinese mainland to save the Kuomintang (KMT) regime there, on January 5, 1950, President Harry S. Truman publicly rejected U.S. intervention to save the KMT regime in Taiwan. One week later, Secretary of State Dean Acheson excluded both Taiwan and South Korea from the U.S. "defense perimeter" in his oft-criticized speech at the National Press Club.

Given the budgetary and political constraints affecting strategy in Washington, the Truman administration's lack of global clarity regarding defense against communism was understandable. But it is also fair to say that the lack of clear commitments to the KMT and South Korea, noted by enemies and friends alike, helped encourage enemy aggression and undercut deterrence in the region in the months leading up to the Korean War. However reckless and aggressive it was for the communists to act on that belief, it was not entirely unreasonable for leaders in Pyongyang, Beijing, and Moscow to conclude that Washington would likely take no direct action to intervene in a conflict on the Korean peninsula or reinsert itself in the Chinese Civil War. Moreover, given the weak U.S. deployments in the region and the challenges that Washington faced in rebuilding Japan, it was not unreasonable for Joseph Stalin, Mao Zedong, and Kim Il-sung to conclude that the United States and its regional partners would have real difficulties even if they were to decide to resist resolutely North Korea's invasion of the South. Stalin and, to a lesser degree, Mao were swayed by Kim Il-sung's hopeful analysis that U.S. efforts would prove to be too little or too late, especially if North Korea were able to launch a massive surprise attack. The realization that the United States would and could intervene quickly and decisively in both Korea and the Taiwan Strait and do so largely on its own shocked Kim Il-sung's allies in both Moscow and Beijing.

The lack of a clear U.S. security commitment toward Korea and the KMT proved costly for the United States, but the alternative strategy toward Japan—a relatively clear commitment—was also not cost-free. The United States was much more resolute and transparent about its defense commitments toward Japan and its desire for more active Japanese burden-sharing in the nascent Cold War. The United States did not yet have a formal alliance with Japan when the Korean War broke out, as it continued to occupy the defeated World War II enemy until 1951, but Japan was included explicitly in Acheson's "defense perimeter." Moreover, in the famous "reverse course" of 1948, Washington demonstrated that it had every intention of making Japan a powerful ally.

Such clarity and resolve, however, created problems of its own. The Soviets and the communists in East Asia did not doubt the U.S. commitment to Japan, even as they underestimated the U.S. commitment to South Korea and Taiwan. Quite to the contrary, they seemed very concerned about long-term trends in U.S.-Japan relations, expecting Japan to play a major role in the anticommunist struggle in the region. As I will discuss, if anything, the communist capitals grossly exaggerated the military importance of the U.S.-Japan alliance to the United States and, more specifically, the willingness and ability of Japan to project military power to the Asian mainland once again. The prospects of future Japanese troop deployments abroad were seen as very high in communist capitals, an odd notion to analysts today looking back on the period. This is why the Sino-Soviet Treaty of February 1950 itself specified Japan and its allies as the target of the treaty. The Soviets were beholden to assist China in instances of "aggression on the part of Japan or any other state that may collaborate with Japan in acts of aggression."[21]

The still-incubating U.S.-Japan security relationship was viewed by the Chinese communists in particular not only as dangerous and provocative over the longer term, but also as a symbol of U.S. regional weakness and global

---

[21] For the full document, see Odd Arne Westad, *Decisive Encounters: The Chinese Civil War: 1946-1950* (Stanford: Stanford University Press, 2003), 314–15.

overstretch in the near term. Perceptions about current conditions and future trends in this alliance therefore simultaneously undercut both legs of successful coercive diplomacy toward the communist states: the credibility of deterrent threats coming from Washington in the near term if military action were taken by the communists to revise the status quo, and the assurances about the long-term security of the DPRK, PRC, and USSR if those states did not adopt belligerent policies in the present. Arguably, Mao and the Chinese communist elite misread the signals coming from the nascent U.S.-led alliance in two separate but equally important ways: exaggerating Japan's postwar assertiveness and the tightness of the U.S.-Japan alliance, while underestimating the ability and willingness of the United States to fight alone in the near term without significant Japanese provision of forces, which Japan did not yet have at the ready. Such logic created both closing "windows of opportunity" and opening "windows of vulnerability" for the East Asian communists and encouraged them to strike hard early and by surprise. If the Korean Civil War was allowed to drag on into the future, then the United States and its anticommunist partners would be more capable of countering communist forces than they were in the early Cold War period.

For its part, the formative process of the communist alliance also promoted dangerous dynamics that increased the chance for war, the likelihood that war would escalate once it started, and the difficulty that the two opposing alliances would experience in reaching an armistice once escalation had occurred. In the months leading to the Korean War, the weakest communist actor in Northeast Asia, Pyongyang, was able to convince its stronger allies—Moscow and Beijing—to support aggression against South Korea. After several failed attempts dating back to spring 1949, in early 1950 Kim was apparently able to persuade Stalin to offer support in principle for the invasion, but only if Kim could also gain approval from Mao as well. Kim approached Mao for approval of the invasion later in the spring with Stalin's conditional approval in hand. Mao found himself in a bind. Like Stalin, Mao underestimated the likelihood of a large-scale direct and immediate U.S. military response, but was apparently even more worried than Stalin and Kim that the United States might initially dispatch Japanese forces into Korea (and then, eventually, U.S. forces), thus blunting the North's invasion and extending the war, perhaps to China. On the other hand, Mao was trying to prove to a skeptical Stalin and other communists around the world that he was a real member of the communist internationalist club, and a leader of East Asian communism in particular: that is to say, not another Tito. Mao did not want to appear weak on internationalism and thereby take the blame for squelching the completion of a neighboring country's revolution. After all, Mao needed Stalin's aid in creating the forces necessary to "liberate" Taiwan and more generally, to build a socialist economy and modern military. Although it took him over a year to do so, in 1949–50 Kim was able to utilize Beijing's and Moscow's perceptions of the threat from the United States and Japan, as well as their concerns and mistrust in the Sino-Soviet relationship, to achieve a final go-ahead from his two most powerful allies.

Once the war began, alliance politics and the lack of alliance coordination in both the communist and anticommunist camps contributed to the escalation of the Korean War in October and November 1950. As demonstrated in chapter 3, alliance politics contributed to at least two major failures of coercive diplomacy in 1950: the communist alliance's failure to deter U.S. crossing of the 38th parallel in October 1950 and U.S. failure to deter the Chinese from entering the Korean War in force, thus precluding the prospect of defeat for North Korean forces and the near-term unification of the peninsula. Mutual mistrust and lack of cohesive planning in the communist camp meant that the alliance sent signals of weakness to the United States and its UN partners in the critical weeks immediately following Gen. Douglas MacArthur's successful Inchon landing, on September 15, 1950. It was in those weeks that the communists might have been able to deter U.S. forces from crossing the 38th parallel, thus reducing the likelihood that war in Korea would escalate into an extended Sino-American conflict. Instead, the delayed response of the poorly coordinated communist alliance suggested to U.S. intelligence officers and strategists that massive Chinese intervention in the war was unlikely despite a major buildup of Chinese forces that they observed in Manchuria. The American strategic perception was that the alliance was well organized and acting in a concerted fashion. Therefore, China's failure to send large number of troops across the Yalu border to assist in North Korean defense just after Inchon, when it would have done the most good militarily and politically, suggested to American analysts that China's forces in Manchuria were placed there to defend the Chinese border, not to fight deep inside Korea.

What was good for the United States in the short run in terms of war-fighting—that being a weaker communist response to Inchon—was not good in terms of coercive diplomacy and, arguably, was not in the longer-term interest of the United States. Available historical evidence suggests that an earlier, more robust, and more carefully coordinated international effort by the communists in Korea could have deterred the United States from crossing the parallel and could have prevented the disastrous escalation of Fall 1950. The communists did attempt to deter U.S. troops from crossing the 38th parallel after MacArthur's fantastically successful Inchon landing, which cut the North Korean forces in half, isolated and then destroyed many North Korean troops south of the 38th parallel, and rendered the remaining North Korean forces north of the parallel vulnerable to quick annihilation. If the communists had been more successful in their late-September and early-October efforts at coercive diplomacy, the war might have ended in 1950 at the 38th parallel instead of in 1953.

As for the U.S. failure of deterrence in this escalatory process, alliance politics in the U.S.-led camp played a big role here as well. The communist camp as a whole was alarmed when Washington reversed publicly stated prewar policies and sent U.S. forces to defend South Korea and dispatched naval forces to the waters near Taiwan to prevent conflict across the Taiwan Strait. From the communists' perspective, this latter move reinserted the United States in the Chinese Civil War. Moreover, after the Korean War broke out, the United States only increased its efforts to get Japan to build its own military power and to reach a bilateral peace treaty with the United States. After communist efforts to deter U.S. crossing of the 38th parallel had failed, Washington's efforts to deter Soviet and Chinese entrance into the Korean War failed in part for reasons related to U.S. policies toward its own allies and security partners in the region. U.S. alliance policies in the first few months of the Korean War had a powerful impact on strategic thinking in Moscow and Beijing about the long-term implications of the military defeat of the North Korean communist regime and the unification of the Korean peninsula under a government friendly to the United States. This prospect helped create the logic of a closing window of opportunity or an opening window of vulnerability in Moscow and Beijing. Such thinking was prevalent in exchanges between Stalin and Mao in October 1950 and was used by a relatively aggressive Mao to convince his worried comrades on the Politburo in Beijing that war with U.S. forces in Korea was necessary even though the young PRC would appear to lack the wherewithal to take on the forces of the world's leading power.

In the period 1950–54, there is also a clear connection between the lack of tight coordination and leadership from the top in the communist alliance and the overall aggressiveness of the international communist movement in Southeast Asia. Mao's decision to supply significant assistance to the Vietnamese communists perhaps displays best how a relatively loosely knit alliance behaved more aggressively than a more tightly united and hierarchical alliance likely would have under the same circumstances. The USSR played a highly influential role in Korea as the power accepting Japanese surrender north of the 38th parallel in 1945, but in July 1949 Stalin tasked the Chinese Communist Party (CCP) to assist in Asian rural revolution more generally. Stalin did so for reasons that had more to do with his desire not to be bothered with such revolutions than it did with his desire to see Mao play an active leadership role in the region.

Beijing took up Stalin's assigned task to lead Asian revolution with a gusto that surprised and sometimes concerned Stalin. The PRC did much more for the Vietnamese communists than we could have expected from the Soviets, who were not only indifferent about distant Southeast Asia on national security grounds, but were also concerned that active support for anti-French forces there would run against Soviet grand strategy in Europe. The Chinese Communists, on the other hand, had real national security stakes in Southeast Asia, few national security concerns in Western Europe, and a desire to use support of Maoist revolution elsewhere as a way of demonstrating Mao's loyalty to and importance in the international communist movement. An alliance still in its formative phases, therefore, arguably behaved more aggressively in Vietnam than a more mature, Soviet-guided alliance would have if given the same set of challenges and opportunities in Indochina.

Of course, Mao's active support for militant revolution in Korea and Vietnam in 1950 grew in large part out of his heartfelt ideological convictions; but Mao's desire to solidify his own party's position within the international communist movement also played an important role in his decision to support fellow Asian communists. This is particularly true when one considers the question of timing. Mao's support for Korean and Vietnamese communist distracted China from economic rebuilding at home and national unification missions in Taiwan, Tibet, and Xinjiang. The Vietnamese were able to garner external support from China despite the distractions in Korea and the continuing relative indifference of the Soviet Union toward peasant rebels in distant Southeast Asia. In this sense again, the communist alliance was, from the perspective of the United States, worse than a monolith.

Chapter 4: The Counterintuitive Benefits of Communist Cohesion and the Dangers of U.S. Alliance Formation in the Mid-1950s

Chapter 4 discusses the mixed record of the mid-1950s. In this period, a relatively well-coordinated and organized communist alliance allowed for more moderation and clearer signaling during the negotiations that ended fighting in the Korean War (at Panmunjom in 1953) and the conflict in Indochina (at Geneva in 1954). A relatively unified allied position on Southeast Asia in Beijing and Moscow served as a restraint on the most aggressive members of the alliance, the local communists involved in civil wars: Kim Il-sung and Ho Chi Minh.

In the case of the Korean War, after the devastating UN counteroffensive of early 1951, Kim Il-sung stood alone among the communists in wanting to escalate the war further to achieve unification of the peninsula. His stronger allies—China and the Soviet Union—refused to adopt such a strategy and settled for limited war and extended negotiations. Once this reality was clear, Kim was the strongest advocate for compromising at the negotiating table to reach a final armistice, but if he had had his druthers and could have once again manipulated a less cohesive alliance in the way he did in 1950, he would have pushed for more, not less, revolutionary violence than his allies preferred. Alliance cohesion among the Soviets and Chinese, however, prevented a replay of early 1950.

The same is true for Indochina in 1954. Although Ho Chi Minh wanted to continue the fight against the French in the First Indochina War by spreading revolution to South Vietnam in the near term, as he would again in the 1960s, in

1954 the Soviets and Chinese provided a united front and restrained Ho, citing the danger of escalation to Vietnam's own revolution and to the international revolutionary movement. Ho's stronger allies were generally satisfied with the Vietnamese communists' achievement to date and feared the costs of escalation, and therefore prevented Ho from pushing his own agenda too hard at their expense. The Vietnamese communists could not implement their preferred and more radical approach unless at least one of their larger allies was willing to lend material and political support for the effort.

But there were also destabilizing aspects of alliance politics in this period. The growing pains of the U.S.-led alliance in Southeast Asia helped spark a major crisis between China and the United States over Taiwan in 1954–55 and a follow-on crisis in 1958. The process by which Washington drew and redrew lines, leaving certain areas ambiguously included in the U.S. defense perimeter, helped undercut deterrence and contributed to tension and conflict in the Taiwan Strait.

Chapters 5 and 6: How Intramural Rivalry Made the Communist Movement Worse Than a Monolith

Chapters 5 and 6 address how Sino-Soviet rivalry catalyzed communist aggression in East Asia and beyond. I argue that the intensifying disillusionment and competition between the Soviets and the Chinese from 1958 to 1969 actually rendered the containment of communism through coercive diplomacy more difficult for the United States, particularly in Indochina, than it would have been if the Soviets had obtained and maintained a firmer grip on Asian communist internationalism during this period. Until Sino-Soviet rivalry reached the point of open Sino-Soviet conflict in 1969, the competition was detrimental, not beneficial, to U.S. security interests. The Soviets and Chinese competed in supporting communist revolution in Southeast Asia and elsewhere, and the big winners of this competition were Third World revolutionary movements, including those led by Ho Chi Minh and by Fidel Castro. The big losers, at least before 1969, were arguably the United States and its allies.

In the early 1960s Ho Chi Minh received a very large amount of material support from the Chinese communists at a time when the Soviets were still relatively aloof toward the revolutionary civil war of their Vietnamese brethren. Chinese support helped give Ho the wherewithal and confidence to pursue his conflict against the U.S.-backed government in Vietnam and, ultimately, against U.S. forces in the region. The Soviets would become more involved in the conflict over time, providing a great deal of support to the Vietnamese communists once the war escalated in late 1964 and early 1965. But it was fairly clear that a major incentive for this Soviet support was a jealous competition with the Chinese communists for the hearts and minds of the Vietnamese communists and other parties around the world, who had been fed a steady diet of criticism about Soviet weakness and accommodation by their proud and more aggressively revolutionary Chinese comrades. Still, despite this additional material support in the mid-1960s, the Soviets feared escalation of the war in Southeast Asia, an area of the world that was still considered peripheral to the core interests of the Soviets in the global struggle with the West. So, the Soviets' preference as early as 1965 was to help broker a peace settlement between the United States and the Vietnamese communists. But the prospect of such an international negotiation process was destroyed before 1968 in part by Chinese pressure on the Vietnamese to reject Soviet advice, combined with Chinese accusations in Hanoi and around the world that the Soviets were revisionists who were colluding with the United States at the expense of local revolutionaries like the Vietnamese.

In 1968–69 intramural rivalry between Beijing and Moscow escalated into direct conflict. Under these conditions, the United States and its anti-Soviet allies were the eventual beneficiaries in the form of Sino-American rapprochement and Chinese acceptance of the U.S.-Japan alliance and an ongoing defense relationship between the United States and Taiwan. This process took time, however, and the major stumbling blocks in the period 1969–72 remained U.S. alliance policies toward Taiwan and Japan.

Chapter 7: Chinese Concern about Taiwan, Japan, and the United States and Sino-American Diplomacy in the 1970s and in the Post–Cold War World

As discussed briefly in chapter 7, the move from rapprochement to normalization of diplomatic relations would take longer still. Continuing differences over Taiwan, together with domestic politics in both countries, delayed the creation of formal U.S.-PRC relations until January 1979. Sino-American rapprochement led to specific acts of cooperation in intelligence and in the Soviet war in Afghanistan, but perhaps more important, the militarization of the Sino-Soviet border would provide a huge drain on Soviet military resources, a drain that almost certainly contributed to the eventual demise of the Soviet Union. From the particular perspective of this book, the period from 1973 to the demise of the Soviet Union in 1991 was relatively uneventful and therefore will receive minimal coverage. Moreover, the period has been covered quite ably elsewhere.[22]

---

[22] For excellent coverage of this period, see Harry Harding, *A Fragile Relationship: The United States and China since 1972* (Washington, D.C.: Brookings, 1992); Robert S. Ross, *Negotiating Cooperation, The United States and China, 1969–1989* (Stanford: Stanford University Press, 1995); James Mann, *About Face: A History of America's Curious Relationship with China, from Nixon to Clinton* (New York, Alfred A. Knopf, 1998); and Alan Romberg, *Rein In at the Brink of the Precipice* (Washington, D.C.: Stimson Center, 2003).

As that literature reveals, China largely sided with the United States against the Soviet Union. The U.S. alliance system, with the partial exception of Washington's defense relationship with Taiwan, was relatively fixed.

The bulk of chapter 7 will instead explore how in the post–Cold War era, U.S. alignment with Taiwan and alliance with Japan again have figured prominently among issues affecting U.S.-China security relations. The recent context is entirely different from that of the 1950s. While they are far from being allies, the United States and China are not enemies either, but rather major economic partners who have also cooperated to some degree in addressing an increasing range of international problems. But there are still security tensions between the two sides over issues such as relations across the Taiwan Strait, and both nations practice coercive diplomacy toward the other, sometimes tacitly, sometimes less so. Especially in the 1990s, Beijing experts exhibited uncertainty and misperceptions regarding the future form and function of U.S. defense relationships after the collapse of the common Soviet enemy. Chapter 7 will discuss how the legacies of these Cold War alliance relationships—particularly the U.S.-Taiwan relationship and the U.S.-Japan security treaty—have affected U.S.-PRC relations since the collapse of the Soviet Union.

Despite this notable and welcome improvement in U.S.-PRC relations, some of the ways in which security relationships between Washington, Taipei, and Tokyo have affected Washington's post–Cold War relations with Beijing were reminiscent of the problems witnessed in the early Cold War and in the early 1970s. In particular, the end of the Cold War created questions and concerns for Beijing about U.S. Cold War alliances and security relationships that mimicked in many respects the problems of alliances in U.S. coercive diplomacy toward China in those earlier periods. The chapter will address why it has sometimes been challenging under these new circumstances for the United States to balance credible threats and credible assurances toward China while, at the same time, addressing problems of alliance burden sharing and military modernization in the region. At times these dynamics have produced real tensions in U.S.-China relations and in the relationship between mainland China and Taiwan, but fortunately not only have the problems been manageable, but they have in fact also been managed well in the past twenty years.

Chapter 8: Review of the Lessons Learned and Potential Applications Elsewhere

The concluding chapter will review the historical and theoretical lessons of the earlier chapters and conclude with a brief discussion of other cases of alliance disunity and conflict escalation to which the theoretical approaches offered here might apply. As a further illustration of how the process of alliance formation can undercut coercive diplomacy by leaving both current threats and future as-surances less than fully credible, I will discuss how Washington's public sug-gestions that the United States would support Georgia's future membership in NATO exacerbated security tensions between Georgia and Russia and increased the likelihood of conflict over South Ossetia and Abkhazia in 2008. The fact that the United States was not allied with Georgia but seemed to be heading in that direction created incentives for Russia to jump through closing windows of opportunity to maximize its influence there through the use of force. Georgia's own behavior, emboldened in part by the prospect of U.S. support, only further increased the chance for conflict.

To illustrate further the second and arguably more novel theoretical ap-proach of the book, I will focus briefly on two cases in which intramural rivalry in a revisionist alliance exacerbated tensions with the alliance's enemy and made reaching a negotiated peace more difficult. I will explore the internal bickering in the pan-Arab movement and its implications for belligerence toward Israel leading up to the 1967 war; and divisions among the Palestinians in Gaza and the West Bank and the challenges those divisions pose for a negotiated settle-ment between Israel and a new Palestinian state. As it is told by scholars of the Middle East, the story in the 1960s of intra-Arab rivalries among Syria, Egypt, Lebanon and the newly formed Palestinian Liberation Organization and its implications for anti-Israeli irredentism and belligerence is rather reminiscent of the Sino-Soviet competition for prestige among Third World revolutionary parties in the 1960s. More recently, to the degree that Israel faces a divided, ir-redentist Palestinian polity today, that movement itself may prove "worse than a monolith" in negotiations both because it cannot send credible signals and because competition for legitimacy among a population sympathetic to argu-ments for irredentism against Israel might lead to more anti-Israeli violence and more opposition to the compromises necessary for peace deals than we would otherwise expect.

These additional cases will not be covered in the same detail as the U.S.-PRC cases. Moreover, they will not be offered as an exhaustive list, but rather as heu-ristic examples of the broader applicability of the theoretical concepts presented in this book.

Chapter 2

GROWING PAINS: ALLIANCE FORMATION AND THE ROAD TO CONFLICT IN KOREA

This chapter will discuss how problems and politics in the nascent alliances and alignments in both the communist and anticommunist camps affected security relations between the two camps in the first years of the Cold War. The uncertain and poorly defined nature of the U.S. commitment to its partners in East Asia undercut the credibility of the nation's near-term threats and long-term assurances in coercive diplomacy. This combined with political maneuverings in the formative stages of the East Asian communist alliance allowed Kim Il-sung to convince Stalin and then Mao to support Pyongyang's attack on South Korea. The communist aggression that started the Korean War was rooted in a combination of communist elites' underestimation of the resolve and power of the United States to counter such an invasion in the near term and an inflated view of Japan's likely future role in the security politics of the region after its full economic and military recovery from World War II. Political dynamics within the communist camp also contributed greatly to the outcome of war in Korea. The internal politics of a still weakly knitted security relationship between Moscow, Beijing, and Pyongyang was an enabling factor for Kim Il-sung's invasion of South Korea. The weakest actor, Pyongyang, was able to gain support from its stronger partners for its aggressive strategy of national unification through force by manipulating Moscow, Beijing, and the relationship between the two.

EARLY U.S. COLD WAR STRATEGIES TOWARD EAST ASIA

Strongpoint Defense and the Drawing of Lines

George Kennan's strongpoint containment strategy was a creative exercise of prudent realpolitik. It reflected a recognition of the limits of U.S. military power and the large number of security concerns that the Truman administration faced in the early–Cold War era. The strategy focused U.S. efforts on the defense and reconstruction of friendly, anti-Soviet centers of industrial power and the prevention of these and other resource-rich or geographically significant areas from falling under Soviet control. By logical association, it eschewed hard-and-fast defense commitments to areas considered to be peripheral to the Cold War effort for either economic or geographic reasons.[1] In East Asia, Japan was included in Kennan's strongpoints, but neither Taiwan nor South Korea was so included, even though Washington cooperated with both in security affairs in the years prior to the Korean War. Of the various problems within the anticommunist camp prior to the outbreak of war in Korea in June 1950, none looms larger than the ambiguous and anemic nature of Washington's security relationship with South Korea and Taiwan that flowed from Kennan's logic. What is less commonly appreciated is that the still-incubating U.S.-Japan security alliance in occupied Japan also contributed to instability by sending unintended signals about both the short-term and long-term implications of that budding alliance for the communist camp.

At the time of Dean Acheson's January 12, 1950, address to the National Press Club, the Truman administration was forced to work with a defense budget that was one-third the size it would become after the Korean War began on June 25 and then escalated into a Sino-American conflict in November of that year. Before the Korean War the American public was relatively reluctant to pay significant costs or take substantial international risks in the realm of national security policy. Regardless of the mobilizing rhetoric of the March 1947 Truman Doctrine regarding the defense of "free peoples everywhere," the United States could not realistically create a global defense commitment to counter all forms of communist aggression in all places around the world. That is the main reason why in April 1950 Truman shelved NSC-68, the National Security Council's dramatic call for a more expensive and assertive U.S. security policy. There was a broad consensus in the administration about the document's prescriptions, but little sense that the requisite defense budgets, foreign aid allotments, and security commitments would be marketable domestically.[2]

Fully in the spirit of Kennan's strongpoint strategy, Acheson's Press Club speech included in the U.S. defense perimeter Japan, the Ryukyus, and the Philippines as an island defense chain off the Asian mainland. Conspicuously absent on his list were the Republic of Korea (South Korea) and the Republic of China (Taiwan), both of which had received Washington's military assistance, including training by U.S. officers, and would receive additional military assistance

---

[1] For the classic coverage of Kennan's strongpoint containment strategy, see John Lewis Gaddis, *Strategies of Containment: A Critical Appraisal of Postwar American National Security Policy* (New York: Oxford University Press, 1982), chs. 3 and 4. Also see David Mayers, *George Kennan and the Dilemmas of U.S. Foreign Policy* (New York: Oxford University Press, 1988), ch. 6.

[2] For the international and domestic political implications of U.S. grand strategy decisions for policy toward China and East Asia, see Robert Blum, *Drawing the Line: The Origins of American Containment Policy in East Asia* (New York: Norton, 1982); and Thomas J. Christensen, *Useful Adversaries: Grand Strategy, Domestic Mobilization, and Sino-American Conflict, 1947–58* (Princeton: Princeton University Press, 1996), chs. 3–5.

in the following months. In the case of South Korea, Acheson would later state that he had no intention of downplaying the U.S. security commitment to Seoul, which he saw as an extension of the United States' obligation to the UN to stem aggression.[3] In the case of Taiwan, the decision to exclude was more clearly intentional. President Truman himself had stated just one week earlier that the United States would not involve itself in the Chinese Civil War across the Strait "at this time" and even excluded future arms sales to the KMT (the United States would reverse itself on the latter pledge in a matter of weeks and then on the former pledge at the onset of the Korean War).[4] Whatever Acheson's intentions were in crafting these two speeches, we now know that in communist capitals and in Seoul itself, the Truman and Acheson speeches were viewed as symbols of waning U.S. resolve in Northeast Asia.

Ambiguous and Weak Security Partnerships: Early U.S. Cold War Strategies toward South Korea before June 1950

Korea was not considered vital to U.S. national security interests at the time. With the exception of a military advisory group, U.S. forces were withdrawn from the Korean peninsula in June 1949. But Acheson was not just making excuses for a failed strategy when he said in retrospect that Korea's exclusion from the Press Club speech did not mean its survival was considered unimportant before June 25, 1950. After the United States removed ground troops from South Korea in June 1949, Seoul continued to receive military and economic aid packages and a substantial military advisory group (KMAG) remained on the ground to train Republic of Korea (ROK) forces.[5] Although Taiwan's security and the Chinese Civil War received much more domestic attention in the United States than did Korea policy, on two occasions administration proposals for assistance to Korea drove Taiwan policy, rather than the reverse. In June 1949 and late January 1950 proposals for aid to Korea led to domestic political storms in Congress over Asia policy, especially relating to the "loss of China." Fearing the threats that these congressional attacks posed to Korea policy and the precedents that they potentially held for legislative approval of other foreign

---

[3] See, for example, Dean Acheson, *Present at the Creation: My Years at the State Department* (New York: Norton, 1969), 355–56.

[4] For discussion of the Truman speech and the Acheson speech, see Blum, *Drawing the Line*, 179–80, 193. The Press Club speech itself was primarily designed to explain why fighting in China against the communists was not in the U.S. national security interest at a time when accusations about the administration being soft on or even sympathetic to communists were becoming more common in the Congress and the media. See Acheson, *Present at the Creation*, 355.

[5] On the political importance of Korea to the Truman administration, see Bruce Cumings, *Origins of the Korean War*, vol. 2 (Ithaca: Cornell University Press, 2004), ch. 2.

policy initiatives, the administration felt compelled to adjust its preferred policies toward the Chinese Civil War.[6]

Despite the importance of Korea as a noncommunist Asian state and as a test for the young UN, Korea was not, as mentioned earlier, included in Kennan's list of strongpoints. In fact, it was as much the nature of the all-out armored invasion of South Korea, together with memories of European and American weakness in the early phases of World War II, as it was the importance of the target that triggered the Munich analogy in U.S. leaders' minds and led to a resolute U.S. military response under UN auspices.[7] In addition to South Korea's limited strategic value and the opportunity costs of a containment line drawn on the Asian mainland, the Truman administration was very concerned about entrapment in an unwanted war by Syngman Rhee's government in Seoul. In fact, one reason for the United States' limited military assistance to Rhee was the fear that Rhee would exploit larger amounts of aid to attack North Korea. This was hardly an unreasonable concern, as Rhee's nationalist goals were not entirely different from those of Kim Il-sung.[8] As William Stueck, author of several authoritative diplomatic histories of the Korean War, makes clear, each of the two Korean leaders wanted national unification under his leadership, but only one had the military might and international backing necessary to launch an invasion in spring 1950. Rhee's government was supplied insufficient parts and ammunition to sustain combat for more than a few days.[9] As a result, the ROK was rendered militarily weaker on the defensive in 1950 precisely at a time when deterrence was weakened diplomatically by other aspects of the U.S. "strongpoint" strategy,

---

[6] In June 1949 a proposal for a security aid package for Seoul required offering a similar package for Chiang Kai-shek and contributed to a decision to forgo the opportunity for a direct meeting between U.S. Ambassador Leighton Stuart and the CCP leadership in what was then called Beiping (later Beijing). After the shocking failure of a Korea aid bill in January 1950, the Truman administration decided to violate its statements of the previous month and offer military assistance to Taiwan. Christensen, *Useful Adversaries*, ch. 4.

[7] For the importance of the type of invasion, as opposed to the geographic location, see Glenn Paige, *The Korea Decision June 24–30, 1950* (New York: Free Press, 1968); Ernest May, *"Lessons" of the Past: The Use and Misuse of History in American Foreign Policy* (New York: Oxford University Press, 1973); and William Whitney Stueck, *The Road to Confrontation: American Policy Toward Korea and China, 1947–1950* (Chapel Hill: University of North Carolina Press, 1981). See also William Whitney Stueck, *The Korean War: An International History* (Princeton: Princeton University Press, 1997), 43.

[8] For an excellent theoretical analysis of U.S. fears of abandonment and the policy of restraining security partners and eventual allies like the Republic of Korea and the Republic of China (Taiwan) in the early Cold War, see Victor Cha, "Powerplay: Origins of the U.S. Alliance System in Asia," *International Security* 34, no. 3 (winter 2009/2010): 158–96. For the Truman administration's concerns in spring 1949, see Kathryn Weathersby, "Should We Fear This? Stalin and the Danger of War with America," Cold War International History Project (CWIHP) Working Paper No. 39 (July 2002): 5.

[9] Stueck, *The Road to Confrontation*, 163–64.

including the limited U.S. military presence in the region and the withdrawal of military forces from the Chinese and Korean Civil War theaters.

Early U.S. Cold War Strategies toward Mainland China and Taiwan

Many leaders in the Truman administration, particularly in the State Department, would have preferred establishing formal relations with the CCP fairly soon after the communist victory in the Chinese Civil War on the mainland and the founding of the People's Republic of China there on October 1, 1949. Beijing's insistence on sole recognition of the PRC, however, would require Washington to abandon a World War II ally, Chiang Kai-shek's Republic of China (ROC), the military and civilian government of which had fled from the mainland to Taiwan. Many favored such a move in Washington and had favored curtailing military assistance and other aid programs offered to Chiang Kai-shek's KMT as that government progressively lost control of the Chinese mainland in 1948–49. One problem with this strategy was that it contradicted the administration's spoken anticommunist justifications for economic and military aid policies elsewhere, and thereby threatened public and congressional backing for core policies in other areas. Complicating matters further, in 1949 Chiang's forces had fled to the relatively easily defended island of Taiwan, and so assisting Chiang did not seem particularly risky or expensive on its face; therefore geography, together with the anticommunist underpinnings for the administration's global strategy, made the abandonment of Taiwan hard to justify to the American public. Despite these problems, until the outbreak of the Korean War, the Truman administration seemed to have every intention of allowing Taiwan to fall to communist forces without American intervention. Although it hardly would have been considered a positive development in and of itself in Washington, Taiwan's fall would have allowed the United States to extract itself permanently from the Chinese Civil War, avoid entrapment by Chiang Kai-shek in a war with the Chinese mainland, and thereby increase the potential for a future Sino-Soviet split along the lines of Marshal Tito's Yugoslavia. Such a chain of events would also have allowed Washington to begin to coordinate more closely its China policy with Great Britain, America's most important global ally and a nation that would recognize the PRC formally in early 1950.[10] In this context, on January 5, 1950, Truman announced his intention to remove the United States fully from the Chinese Civil War across the Taiwan Strait.

Before June 25, 1950, the Joint Chiefs of Staff and the National Security Council concluded that Taiwan was of considerable strategic importance but

[10] For the classic study, see Nancy Bernkopf Tucker, *Patterns in the Dust: Chinese-American Relations and the Recognition Controversy, 1949–50* (New York: Columbia University Press, 1983).

was not "vital" to U.S. national security interests.[11] This designation of Taiwan as "less than vital" allowed Acheson to push through his still-controversial position that defending the island at the expense of long-term strategy toward mainland China and the region was not worth the candle. Despite Acheson's policy preference, the United States continued to recognize the regime in Taipei as the sole legitimate government of all of China and continued to grant it economic and military assistance during the first half of 1950. Some of these continued aid packages and policy initiatives were rooted in domestic politics of the early Cold War.[12] In addition, Taiwan was still considered of some strategic value and nobody truly relished its collapse, even as it was anticipated.[13]

After the signing of the Sino-Soviet defense pact in February 1950, some in the U.S. government argued that the value of Taiwan to the United States had risen. Advocates of this position saw the island as a geostrategic base from which to counter the perceived Sino-Soviet monolith and as a symbol of U.S. resolve that could reassure Japan and others about the value of friendship with the United States during the Cold War. There is no evidence that these arguments won the day before the Korean War broke out, while there is much evidence that they did not.[14]

The United States did indeed care about Taiwan's ability to fend off absorption from the mainland but was unwilling to make a strong and clear alliance commitment to the island in 1950. In addition to the fear of permanently alienating the Chinese communists by such a commitment, there was a legitimate concern that Chiang Kai-shek's government would attempt to drag the United States into an unwanted war with mainland China. Here we see a parallel between U.S. leaders' concerns about being dragged into war by a security partner and future ally in South Korea and their concerns about becoming embroiled in the aggressive policies of Chiang Kai-shek, another anticommunist irreden-

---

[11] This position was summed up in NSC-37/3 Feb. 10, 1949. See Documents of the National Security Council, microfiche, film 438, reel 1. Also see "Memorandum of the Secretary of Defense," November 24, 1948, in ibid.; and James Schnabel and Robert J. Watson, *History of the Joint Chiefs of Staff, vol. 3, The Korean War, Part 1* (Wilmington, Del.: Glazier, 1979), 30–31.

[12] Christensen, *Useful Adversaries*, ch. 4.

[13] Its continued importance was demonstrated by the fact that the United States allowed itself the ability to reverse course on Taiwan if international conditions changed. The words "at this time" were inserted in Truman's January speech outlining the U.S. extrication strategy from Taiwan to preserve strategic flexibility. On this issue, see Stueck, *The Korean War*, 30.

[14] For a fuller discussion of these issues, see Alan Romberg, *Rein in at the Brink of the Precipice* (Washington, D.C.: Stimson Center, 2003), ch. 1; and David Finkelstein, *Washington's Taiwan Dilemma: From Abandonment to Salvation* (Fairfax, Va.: George Mason University Press, 1993); and Christensen, *Useful Adversaries*, 128–33. For an important document that suggests that U.S. Taiwan policy had not been adjusted in the days just before the Korean War broke out, despite strong recommendations for such an adjustment by key advisors like Dulles and Rusk, see Clubb to Rusk, June 16, 1950 Decimal File 794A.00/6-1650, box 4254, National Archives.

tist involved in a bitter civil war.[15] ROC forces had continued to bomb main-land China's urban targets, often with American planes showing remnant U.S. markings, and to harass mainland ports with navy vessels transferred from the United States. This was a source of great frustration to the United States, espe-cially to Dean Acheson, who was trying to get his country out of the Chinese Civil War, while Chiang was apparently trying to drag it back in. Chiang was simply ignoring State Department warnings and, according to Acheson, was "telling us to go to hell."[16]

American strategy toward Taiwan would fundamentally change once the Ko-rean War started and the U.S. Seventh Fleet symbolically entered the Taiwan Strait with the purpose of preventing aggression from either side. After the Ko-rean War escalated into a direct Sino-American conflict later in the year, these changes became rather fixed. Korean War escalation destroyed any subsequent prospect for U.S. rapprochement with Beijing or distancing from Taipei.

As we will see later in the discussion of communist calculations, U.S. fear of entrapment and a desire to encourage divisions within the communist alli-ance over time led Washington to signal weakness about its support for Asian anticommunist forces, particularly those facing compatriots on mainland Asia in China and Korea.

Early U.S. Cold War Strategies toward Japan:
America's Reverse Course of 1948–51

U.S. policies toward Seoul and Taipei in the months leading to the Korean War have understandably received the lion's share of scholarly attention in analyses of deterrence failure on June 25, 1950. A less-appreciated factor in the breakdown of coercive diplomacy is the destabilizing effect that signals sent from the bud-ding U.S.-Japan alliance had on Soviet and Chinese calculations related to Korea during 1949–50.

Japan was always considered an important industrial power in the early Cold War years by U.S. grand strategists such as George Kennan, George Marshall, and Dean Acheson. However, in the first year of the Cold War (1947–48), a concerted effort to rebuild Japanese power was delayed. Since V-J Day in August 1945, Washington had given priority to General MacArthur's efforts to restruc-ture Japanese politics and economics so as to prevent either a return to militarism or a turn toward socialism. For this purpose, MacArthur's occupation regime purged from politics many wartime leaders, instituted land reform, and began

---

[15] Cha, "Powerplay."

[16] Hackler to Merchant, February 17, 1950, China: Internal Affairs, LM 152, reel 1, frame 886, National Archives.

breaking down Japan's large corporate conglomerates (*zaibatsu* or *keiretsu*). This strategy would begin to change in late 1947 and early 1948 with what diplomatic historians label "the reverse course." As fears of the Soviet Union grew in Washington, concerns also grew that MacArthur's reform program was making Japan politically unstable, economically weak, vulnerable to subversive political infiltration, and militarily unable to defend itself or assist in anticommunist efforts elsewhere. Washington, therefore, adopted a more politically and economically conservative program designed to stabilize the Japanese political economy. Many, especially officials in the Department of Defense, hoped that this would set the stage for increased Japanese military strength and burden-sharing in the future. Japanese national power, it was argued, would assist American efforts in countering international communism in Japan and the rest of East Asia.[17] Especially after the outbreak of the Korean War, this view would win the day in Washington, and it would be presented in Tokyo by Truman's special envoy to Japan's government, John Foster Dulles, in preparation for the turnover of sovereignty.

Despite concerted efforts to get Japan to do more on the military side, the United States agreed to compromise with Tokyo and accept much smaller Japanese military contributions to the alliance than were preferred in Washington. In the 1950s, the Truman and Eisenhower administrations both believed that Japanese discrimination against certain American imports, the systematically undervalued yen, and the reliance on American military protection were necessary to strengthen Japan domestically and internationally as an anticommunist ally. After much persuasion by Prime Minister Yoshida Shigeru, American leaders also decided that pushing the Japanese government to build up its military

[17] On the reverse course, the demise of the 1947 guidelines, FEC 230, and the adoption of NSC-13/2, see Robert A. Pollard, *Economic Security and the Origins of the Cold War* (New York: Columbia University Press, 1985), 182–87; Andrew Rotter, *The Path to Vietnam: Origins of the American Commitment to Southeast Asia* (Ithaca: Cornell University Press, 1987), ch. 2; also see John Dower, *Embracing Defeat: Japan in the Wake of World War* II (New York: Norton, 1999), 272, 511, 525–527. Roger Buckley argues that there was no reverse course in 1948, at least not one based on security calculations. See his *U.S.-Japan Alliance Diplomacy, 1945–90* (Cambridge: Cambridge Studies in International Relations, 1992), ch. 2. The decisions to adopt the reverse course were made in Washington, which of course meant that they were treated as highly suspect by MacArthur, who had all but set himself up as the emperor of Japan and who had made a postwar reputation criticizing the State Department for shortcomings ranging from Eurocentrism to excessive meddling in the Pacific. He would not only argue against denuding the antizaibatsu program, but perhaps more surprisingly, against strengthening the Japanese military. The general argued that economic growth and a stable, liberal political order were the most important weapons in the struggle for containment in Japan, not the creation of military might. Nobody doubted the general's argument about the importance of economic strength and political stability, but many at the Defense Department and some at the State Department insisted that they wanted Japanese military strengthening as well. For differences between the State Department and the Defense Department, see Buckley, *U.S.-Japan Alliance Diplomacy*, ch. 2.

significantly and thereby reduce the American defense burden could lead to a popular backlash in Japan that might threaten the budding alliance and, by association, the maintenance of American bases in Japan.[18] The Japanese authorities thereby resisted American pressure. The biggest source of frustration in Washington was Tokyo's refusal to meet American targets for Japanese troop strength. In his advisory capacity, in 1950 Dulles had argued to State Department officials for a significant Japanese military buildup.[19] Japanese leaders seemed to be able to fend off American demands for a larger and more active military by pointing out that socialists and communists waited in the wings to overthrow the conservative postwar coalition. The gap between American desires and Japanese realities were marked. For example, in 1951 Japan agreed to supply only about one third of the troops requested by Washington for its euphemistic "National Police Reserve."[20] In the 1954 Mutual Security Agreement, Japan agreed to only one half of the total self-defense forces prescribed by Washington.[21] This is not to imply that the U.S.-Japan security relationship was all a one-way street. The United States gained much from the Cold War relationship with Japan, as it does from its cooperation with Japan today.[22] The most obvious U.S. benefit has been Japan's willingness to serve as an invaluable base for U.S. power projection forces. This arrangement has cost relatively little to the U.S. taxpayer.

Especially given the legacies of the recent Pacific War and the generous American aid, trade, and military policies that Washington had already granted to Tokyo, the Truman administration was concerned about the domestic marketability of the budding U.S.-Japan alliance in the United States. Yoshida, therefore, would be pushed to accept Washington's pro-Taiwan and tough anti-CCP China policies. American elites worried that if Yoshida diverged too sharply from the anticommunist strategies being advocated by the United States, the American Congress and public would demand a fundamental reconsidera-

---

[18] On Japanese public opinion on defense issues during the peace treaty and alliance treaty negotiations, see Buckley, *U.S.-Japan Alliance Diplomacy*, chs. 2–3.

[19] See Makato Momoi, "Basic Trends in Japanese Security Policies," pp. 343–45 in Robert Scalapino, ed., *The Foreign Policy of Modern Japan* (Berkeley: University of California Press, 1977). Momoi points out that on September 15, 1950, Dulles argued to the State Department that there should be no limits on Japanese military power.

[20] Michael Schaller, *Altered States: The United States and Japan since the Occupation* (New York: Oxford University Press, 1997), 46.

[21] The United States had requested that Japan keep 325,000 to 350,000 troops under arms, and Tokyo agreed to only 165,000. Buckley, *U.S.-Japan Alliance Diplomacy*, 56–57. Momoi, "Basic Trends in Japanese Security Policies," offers somewhat different numbers, an American request of 325,000 and a Japanese agreement to 180,000.

[22] As Richard Samuels argues very clearly, the original U.S.-Japan security arrangement was a bargain benefitting both sides. Richard Samuels, *Securing Japan: Tokyo's Grand Strategy and the Future of East Asia* (Ithaca: Cornell University Press, 2007), ch. 2.

tion of the already controversial relationship.[23] This concern was not unique to America's Japan policy. Washington presented the same line to foreign leaders around the world as a reason for them to reconsider independent policy lines toward the communist world.[24]

Judging from a careful analysis of the rather fickle nature of American public opinion on Cold War issues during 1947–50, the Truman administration's argument to foreign leaders was not just a bill of goods. Although principled isolationism was relatively rare after Pearl Harbor, popular attitudes about the Cold War effort were hardly steady, swinging from dangerous levels of apathy toward international affairs, to dangerous levels of righteous anger toward the communist bloc and anyone in the domestic and international arena who seemed "soft" on communism.[25]

American domestic politics in 1950 did much more than constrain the United States in its China policy. It made it impossible for the Truman administration to support any new apparent accommodation of the Chinese communists on the part of America's key allies, including Britain and Japan. For example, in early December 1950, Acheson and Truman would shoot down British proposals for London to promote early peace negotiations in Korea, arguing that any accommodation by the United States of the Chinese communists that appeared to result from pressure by America's allies would undercut American popular support for the global containment effort under NSC-68 and would lead the American public to reject further security assistance to all allies that broke ranks with the United States in its struggle against Asian communism.[26]

In 1950–51, John Foster Dulles, as President Truman's special envoy to Japan, would apply the same logic to U.S. dealings with the Japanese leader, Yoshida. Dulles made various arguments why Japan should reject Beijing as a diplomatic partner, recognize Chiang Kai-shek's regime on Taiwan as the sole legitimate

---

[23] Schaller, *Altered States*, ch. 2.; Walter LaFeber, *The Clash: U.S.-Japan Relations Throughout History* (New York: Norton, 1997), 280; and Michael M. Yoshitsu, *Japan and the San Francisco Peace Settlement* (New York: Columbia University Press, 1983), ch. 4.

[24] See, for example, the minutes of the summit meetings between Prime Minister Attlee of Great Britain and President Truman in early December 1950 in U.S. Department of States, *Foreign Relations of the United States, 1950*, vol. 3: 1698–787; related memoranda regarding the Truman-Attlee Talks, in Memoranda of Conversations, Acheson Papers, Box 65, Harry S. Truman Library (HSTL). Truman and Acheson would continually return to the theme that anything short of British-American consensus in East Asia would threaten American popular support for assistance to Britain in Europe.

[25] Christensen, *Useful Adversaries*, chs. 4–5.

[26] So, just after the escalation of the Korean War in late 1950, Dean Acheson rejected British Prime Minister Attlee's proposal for peace talks and reconciliation with China: "If we surrender in the Far East, especially if this results from the actions of our Allies, American people will be against help in the West to those who brought about the collapse." See "Truman-Attlee Talks," December 4, 1950, in Memoranda of Conversations, Acheson Papers, box 65, Harry S. Truman Library.

government of all of China, and sign a peace treaty with the ROC, rather than the PRC. Dulles also sought Yoshida's general compliance with American limits on trade contacts with the PRC. Like most Japanese elites, Yoshida was very anticommunist. But as a practical matter, Japan wanted diplomatic ties with Beijing and much more extensive trade relations than Dulles's preferred scenario would allow. As Yoshida bluntly put it, "I don't care whether China is red or green. China is a natural market, and it has become necessary for Japan to think about markets."[27] In his effort to persuade Japanese leaders to toe the U.S. policy line, Dulles's trump card was not a straightforward geostrategic argument but a domestic political one with geostrategic implications. He emphasized that, if Japan did not comply with America's general Cold War strategy, the American Congress and the American public would not support the Peace Treaty with Japan. Under such conditions, the military protection of Japan by American forces would become more controversial domestically, as would economic aid and Japan's preferential trade and financial arrangements. It appears that it was this domestic political argument, above all others, that convinced the reluctant Japanese that questioning the American leadership role in the Cold War in Asia could carry devastating results for the recovering nation's security and economic interests.[28] Dulles would return to this tried-and-true bargaining tactic again as President Eisenhower's Secretary of State in order to prevent Japan from establishing politically significant trade offices in China.[29]

To understand the sacrifice that Tokyo had to make in order to grant the United States a firm leadership role on the budding alliance's China policy, it is critical to note just how important the Chinese economy had been to Japan in the recent past. As Michael Barnhart points out, it was partially the perceived need in Tokyo for a secure economic relationship with China that fueled Japanese aggression on the mainland in the 1930s. According to Barnhart, the interwar Japanese leadership was actually obsessed first and foremost with the threat from the Soviet Union and the lessons of World War I about the need for an autarkic economy to provide staying power in war. The quest for Japanese autarky on the Asian mainland helped drive Japan deeper and deeper into a quagmire in China and, eventually, into war with the United States.[30] In the 1920s and for most of the 1930s, China (including Manchuria) was by far Japan's biggest export market and import provider in the region.[31]

[27] See LaFeber, *The Clash*, 280.

[28] For an excellent discussion of Dulles's strategy toward Japan in these months, see Schaller, *Altered States*, ch. 2. Also see Yoshitsu, *Japan and the San Francisco Peace Settlement*, ch. 4; and Shimizu Sayuri, "Perennial Anxiety: Japan-U.S. Controversy over Recognition of the PRC, 1952–58," *Journal of American-East Asian Relations* 4, no. 3 (fall 1995): 223.

[29] Shimizu, "Perennial Anxiety," 238.

[30] Michael Barnhart, *Japan Prepares for Total War* (Ithaca: Cornell University Press, 1987).

[31] See *Hundred Year Statistics of the Japanese Economy* (Tokyo: Bank of Japan, Statistical Department, July 1966), 292–93.

Not only would normal relations between Tokyo and Beijing facilitate business contacts, which were much more likely to be viewed as separate from politics by Japanese leaders than American ones, but also the arrangement would allow the new postwar Japan to seem less abrasive among the new Asian postcolonial nationalist movements that Japan's militarist predecessors helped create.[32] Yoshida's desire to establish relations with Beijing after the founding of the PRC did not necessarily mean, however, that he was eager to break relations with Taiwan or that he would have been indifferent to a PRC invasion and takeover of Taiwan. In fact, in spring 1950 when Dulles, then a Republican consultant to Secretary Acheson, and Dean Rusk, then Assistant Secretary of State for Far Eastern Affairs, were trying to convince Truman and Acheson to reverse their policy of aloofness toward the Chinese Civil War and to agree to defend Taiwan, one of the most powerful sections of their appeal was a reference to an early May 1950 discussion between Japan's Finance Minister Hayato Ikeda and Joseph Dodge, an American official in the Department of the Army. Ikeda, claiming that he was representing Prime Minister Yoshida, emphasized that many in Japan had come to doubt that the United States would defend Japan. One reason for this skepticism was the apparent lack of commitment of the United States to Korea and to Taiwan. In the Taiwan case in particular, the arguments within the American government about the island's limited strategic importance seemed to carry direct implications for Japanese observers about the strategic status of the island nation of Japan in the eyes of the Americans. Ikeda was clearly asking the United States to take a firm position somewhere in East Asia so as to reassure Japan that America's commitment to his recovering nation was a firm one.[33]

Within the U.S. government there were two very different interpretations of the relationship between Taiwan and Japan's security. To some U.S. elites, including Dulles and Rusk, Ikeda's statement signaled that Japan desperately wanted Taiwan to be defended by the United States. If Taiwan were not defended, they believed, Japan would lose faith in America's resolve in the region. Dean Rusk presented this position to Dean Acheson in an important policy memo on Taiwan.[34]

But this was not the only reading of the Japanese position. On May 19, the State Department Office of Intelligence Research presented Acheson with quite an opposite reading of the Japanese attitude about Taiwan's importance. The

[32] See Shimizu, "Perennial Anxiety," 223; Schaller, *Altered States*, ch. 1; LaFeber, *The Clash*, ch. 9.

[33] For the presentation of these issues to Secretary Acheson by Assistant Secretary Rusk as part of the joint effort by Dulles, Rusk, and Secretary of Defense Louis Johnson to cause a reversal on Taiwan policy, see "Rusk to Dulles, U.S. Policy Toward Formosa, May 30, 1950," Office of Chinese Affairs, film C0012, reel 15, frames 696–710, National Archives. For the argument that Ikeda's statements meant Japan clearly wanted a U.S. commitment to the defense of Taiwan and South Korea, see Schaller, *Altered States*, 27.

[34] "Rusk to Dulles, U.S. Policy toward Formosa, May 30, 1950," Office of Chinese Affairs, film C0012, reel 15, frames 696–710, National Archives.

analysts argued that the fall of Taiwan would have a minimal impact on Japan's economy, and would open the door for Japan to restore critical trade relations with mainland China. More important, the document states, "The Japanese appear to accept what they interpret as the American appraisal of the strategic value of Taiwan [that it is not vitally important]. The fact that the U.S. officially disclaimed any intention of defending Taiwan precludes loss of U.S. prestige in Japan as a result of Taiwan's fall."[35]

It is possible to explain why two different U.S. government documents could draw such different conclusions about the Yoshida government's position in May 1950. One group of analysts believed that Ikeda's statement about the need for a tougher American posture meant that the fall of Taiwan would be viewed in Tokyo as a sign of failed American resolve. Another group may have seen the same statements only as a concern that the same measure of strategic significance given to Taiwan (less than vital to American security) could also be applied to the islands of Japan. For this latter group the solution was not for the United States to reverse its decision on Taiwan's importance and decide to defend it. Rather the United States needed to reassert its commitment to Japan, and to demonstrate, through some concrete action, that Japan would not be labeled peripheral and abandoned to communist pressure. Supporting this interpretation is evidence of Yoshida's own rather fatalistic attitudes about the future of Taiwan at this time and his general desire to prevent Japan's Taiwan policy from undercutting efforts to improve relations with the mainland.[36]

Japan's acquiescence to American demands resulted in the December 1951 Yoshida Letter and subsequent bilateral Peace Treaty negotiations with Taipei in 1952, which locked Japan into a pro-Taiwan, anti-Beijing diplomatic posture for the next twenty years. With Japanese acceptance of America's harsh economic sanctions regime against China, the small-scale but promising trade between Japan and the PRC allowed by the United States in 1949–50 practically disappeared.[37]

From the perspective of alliance maintenance, the burden-sharing arrangement between Washington and Tokyo made a great deal of sense. But from the point of view of international strategy toward the enemy camp, the situation was hardly ideal. Obviously, having Tokyo and Washington on the same page made it more difficult for the communists to drive a wedge in the budding alliance

[35] See "Consequences of Fall of Taiwan to Chinese Communists in 1950," Office of Intelligence Research, Intelligence Estimate No. 5, May 19, 1950, Office of Chinese Affairs, film C0012, reel 15, frames 711–720, National Archives. Words in parentheses added by author.

[36] Yoshitsu, *Japan and the San Francisco Peace Settlement*, 66–69.

[37] Sadako Ogata, "The Business Community and Japanese Foreign Policy: Normalization of Relations with the People's Republic of China," in Robert Scalapino, ed., *The Foreign Policy of Modern Japan* (Berkeley: University of California Press, 1977), 179; and Shimizu, "Perennial Anxiety."

and made it easier for U.S. administrations to sell at home the domestically controversial economic and alliance policies with Japan. But there were also large costs. Beijing noticed Japan's hostile diplomatic stance but, as noted later, CCP elites were not particularly reassured by Japan's unwillingness to play a larger military role. For much of the Cold War, Japan's restraint on military security affairs was generally viewed as temporary and mercurial in Beijing. The prospect of future Japanese military assertiveness in Asia in alignment with the United States, Taiwan, and others was a fairly consistent concern of PRC strategists, even if on several occasions Beijing took some comfort in its overestimation of U.S. dependence on allies like Japan and Germany.

To be fair to the Asian communist elites on this one score, in 1949–50 it would have been nearly impossible to predict the persistence of the rather unusual U.S.-Japan defense relationship that we have witnessed since the early 1950s. But for our purposes, the United States' bargain with Japan produced a bad mix of factors for coercive diplomacy. As we will see in the following sections, U.S. efforts to encourage Japanese industrial reconstruction and eventual military strengthening and Washington's efforts to limit any rapprochement between Japan and the Chinese communists exacerbated fears in the communist world of an aggressive, anticommunist Japanese military presence in Asia in the future (especially in Korea and Taiwan). This impression in Moscow, Beijing, and Pyongyang would help tighten the Sino-Soviet alliance, and encourage Moscow and Beijing to give robust military aid to the North Koreans before the Korean War. While it fed such destabilizing reactions, the U.S. "reverse course" did not produce much in the way of actual near-term Japanese military power. So, the U.S.-Japan security relationship in 1949–50 was therefore quite poor from the perspective of bolstering U.S. coercive diplomacy in the region. Japan's weakness and reluctance to play a larger military role early in the Cold War meant that the U.S.-Japan proto-alliance did not significantly enhance the deterrent threat of effective anticommunist intervention in Korea or Taiwan in the short run.[38] At the same time, from a longer-term perspective, the U.S.-Japan relationship served to undercut reassurances to the enemy camp about the future, creating fears of potential South Korean offensives and the growth of a powerful and aggressive U.S.-Japan-Korea alliance. These factors contributed to the communists' decision to strengthen Kim Il-sung's military in 1949, largely for defensive purposes. Then, when the balance seemed to turn in Kim's favor in 1950, concerns about Japan's future role in Asia made the Soviets and mainland Chinese more willing to agree with Kim that he should finish the job of unifying Korea on his terms before the regional balance of power turned against the communists.

---

[38] Shimizu, "Perennial Anxiety, 223; Schaller, *Altered States*; and LaFeber, *The Clash*.

THE FORMATIVE YEARS OF THE COMMUNIST ALLIANCE SYSTEM IN ASIA

Mao and Stalin: China Crosses the Yangzi (Yangtze)

After victories throughout the northern half of mainland China, in early 1949 forces of the Chinese People's Liberation Army (PLA) were preparing to cross the Yangzi River to seize the KMT capital at Nanjing and to then unify the country by force. The more globally oriented Soviets were quite nervous about the CCP launching such a campaign, fearing the possibility of U.S. intervention and the overextension of PLA forces. The big fear in Moscow was that the Chinese Civil War might escalate, endangering the CCP's victories to date and, perhaps, eventually dragging the Soviet Union into the Chinese Civil War.

The Soviets had much to lose by early 1949, having already insured a friendly government in northern China, an area where Moscow coveted raw materials and warm-water port access near the Soviet border. One might speculate that Stalin preferred a divided China to one unified under communist rule, but there is little evidence to support such a position. It seems more likely that Stalin's cautious attitude in spring 1949 was underpinned instead by a combination of risk aversion, satisfaction with what the CCP had secured to date, and remnant condescension toward his Chinese communist brethren. In any case, in January 1949, Stalin and Mao exchanged rather frank messages debating whether or not the CCP should agree to negotiate with the KMT or simply drive south and finish the job by force. Mao rejected the idea of negotiations with the KMT and resented the Soviets' offer at mediation in the Chinese Civil War. In addition, Mao clearly decided on his own to go forward with the crossing of the Yangzi, quickly seizing Nanjing in April and unifying most of the eastern portion of what is now the PRC by summer's end.[39] As would become a pattern, local communists seeking unification of their nations under their own rule were more aggressive than the more cautious and satisfied Soviets, who were giving Mao's followers advice from a safe distance and from a less parochial perspective.

---

[39] For excellent coverage of these events, see Westad, *Decisive Encounters,* ch. 7. Also see Sergei N. Goncharov, John W. Lewis, and Xue Litai, *Uncertain Partners: Stalin, Mao and the Korean War* (Stanford: Stanford University Press, 1993). For a dissenting view, see Michael Sheng, *Battling Western Imperialism: Mao, Stalin, and the United States* (Princeton: Princeton University Press, 1997), ch. 8. Sheng at times seems to agree with the thesis that Mao was more aggressive in the civil war period toward the U.S. and its allies than was Stalin, claiming that "Mao was more of a radical revolutionary, inclined to simple solutions, while Stalin was more of a tactful statesman"; at other times Sheng discusses Mao's slavish following of Stalin's lead. The quotation is from p. 167. For evidence of Mao's rejection of Soviet mediation in January 1949, see "Cable, Terebin to Stalin [via Kuznetsov]," January 13, 1949, in *Cold War International History Project Bulletin (CWIHPB)* no. 16 (fall 2007/winter 2008): 129.

This difference of strategy between Stalin and Mao colored not only their future relations but also Mao's own attitudes about his responsibilities as an international revolutionary to support leftist national "liberation" struggles, particularly those movements seeking to avoid permanent division of their countries. Mao's experience in the first half of 1949 would almost certainly influence his later attitudes about the Korean War. For his part, Stalin also seemed to learn dangerous lessons from the Chinese crossing of the Yangzi. Up to that point he had clearly been overly passive in his support for the Chinese communists' efforts at national unification and had overestimated the danger of U.S. intervention and anticommunist resistance. In the months immediately after the crossing, Stalin would promise more energetic assistance to and deeper political relations with the Chinese communists in the future. For example, Stalin volunteered assistance in certain military areas, such as Soviet help in the creation of a Chinese communist air force.[40] A CCP entourage headed by the high-ranking Communist Party leader Liu Shaoqi secured these promises from Stalin on a trip to Moscow in early summer 1949. During Liu's trip, Mao would announce in Beiping (later, Beijing) his "Lean to One Side" policy of joining the Soviet camp in the Cold War. Although a formal alliance between the CCP and the USSR was out of the question until the CCP finished the Civil War against the KMT and founded a communist state in China (which would occur on October 1, 1949), the Liu visit certainly set the CCP on course toward the December-February 1950 treaty negotiations between Beijing and Moscow.

During the visit of Liu's entourage Stalin made clear his position that Mao's CCP should support and guide Asian revolution.[41] As events in Korea would demonstrate, despite Stalin's language of putting the CCP in charge of East Asia, Stalin's gesture had little to do with Stalin's sincere desire to cede leadership to Mao. Rather Stalin was most likely interested in limiting the burden on the Soviet Union in an area of marginal importance to Moscow by passing the buck to his Chinese "comrades."[42] In a sense, Stalin did not have a tremendous amount of choice in the matter, as the Soviet Union lacked power projection capability in the Pacific until the 1970s and was busy rebuilding its economic heartland in Europe after World War II. Outside of the nearby Korean peninsula, Stalin really required Mao to pick up a share of the burden of supporting Asian revolutionaries loyal to the international movement that he led.

[40] Westad, *Decisive Encounters*, 265–70.
[41] Goncharov et al., *Uncertain Partners*, 71–72; Westad, *Decisive Encounters*, 317–18.
[42] Chen Jian, "China and the First Indo-China War, 1950-54," *China Quarterly* (1993): 88–89.

Mao's and Stalin's Attitudes toward the
United States, Japan, and Korea

For the purpose of understanding the interactions between the U.S.-led system in East Asia and the communist alliance, it is most important to note how Mao's and Stalin's attitudes toward the U.S. presence in the region evolved in 1949–50. These attitudes would affect their views on whether or not Moscow and Beijing should support Kim Il-sung in his repeated proposals to unify the Korean peninsula by force. Such a decision would be based on what opportunity costs a war in Korea might carry for the regional communist movement.

For his part, Mao grew increasingly confident throughout 1949 that the threat of U.S. intervention in the Chinese Civil War was declining with each CCP victory and each KMT defeat. In January 1949, he warned the CCP Central Committee against "the mistaken viewpoint which overestimates the power of American imperialism." While not dismissing the threat of U.S. intervention in the civil war, Mao emphasized that "as the Chinese people's revolutionary strength increases and becomes more resolute, the possibility that the United States will carry out a direct military intervention also decreases, and moreover, in the same vein, the American involvement in financial and military assistance to the KMT may also decrease."[43] In a telegram to Stalin in the same month Mao also called into question U.S. staying power in its support of the KMT.[44] In April, after crossing the Yangzi, Mao anticipated that the United States would not only distance itself from the KMT, but would also begin to consider ways to establish diplomatic relations with the CCP regime. Although he was willing to consider such relations with the United States in spring 1949, he would offer no concessions and would demand a total break between Washington and the KMT as a prerequisite, in line with Stalin's own suggestions to Mao on the matter.[45]

Of course, in negotiations with Stalin for assistance, the CCP would emphasize the need for Soviet help in guarding against the prospect of future U.S.-led imperialism in the region. A remilitarized Japan would play a central role in such scenarios. For example, in early 1949, Liu Shaoqi expressed concern to

[43] "Talk to the Central Committee on Tasks Related to Completion of the Civil War," *Mao Zedong Junshi Wenxuan* [Selections of Mao Zedong's Writings on Military Affairs] (Beijing: Liberation Army Soldiers Press, 1981) 326–32; internally circulated.

[44] "Mao's Letter to Stalin on Negotiations with the GMD, January 1949," *Mao Zedong Junshi Wenxuan*, 218–19.

[45] See for example "CCP Military Commission Instructions regarding English, French, and Foreign Nationals and Diplomatic Personnel," April 28, 1949, in *Dang de Wenxian* (Party Documents) 1989, no. 4, p. 43. For a discussion of Stalin's and Mao's calculations on recognition in spring 1949, see Yafeng Xia, *Negotiating with the Enemy: U.S.-China Talks during the Cold War, 1949–1972* (Bloomington: Indiana University Press, 2006), 22–34. For further coverage of the recognition issue, see Gordon Chang, *Friends and Enemies: The United States, China, and the Soviet Union, 1948-72* (Stanford: Stanford University Press, 1990), chs. 1–2; Christensen, *Useful Adversaries*, 138–47.

Stalin's China expert, Ivan Kovalev, that the Chinese communists might face a major offensive by the United States, Japan, and Chiang Kai-shek's KMT.[46] But Mao was actually more confident than Stalin that the tide in China was with the CCP and against the KMT and the United States. Washington offered no help to the KMT when the CCP crossed the Yangzi and seized Nanjing and Shanghai. Moreover, the United States and Britain even largely ignored the PLA's attacks on the British naval ship the HMS *Amethyst*, during the Nanjing operation.

What was unfortunate from a South Korean perspective was that the KMT was collapsing without U.S. intervention precisely at the time that the United States was preparing to pull troops out of South Korea as well. In April and May 1949 the Soviets were receiving intelligence from South Korea about the U.S. preparations to withdraw in the coming months.[47] The U.S. reputation for resolve to intervene in Asian civil wars was, therefore, weakening in Beijing and Moscow, with dangerous implications for Seoul. This perception of U.S. lack of resolve would only be reinforced in January 1950 with the delivery of President Truman's speech on U.S. policy toward the Chinese Civil War and Acheson's National Press Club speech, which seemed to indicate that any U.S. commitment to South Korea or Taiwan had been abandoned.

What was ironic about Soviet, Chinese, and North Korean estimations of U.S. withdrawal from Korea in 1949 was that, in the near term, the event did not reduce their sense of threat from their enemies. In important ways it increased that sense of threat. Given the perceived weaknesses of North Korea at the time, the apparent aggressiveness of the South Korean deployments along the 38th parallel, and persistent fears of Japan's reentry into the Korean theater at a later date, the communist capitals did not take great comfort in the prospect of U.S. withdrawal, at least not at first.

All three communist capitals seemed concerned that a U.S. exit from Korea would unleash Syngman Rhee's government in South Korea to launch offensives against the North, perhaps with eventual assistance from Japan. As stated earlier, this could not have been further from an accurate read of the U.S. intention in the withdrawal. Declassified telegrams in the Soviet Archives demonstrate the concern for North Korean vulnerability in both Moscow and Beijing at the time. North Korean leaders told their quasi-allies that they feared attack from South

---

[46] Goncharov et al. *Uncertain Partners*, 59.

[47] Kathryn Weathersby, "Should We Fear This?"4–5. For the original document, see "Weixinsiji Guanyu Heshi Meijun Tuichu Nan Chaoxian Deng Wenti Zhi Shitekefu Dian" [Vyshinsky's Telegram to Shtykov Regarding Verification of the Issue of U.S. Withdrawal from South Korea, etc.], April 17, 1949, in Shen Zhihua, ed., *Chaoxian Zhanzheng: Eguo Danganguan de Jiemi Wenjian* [The Korean War: Declassified Documents from the Russian Archives] (Taipei, Academia Sincai, Institute of Modern History, Historical Material Collection No. 48, 2003), 1:170.

Korea and Japan after U.S. withdrawal.[48] Kim Il-sung played on Stalin's concern regarding Japan when emphasizing the dangers posed by the U.S.-backed South Korean government and the need, with Soviet help, to unify the peninsula under communist control. According to Kathryn Weathersby, Stalin in particular believed that Japan, with U.S. backing, would rearm and eventually use a base in Korea as a threat to the Soviet Far East. Only Stalin's remaining concern about direct conflict with the United States in the near term would prevent him from accepting Kim's bait by approving such an invasion.[49] Stalin apparently took Kim's prognosis very seriously, viewing June 1949 as a period of high danger for such a South Korean invasion of the North.[50] Beijing in particular was very concerned about the threat of large-scale Japanese intervention in a protracted Korean conflict. Soviet reports from Pyongyang indicate that in a meeting with an emissary of Kim Il-sung in Beijing in early May, Mao said the following:

> A war in Korea can be one of quick resolution, or it can be protracted. A protracted war is not in your [North Korea's] interest; because in that kind [of war] Japan can intervene and help South Korea. But you need not worry, because you have the Soviet Union beside you and us in Manchuria. When needed, we can stealthily send you Chinese troops. Mao added, we [Koreans and Chinese] all have black hair, nobody can tell us apart [*shei ye fen bu qing*].[51]

The U.S. alliance policy of withdrawing troops in 1949 from South Korea only stoked Beijing's and Moscow's concerns about the threat to North Korea, because it would free up local actors for aggression. Kim was able to parlay these concerns into increased assistance from his foreign communist supporters.

Pyongyang would procure most of its military equipment and training from the Soviets, who also kept over 1,000 advisors in North Korea after Soviet troops left in early 1949.[52] North Korea also requested from China the transfer or repatriation of ethnic Korean troops fighting in the Civil War along with their weapons and ammunition. Mao would agree to the transfer of two of these three divisions in May 1949 (the third was already engaged in offensive operations in

[48] "Weixinsiji Guanyu Heshi Meijun Tuichu Nan Chaoxian Deng Wenti Zhi Shitekefu Dian," [Telegram to Shtykov from Vyshinsky Regarding Verification of U.S. Withdrawal from South Korea, etc], April 17, 1949, in Shen, ed., *Chaoxian Zhanzheng*, 1:170; and "Huaxilifusiji he Shenjiemianke Guanyu Sanba Xian Xingshi Gei Shi Dalin de Baogao" [Huaxilifusiji and Shenjiemianke's report to Stalin Regarding the Situation at the 38th Parallel], April 20, 1949, in ibid., 171.

[49] Weathersby, "Should We Fear This?"

[50] Ibid., 5.

[51] "Shitekefu Guanyu Jin Richeng Tongbao Jin Yi Zai Beiping Tanpan Qingkuang Zhi Wei Xin Si Ji Dian" [Shytkov (later ambassador to NK) Telegram to Vishinski Regarding Kim Il-sung's Report on Kim Il's Discussions in Beiping of the Situation], May 15, 1949, in Shen, ed., *Chaoxian Zhanzheng*, 1:187–88.

[52] Xu Yan, *Mao Zedong yu KangMei YuanChao Zhanzheng: Zhengque er Huihuang de Yunchou Weiwo* [Mao Zedong and the Resist America, Aid Korea War: A Correct and Glorious Mapping Out of Strategy] (Beijing: PLA Publishing House, December 2003), 41.

southern China and was, therefore, unavailable at the time).[53] In early January 1950 while Mao was in Moscow negotiating the Sino-Soviet Treaty, Beijing and Pyongyang would agree to the transfer of an enlarged third division of ethnically Korean troops (some 14,000–16,000), the most experienced and battle-hardened of the three divisions to be sent to Korea.[54] Chinese military historian Xu Yan and diplomatic historian Shen Zhihua argue that these transfers in 1949 and 1950 were ordered by Mao to help the North Koreans defend against future invasion from South Korea and Japan, not to encourage the North to launch an attack themselves.[55] This seems quite plausible, at least for the 1949 transfers. In the end, however, this assistance would not be used for defense of the North against invaders, but in the North's invasion of South Korea in June 1950. The three ethnically Korean divisions transferred over the past year would eventually spearhead the invasion.[56] Moreover, the changed intrapeninsular balance between spring 1949 and January 1950 seemed an important element in Stalin's decision to reconsider his opposition to Kim Il-sung's disastrous plan to attack the South.[57] Despite the significance of this Chinese transfer of troops, the more

[53] "Shitekefu Guanyu Jin Richeng Tongbao Jin Yi Zai Beiping Tanpan Qingkuang Zhi Wei Xin Si Ji Dian" [Shtykov Telegram to Vishinski Regarding Kim Il-sung's Report on Kim Il's Discussions in Beiping of the Situation], May 15, 1949, in Shen, ed., *Chaoxian Zhanzheng*, 1:187–88; and Xu, *Mao Zedong yu KangMei YuanChao Zhanzheng*, 44–45.

[54] For relevant documents, see "Guanyu Zhongguo Renmin Jiefang Jun Zhong Chaoxianzu Ren Huiguo Wenti Zhi Shitekefu Dian" [Telegram to Shtykov Concerning the Repatriation of Ethnically Korean Troops in China's PLA], January 8, 1950, and "Shitekefu Guanyu Chaoxian Tongyi Jieshou Zhongguo Renmin Jiefangjun Zhong Chaoxianzu Ren Dian" [Shtykov's Telegram Regarding Korea's Acceptance of the Ethnically Korean Troops in the Chinese PLA], January 11, 1950 in Shen, ed., *Chaoxian Zhanzheng*, 1:280–81.

[55] Xu, *Mao Zedong yu KangMei YuanChao Zhanzheng*, 45; also see Shen Zhihua, *Mao Zedong, Si Dalin yu Chao Zhan: Zhong Su Zui Gao Jimi Dangan* [Mao Zedong, Stalin and the Korean War: The Top Secret Sino-Soviet Archives] (Hong Kong: Cosmos Books, 1998), 210–15. Shen argues that the return of the three divisions of Korean soldiers in early 1950 was not a sign that Mao wanted to assist in an invasion of the South but rather that Mao wanted to assist in the defense of the North. The argument is persuasive in that Mao learned of and approved of Kim's final plan for invasion only in mid-May during the latter's secret trip to Beijing. That trip followed Kim's trip to Moscow, where he first convinced Stalin to offer his conditional support.

[56] Goncharov et al., *Uncertain Partners*, 140–41. For relevant documentation, see the three telegrams under the heading, "JunWei Tongyi Di Si Yezhan Jun Zhong Chaoxian Guan Bing Hui Chaoxian de Dianbao" [Telegrams (Regarding) the Central Military Commission's Agreement to the Return to Korea of Korean Officers and Soldiers in the Fourth Guerrilla Army], January 11 and January 28, 1950, in *Jianguo Yilai Liu Shaoqi Wengao* [The Manuscripts of Liu Shaoqi Since the Founding of the Nation], vol. 1, July 1949–March 1950 (Beijing: Central Documents Publishing House, November 1998), 249–50. For discussion of the importance of the transfer of Korean forces, see William N. Stokes, "War in Korea," in Marshall Green, John H. Holdridge, and William N. Stokes, eds., *War and Peace in China* (Bethesda, Md.: DACOR Press, 1994), 26–27.

[57] For Soviet documents on Kim's appeal to Stalin and the importance of the January 1950 assessment of the balance of power in the final decision to support the invasion, see Kathryn Weathersby, "Soviet Aims in Korea and the Origin of the Korean War, 1945–50: New Evidence From Russian Archives," Cold War International History Project (CWIHP) Working Paper No. 8, pp. 23–26; and Song Liansheng, *KangMei YuanChao Zai Huishou* [Looking Back on the War to Resist

powerful Soviets provided much more assistance to the North than did Beijing and clearly had the stronger influence on Pyongyang.

On at least three occasions (March 1949, August–September 1949, and January 1950) Kim Il-sung had requested from Stalin approval and support for an attack on South Korea (and on at least two occasions Pyongyang also approached Mao). Until late January 1950 Stalin refused to back such an adventure fearing both local failure against anticommunist forces that he perceived as superior and the potential for escalation of the war to involve the United States. The more cautious Stalin encouraged North Korea instead to adopt defensive measures and build up its military to improve the local balance of forces in its favor.[58]

In March 1949 Kim Il-sung initially approached Stalin with a request for help in a military offensive to unify the peninsula under his leadership. This notion was rejected almost out of hand by Stalin as the peninsular balance of power seemed to favor the South and the United States had not yet withdrawn forces from South Korea. For his part, Mao rejected the idea of a North Korean invasion of South Korea in May 1949, citing his concern about the North's inability to win relatively quickly. Mao was worried that an attack in 1949 would be premature given the international situation at the time. Mao apparently believed that the South still had the upper hand in the peninsular balance of power, particularly when one factored the prospect of Japanese reinforcements into the equation, as Mao consistently did.[59] A document in the Soviet archives suggests that Mao shared Kim's concern that Japanese forces might replace U.S. forces in the South following the expected U.S withdrawal later that spring. Mao believed that Japanese forces could then assist in an invasion of the North. Mao advised that North Korea prepare to counterattack such an invading army. But he added that the North Koreans would need to be cautious especially if the Japanese were involved directly in the future invasion. In that case, Pyongyang should be willing to sacrifice some land to the enemy in order to gain a better military position, surrounding the enemy troops and annihilating them later. Mao stated that Chinese troops could then eventually help the North Koreans defeat the Japanese army.[60] A Soviet document quotes Mao as saying, "Even under condi-

America and Assist Korea] (Yunnan People's Press, January 2002), 41. Soviet internal historical documents discuss how by January 1950, when Stalin had a change of heart on the Korean invasion, the North had amassed an army of 110,000 troops that was larger and more powerful than the army of the South. See Weathersby, "New Findings on the Korean War," *Cold War International History Project Bulletin (CWIHPB)* 3 (fall 1993): appendix one, document dated 9 August 1966, "On the Korean War, 1950–53."

[58] Xu, *Mao Zedong yu KangMei YuanChao Zhanzheng*, 41.

[59] On the balance on the peninsula in spring 1949, see Xu, *Mao Zedong yu KangMei YuanChao Zhanzheng*, 43.

[60] See, for example, "Kewaloufu Guanyu Mao Zedong Tongbao yu Jin Yi Huitan de Qingkuang Zhi Shi Dalin Dian," [Kovalev's Telegram to Stalin Concerning Mao Zedong's Report on a Meeting

tions where the Americans go and the Japanese do not come [to replace them], we still advise the Korean comrades not to launch offensives against the South but to wait for a more propitious situation."[61] The time was not ripe for a North Korean attack in the minds of the leadership in Moscow and Beijing. One key reason was that CCP elites, still caught up in their own civil war, believed that the Chinese communists would have great difficulty lending support to North Korea as quickly as the United States could support South Korea, perhaps by sending in divisions of Japanese soldiers to fight there.[62]

In general, Mao believed that North Korea and China in combination could defeat a South Korean and Japanese condominium on the peninsula, but he was not itching for a fight and did not believe in May 1949 that the balance of forces favored the communists.[63] It was in this more defensive context that Mao agreed to the transfer of experienced troops and their equipment from China to North Korea. Ominously for the anticommunist forces in the region, in spring 1949 Mao suggested that by early 1950 the situation might have changed for the better from the communists' point of view and the communists could revisit this issue.[64]

In August and September 1949, the Soviets again would reject the idea of a North Korean invasion of South Korea, which was apparently raised by Kim Il-sung on August 12. In late August the Soviet ambassador to Pyongang, Terentii F. Shtykov, cited South Korea's perceived relative military strength, the danger that the United States might send weapons to the South and dispatch Japanese forces there, and, rather presciently, the danger that the United States might use the occasion of a major North Korean offensive to launch a major international campaign against the USSR.[65] In draft instructions to the embassy in late September, Molotov stated that the Soviet ambassador to North Korea should

with Kim Il on the Situation], May 18, 1949, in Shen, ed., *Chaoxian Zhanzheng*, 1: 189–90; "Shitekefu Guanyu Jin Richeng Tongbao Jin Yi Zai Beiping Tanpan Qingkuang Zhi Weixinsiji Dian," [Shtykov Telegram to Vyshinski Regarding Kim Il-sung's Report on Kim Il's Discussions in Beiping of the Situation], May 15, 1949, in ibid., 1:187–88.

[61] "Kewaloufu Guanyu Mao Zedong Tongbao yu Jin Yi Huitan de Qingkuang Zhi Shi Dalin Dian," [Kovalev's Telegram to Stalin Concerning Mao Zedong's Report on a Meeting with Kim on the Situation], May 18, 1949, in Shen, ed., *Chaoxian Zhanzheng*, 1: 190.

[62] Song, *KangMei YuanChao Zai Huishou*, 39–40.

[63] See, for example, "Kewaloufu Guanyu Mao Zedong Tongbao yu Kim Yi Huitan de Qingkuang Zhi Shi Dalin Dian," [Kovalev's Telegram to Stalin Concerning Mao Zedong's Report on a Meeting with Kim Il on the Kang Mei: Situation], May 18, 1949, in Shen, ed., *Chaoxian Zhanzheng*, 1:189–90; and Xu, *Mao Zedong yu Kang Mei Yuanchao*, 43–45.

[64] See, for example, "Kewaloufu Guanyu Mao Zedong Tongbao yu Kim Yi Huitan de Qingkuang Zhi Shi Dalin Dian," [Kovalev's Telegram to Stalin Concerning Mao Zedong's Report on a Meeting with Kim Il on the Situation], May 18, 1949, in Shen, ed., *Chaoxian Zhanzheng*, 1:189–90. For a Chinese history that supports this interpretation, see Song, *KangMei YuanChao Zai Huishou*, 40–45.

[65] Weathersby, "Should We Fear This?" 7.

"sternly and seriously" relay the message that the situation on the peninsula was neither politically nor militarily ripe for such an attack. The North still did not have "absolute superiority" on the battlefield and, combined with the lack of political preparation, would face a long war involving the United States. In an incisive analysis, Molotov surmised that an unsuccessful North Korean invasion could lead to a long-term U.S. military presence in South Korea under UN auspices.[66] In mid-October 1949, two weeks after the founding of the PRC, Mao and Stalin apparently exchanged notes agreeing that a North Korean attack on the South would be unwise. The North was advised to use guerrilla operations to weaken and destabilize South Korea and to continue with its own military strengthening at home.[67]

The Tail Wags the Dog: Stalin's Reversal on the North Korean Invasion

On January 17, 1950, at a luncheon in Pyongyang attended by the Soviet ambassador, Kim Il-sung again requested Stalin's support and approval for an attack on South Korea. Kim emphasized that he had an opportunity to exploit South Korean popular opposition to the government there and garner backing in the South for himself. He requested a meeting with Stalin. He said that, as Stalin had suggested, he had waited for a South Korean attack on the North to provide a legitimizing pretext for a North Korean "counterattack," but the South had not obliged him. Kim played well on Stalin's combination of arrogance and insecurity, saying that, as a communist, Kim could not launch an attack by himself and that "Stalin's directives were law," in his mind. However, he added that if Stalin could not meet with him to discuss this matter, he, Kim, could meet with Mao in China after the latter's return from the Soviet Union (Mao was in Moscow negotiating the Sino-Soviet defense treaty). Kim pointed out that based on what Mao had said in the past, he believed that the PRC leader would support such a plan after the completion of the war in China. Kim said that he had other issues to discuss with Mao in any case, such as the creation of an Asian Cominform. He added that Mao would have received all the appropriate directives in Moscow and could discuss the Korean proposal for an invasion with him. The North Korean president finished his discussion with a request for Stalin's appraisal of the situation in South Korea.[68]

---

[66] "Maloutuofu Chengbao de Ni Dafu Jin Richeng de Zhishi Gao," [Molotov's Draft Plan for Instructions on Responding to Kim Il-sung], September 23, 1949, in Shen, ed., *Chaoxian Zhanzheng*, 1:255–59. The document offers three separate drafts of the reply to Kim. Corroborating the accuracy of this document is another similar draft Politburo decision of the same date cited by Kathryn Weathersby and attributed by her to Gromyko and Bulganin. See Weathersby, "Should We Fear This?" 7–8.

[67] Shen, *Mao Zedong, Si Dalin yu Chao Zhan*, 216.

[68] "Shitekefu Guanyu Jin Richeng Tichu Xiang Nanfang Fadong Jingong Wenti Zhi Weixinsiji

Whether it was intended to play such a role or not, such talk from Kim could only have helped feed Stalin's well-documented fears regarding Mao's influence in Asia as an actor independent of Stalin's control. Stalin had treated Mao with great caution when the latter arrived in Moscow to negotiate a defense treaty, delaying substantive talks for weeks. According to Odd Arne Westad, Stalin wanted to test Mao's dedication to "proletarian internationalism" and to ensure Soviet territorial claims in the region.[69] Although the causal connection is not clear, once Stalin agreed to discuss the treaty in early January, Mao made decisions in Moscow that were consistent with an effort to demonstrate his revolutionary credentials to Stalin.[70] On January 8, Mao agreed to send the third division of ethnically Korean forces in the PLA to North Korea.[71] Two days after Acheson's January 12 "defense perimeter" speech, which was viewed in Moscow as designed to create a wedge between Mao and Stalin, Mao ordered the seizure of the U.S. consular property in China.[72] On January 17–18 Mao ordered Beijing to recognize Ho Chi Minh's government in Vietnam.[73] Whether these actions fully convinced Stalin of Mao's dedication to proletarian internationalism is unclear. We do know from Russian sources that Stalin remained very concerned about Mao's loyalty to Soviet leadership.[74] Given this concern, Stalin could not have been pleased with the idea of Kim Il-sung turning to Mao for leadership after Stalin had refused Kim's requests regarding military unification twice in the past. Given the conspiratorial and zero-sum nature of revolutions, it is perhaps not surprising that revolutionary leaders in the Soviet Union and China would view each other with such deep mistrust, particularly in the early stages of forming strategic partnerships. Moreover, on an individual level, both Stalin and Mao were arguably unusually untrusting of comrades at home and abroad, even by communist revolutionary standards.

After determining that the international situation had changed over the previous months and after receiving some recent intelligence from South Korea

---

Dian," [Shtykov's Telegram to Vishinski Regarding Kim Il-sung's Raising of the Issue of Launching an Attack on the South], January 19, 1950, in Shen, ed., *Chaoxian Zhanzheng*, 1: 305.

[69] Westad, *Decisive Encounters*, 311–12.

[70] For this interpretation of events, see John Garver, "Paradigms, Polemics, Responsibility, and the Origins of the U.S.-PRC Confrontation in the 1950s," *Journal of American-East Asian Relations* (spring 1994): 9.

[71] See "Guanyu Zhongguo Renmin Jiefang Jun Zhong Chaoxianzu Ren Huiguo Wenti Zhi Shitekefu Dian," [Telegram to Shtykov Regarding the Repatriation of Ethnic Koreans in the Chinese People's Liberation Army], January 8, 1950, in Shen, ed., *Chaoxian Zhanzheng*, 1:280.

[72] Goncharov et al., *Uncertain Partners*, 102.

[73] "Guanyu Tongyi yu Yuenan Zhengfu Jianli Waijiao Guanxi Gei Liu Shaoqi de Dianbao" [Telegram to Liu Shaoqi Regarding Agreement to Form Diplomatic Relations with Vietnam's Government], January 17 and 18, 1950 in *Jianguo Yilai Mao Zedong Wengao*, vol. 1: 1949–50 (Beijing: Central Document Publishing House, November1987), 238–39.

[74] Goncharov et al., *Uncertain Partners*, ch. 3.

about political instability and the low state of morale there, on January 30, 1950, Stalin decided to reconsider his earlier opposition to a North Korean invasion of the South and to discuss this issue directly with Kim in Moscow at an appropriate time. The Soviet leader's reconsideration of the issue occurred while Mao and Zhou Enlai were still in Moscow negotiating the February 1950 Sino-Soviet defense treaty. But all indicators suggest that Stalin kept Mao in the dark about his changed plans for Korea. According to documents cited by Weathersby, and consistent with the foregoing analysis, on February 2, 1950, Stalin asked Kim not to discuss his forthcoming trip with his own colleagues or with the Chinese communists. Stalin cited security concerns and the sensitivity of the topic.[75] Kim arrived in Moscow on March 30 and would stay until April 25 to discuss his plan and secure a final blessing from Stalin.

What changed in Stalin's thinking from September to January? Scholars studying the Russian and Chinese archives cite various factors. One very important factor was the U.S. withdrawal from the Chinese Civil War. Weathersby and others cite Stalin's statements to Kim in March 1950 that he was reassured by the CCP victory in China and the lack of U.S. intervention during the KMT's collapse.[76] A Party document from the Central Committee that is presented by Weathersby states the following:

> Comrade Stalin confirmed to Kim Il-sung that the international environment has sufficiently changed to permit a more active stance on the unification of Korea.
>
> Internationally, the Chinese Communist Party's victory over the Guomindang [KMT] has improved the environment for actions in Korea. China is no longer busy with internal fighting and can devote its attention and energy to the assistance of Korea. If necessary, China has at its disposal troops which can be utilized in Korea without any harm to the other needs of China. The Chinese victory is also important psychologically. It has proved the strength of Asian revolutionaries, and shown the weakness of Asian reactionaries and their mentors in the West, in America. Americans left China and did not dare to challenge the new Chinese authorities militarily.
>
> Now that China has signed a treaty of alliance with the USSR, Americans will be even more hesitant to challenge the Communists in Asia. According to the information coming from the United States, it is really so.[77]

Stalin then placed two critically important conditions on Kim's invasion: "However, we have to weigh once again all the 'pros' and 'cons' of the liberation. First

---

[75] Weathersby, "Should We Fear This?" 9.

[76] Ibid., 9–11.

[77] Report on Kim Il-sung's visit to the USSR, March 30–April 25, 1950, Prepared by the International Department of the Central Committee of the All-Union Communist Party (Bolshevik) as presented in Weathersby, "Should We Fear This?" 9–11.

of all, will Americans interfere or not? Second, liberation can only be started if the Chinese leadership endorses it."[78]

Weathersby argues that, in addition to the signaling of U.S. weakness inherent in its withdrawal from the Chinese Civil War, a secret U.S. policy document on East Asia, NSC-48, had been passed to the Soviets by British spy Donald McLean.[79] There is little doubt that Stalin had a spy network in the United States, Europe, and Asia, but in this case access to NSC-48 might not have been necessary. The Truman administration was publicizing the key elements of its strategy. In a speech on January 5, 1950 Truman for all intents and purposes announced a permanent U.S. withdrawal from the Chinese Civil War, including the cessation of military aid. On the following day, Great Britain recognized the PRC. The issue of the importance of U.S. acquiescence to Chinese communist victory as a precedent in Korea is evident in the Soviet documentation and in the secondary histories.[80] Moreover, as noted earlier, Acheson's National Press Club speech on January 12, 1950, excluded both South Korea and Taiwan from the U.S. defense perimeter. That speech revealed to the public the most important aspects of NSC-48 regarding U.S. intentions toward South Korea. Although the full weight of its importance in Stalin's mind is not entirely clear, we know that Stalin and Mao discussed the speech during the latter's stay in the Soviet Union to negotiate the Sino-Soviet Treaty.[81] According to diplomatic historians studying both the PRC and the USSR, Acheson's speech had an impact on Stalin and Mao and, along with the earlier Truman speech on Taiwan, may have contributed to Stalin's conclusion that the likelihood of U.S. intervention in Korea was limited. Stalin apparently mentioned the speech to Mao and his entourage, characterizing it as an effort in Washington to drive a wedge between Moscow and Beijing and as part of a longer-term strategy to deny Taiwan to the PRC.[82]

[78] Ibid., 9.

[79] Ibid., 11; for the original documents, see National Security Council 48/1 and 48/2, December 23, 1949 and December 30, 1949, in *Documents of the National Security Council*, film A 438, reel 2.

[80] See, for example, Xu, *Mao Zedong yu KangMei YuanChao Zhanzheng*, 42.

[81] Marc J. Selverstone, *Constructing the Monolith: The United States, Great Britain, and International Communism, 1945–50* (Cambridge: Harvard University Press, 2009), 213.

[82] Goncharov et al. argue that the Acheson speech was an important data point in Stalin's thinking, though they emphasize Stalin's fear that Acheson's apparent decision to withdraw from Asian civil wars was designed to place a wedge between the Soviets and the Chinese. See Goncharov et al., *Uncertain Partners*, 101. On the Truman and Acheson speeches' effect on Kim and Stalin, see Song, *KangMei Yuanchao Zai Huishou*, 30. Shen, *Mao Zedong, Si Dalin yu Chaozhan*, 217, states that on January 22 Stalin and Mao had a special meeting to discuss the Acheson speech, only one of three formal, substantive meetings between the two top communist leaders. Also see Xu, *Mao Zedong yu KangMei YuanChao Zhanzheng*, 41; also see Elizabeth A. Stanley, *Paths to Peace: Domestic Coalition Shifts, War Termination and the Korean War* (Stanford: Stanford University Press, 2009), 98.

Finally, we now know from declassified Soviet documentation that in January 1950 Stalin had relevant intelligence from his spies in South Korea. These intelligence reports suggest that leaders in Rhee's government also believed that the United States would not intervene to save them if they were to become embroiled in a war of unification. According to the January 28, 1950, Soviet intelligence report, South Korean elites were very concerned about the state of their economy, politics, and international backing. On the international front, ROK leaders listed three important factors: 1) South Korea was extremely unlikely to get help from the United States; 2) the United States would not help Chiang Kai-shek defend Formosa (Taiwan); and 3) Great Britain had recognized the PRC. They believed that the ROK would receive the same treatment as the KMT, that Britain would recognize North Korea next, and that the United States would follow suit in China and North Korea. Even before the Acheson speech, Syngman Rhee's cabinet concluded that the United States was not willing to fight for South Korean interests. Rhee did not believe that the United States would write off South Korea entirely (as he believed Washington had just done to Taiwan) until after the issue of Japan was resolved, but he did not believe that the United States would fight alongside the South. His prescription was a diplomatic initiative to gain tighter coordination between South Korea, Japan, and the United States in a united anticommunist front.[83]

Stalin received this information regarding South Korea just two days before his January 30 policy reversal on North Korea's invasion of the South. The mixed message he received from Seoul about U.S. alliance policy could not have been worse from a deterrence point of view. On the one hand, its strategic decisions about where to defend and with whom to ally were making the United States seem weak in the near term in areas like South Korea. Moreover, South Korean morale appeared shaken at a time when the balance on the peninsula was moving in the favor of the North. On the other hand, while Japan was not yet ready to contribute greatly to South Korean security, the South Korean leader was discussing the need for closer collaboration with Japan in the future as a solution to the potential for abandonment by the United States. Rhee's alignment strategy, which is in keeping with the theoretical work of Victor Cha on the later ROK-U.S.-Japan relationship, must have helped undercut both sides of a standard deterrence relationship for Stalin.[84] The anticommunist allies appeared temporarily weak and in disarray, inviting attack in the near term, but Seoul's prescription for healing its woes played into Stalin's fears about the future: the

---

[83] See the January 28, 1950, Soviet intelligence report on the January 6, 1950, meeting of the ROK State Council in Shen, ed., *Chaoxian Zhanzheng*, 1:307–308.

[84] Victor D. Cha, *Alignment Despite Antagonism: The U.S.-Korea-Japan Security Triangle* (Stanford: Stanford University Press, 1999).

formation of a tight Korea-Japan-U.S. condominium after Japan had become politically independent and militarily more powerful.

Elizabeth Stanley argues that in early 1950 hawkish advisors in Moscow had been pushing exactly the same line as Kim regarding the future activities of the U.S. security partners in the region. She writes that

> the hardliners argued the United States was preparing South Korea and Japan as 'military springboards' for aggression aimed at the Soviet Union and Asian 'national liberation movements.' Now the hardliners imputed a U.S.-Japanese threat to Soviet interests in Asia to justify their plans for Korean unification. Obviously this line of argument contradicts the previous argument [made by the same advisors] about U.S. non-intervention if North Korea invaded the south. The hardliners were arguing it both ways: while they asserted the U.S. position in Asia threatened Soviet interests, demanding aggressive action, they simultaneously implied the United States was indecisive and could easily be forced out of Korea.[85]

The uncertainties of U.S. alliance politics in the early Cold War managed to undercut both portions of a successful attempt at coercive diplomacy: credible threats and assurances. There were both insufficient threats to deter communist aggression in the near term, and insufficient assurances that near-term peace would provide longer-term security from a Soviet perspective.

In the first months of 1950, Stalin still had major concerns about the operation. He was simply not as confident as Kim that the victory would be quick or that the chances of effective U.S. intervention were quite so small. Stalin insisted on careful preparation, secrecy, and, eventually, securing of Chinese approval of the invasion after his meetings with Kim and Pak Hon-yong in Moscow. During their meetings, Stalin insisted to Kim that "liberation can be started only if the Chinese leadership endorses it."[86] He also told Kim that Soviet forces would not be used in Korea if the invasion went awry. North Korea would have to rely on its own efforts and Chinese assistance. Kim replied by asserting that that the United States would not get involved; that Mao was fully supportive and had offered to help in the past; and that because the United States would not get involved, Chinese help would not even be necessary. Stalin reiterated that Kim should get Chinese approval, stating that, especially if the United States were to get involved in the war, China, not the Soviet Union would have to provide direct assistance. Kim insisted that, by the time the United States could digest the invasion and reverse course on its South Korea policy, the war would be over. Kim cited not only a distribution of military power on the peninsula that by spring 1950 would clearly favor the North, but he

---

[85] Stanley, *Paths to Peace*, 99.
[86] Weathersby, "Should We Fear This?" 9.

and Pak also predicted a massive uprising in the South in support of the North. They therefore posited that the war would end within a few days.[87]

U.S. policies toward Japan, Chiang Kai-shek's Republic of China, and South Korea all contributed to Stalin's false sense in 1950 that a North Korean invasion of the South would be much wiser than it had seemed in 1949. But Stalin still appeared concerned about the potential risk of North Korean failure and the danger of escalation involving the United States. So, why did he approve of the invasion despite those concerns?

Judging from the writings of leading diplomatic historians in the United States, Russia, and China, Stalin's suspicions about Mao's potential Titoism or Asian Leninism assisted Kim Il-sung in his efforts to be the tail that could wag the communist international dog. For example, historians Vladimir Zubok and Constantine Pleshakov argue that knowledge of Mao's likely future independence as a communist leader is "critical for explaining why Kim succeeded in getting Stalin to consent to the invasion of South Korea." They write,

> Had Stalin said no to North Korea, it would have looked as if again, as during the civil war in China, he were putting the brakes on the revolutionary process in the Far East. And Mao Zedong was autonomous and unpredictable. The Chinese could start supporting Kim without the sanction of Moscow, in the same way that Tito's Yugoslavia had supported the Albanians and the Greek guerrillas, ignoring Moscow's objections. Taking issue with the PRC just months after the much-trumpeted conclusion of the Sino-Soviet treaty in Moscow would be unacceptable and ruinous. Equally so would be the recognition of Mao's revolutionary supremacy in Asia. ... When, in early April 1950, Stalin supported Kim's invasion plan, he believed that he was preventing both of these developments.[88]

In an article based on Soviet archives, Weathersby speculates that Stalin approved of the North's attack on the South in large part to guarantee that the PRC would not stray from Soviet leadership and in part to guarantee the Soviets' relative prestige in the international communist movement.[89] This makes sense. By offering conditional approval for Kim's invasion, Stalin was pushing Mao into backing an initiative for which Stalin could still receive the credit. If Mao refused to back Kim's plan, Mao rather than Stalin could be blamed for balking and refusing to back the invasion.

A leading Chinese expert of Soviet diplomatic history, Shen Zhihua draws a somewhat similar conclusion about Stalin's motives. He argues that Stalin in early 1950 did not trust that Mao would play the role of the vanguard (*xianfeng*) in

---

[87] Ibid., 8–10.

[88] Vladimir Zubok and Constantine Pleshakov, *Inside the Kremlin's Cold War: From Stalin to Khrushchev* (Cambridge: Harvard University Press, 1996), 63.

[89] Weathersby, "New Findings on the Korean War," 14.

allying with the Soviet Union and countering the United States. Moreover, Stalin feared being dragged into a war with the United States by any ally, especially over an issue in Asia, a theater of secondary importance to Moscow. Pushing for armed unification of Korea before he allowed China to take on Taiwan and doing so through conditional approval served both of Stalin's purposes: avoiding World War III between the United States and the Soviet Union and assuring China's allegiance and second-ranking status in the movement that he led. According to Shen, if the Chinese recovery of Taiwan were to have preceded the unification of Korea, Stalin would have had two worries. First, as a sea-and-air battle, where the Chinese were weak and the United States was strong, a fight over Taiwan was more likely to drag the Soviet Union into a war with the United States than a fight over Korea, in which Stalin could hope to pass the buck to the Koreans and Chinese fighting on land by merely supplying weapons. Second, China was not yet considered a trustworthy ally by Stalin in Northeast Asia, and if it were able to solve the Taiwan problem in the near term—before Korean unification—China would be more likely to go its own way, refusing to play the role of the Soviets' main ally in Northeast Asia and potentially even becoming a threat to Soviet interests in East Asia. So, according to Shen, the conditional approval of a Korean invasion in May fit Stalin's strategy perfectly. Such a formula allowed Stalin to avoid direct U.S.-USSR conflict, pushing China into an internationalist position, distracting it from Taiwan and thereby preserving China's dependence on the Soviets for assistance on that issue. A Korean conflict in the near term solidified the Sino-Soviet alliance, and thereby better guaranteed Chinese loyalty to the Soviet-led movement. As Shen puts it, the outbreak of the Korean War was the first "great test" (*da kaoyan*) of the Sino-Soviet alliance, and the process leading up to it exemplified Stalin's prewar lack of trust in Mao.[90]

Shen's account is consistent with other studies of Stalin's thinking in early 1950 by Russian diplomatic historians. In the first half of January Stalin was allegedly quite concerned that the United States would settle its account with Beijing by abandoning Taiwan and thereby undercut the budding Sino-Soviet alliance. In fact, the speeches by Truman and Acheson were hardly all good news for Stalin on this score. By suggesting the prospect of peace between the United States and his potentially Titoist Chinese comrades even before a Sino-Soviet alliance was signed, Stalin reportedly worried that China might not stay on board with him.[91] Mao's own later recollections of Stalin's attitudes at the time confirm this analysis. Mao said that Stalin did not fully trust him to be a real revolutionary until after China's massive intervention in the Korean War later in the year.[92]

[90] Shen, *Mao Zedong, Si Dalin yu Chao Zhan*, 209–21.

[91] Sergei N. Goncharov, "Stalin's Dialogue with Mao Zedong," *Journal of Northeast Asian Studies* (winter 1991–92): 45–76; Goncharov et al., *Uncertain Partners*, ch. 4.

[92] See "Guanyu Guoji Xingshi de Jianghua Tigang" [Outline of a talk on the International

Mao's Alliance Dilemma: To Support or Veto?

Having gained Stalin's conditional approval in April, Kim then traveled to Beijing in mid-May to seek Mao's approval. Mao had apparently not been informed of any of the deliberations between Stalin and Kim, and the Kim proposal for an invasion came as a surprise. According to a document in the Soviet archives, in preparing for Kim's trip to Beijing in May, Mao had told the North Korean Ambassador to China that Kim should come secretly if he was planning to attack the South in the relatively near future. Mao told the DPRK representative that, in general, he did not believe that Korea could be unified peacefully and he thought it impossible that the United States would "launch World War III over such a small piece of territory [as South Korea]."[93] It is important to treat this document's contents with some suspicion, especially the report of Mao's casual confidence about nonintervention by the United States. Kim was hardly beyond embellishing these points for Stalin's ears at a time when the latter was still somewhat concerned about the threat of war with the United States and quite insistent that Kim receive the approval of Mao before he move forward. The portrayal of Mao that Kim's ambassador gave to Stalin fit Kim's plan perfectly: Mao agrees that we must use force to achieve our goals; he believes it will be absolutely safe to do so; and we have Mao's full backing. This was likely all part of Kim's quite successful plan to be the tail that wagged the communist alliance's dog.

Along the same lines, judging from the Russian documentation, when Kim arrived in Beijing he strategically avoided mentioning to Mao the conditionality of Stalin's earlier approval of Kim's plan. He apparently did not inform Mao that Stalin would not back the plan unless Mao did as well, instead telling the CCP leadership that Stalin had agreed to Kim's plan but simply wanted Kim to discuss the issue with Mao in person. Appropriately distrusting his Korean quasi-ally, Mao urgently sought clarification from Moscow. On May 13, 1950, Zhou Enlai visited Soviet Ambassador to China N. V. Roschin and asked him to cable Moscow immediately for clarification. The description of what Kim had told Mao and Zhou is telling and important, as is Stalin's reply. On May 13, Roschin wrote, "This evening Comrade Mao Zedong met with their [DPRK leaders'] entourage. In discussions with Comrade Mao Zedong, the Korean Comrades notified [him] of the following directive from Filipov [Stalin]: at present the situation is different than it was in the past, North Korea may begin to take action;

Situation, December 1959], vol. 8 in *Jianguo Yilai Mao Zedong Wengao* [The Manuscripts of Mao Zedong since the Founding of the Nation] (Beijing: Central Documents Publishing Company), 599–603.

[93] "Shitekefu Guanyu Jin Richeng Fang Hua Jihua Zhi Weixinsiji Dian" [Shtykov's Telgram to Vyshinsky Regarding Kim Il-sung's Planned Trip to China], May 12, 1950, in Shen, ed., *Chaoxian Zhanzheng*, 1:381.

but, this issue must be discussed with Chinese Comrades and Comrade Mao Zedong himself."[94] The following was Stalin's reply the next day: "In meetings with Korean comrades, Filipov [Stalin] and his friends raised [the following], in light of [the fact] that the international situation has already changed, they agree with the Koreans' proposal to realize unification. At the same time they added a point, this question must be decided in the end by the Chinese and Korean comrades in common. If Chinese Comrades do not agree, then we have to reconsider how to resolve this question. The details of the discussion can be related to you by the Korean Comrades."[95] Stalin's reply underscored that he was offering much more decision-making authority to Mao than Kim had initially reported to the Chinese leader.

In what is a major breakthrough in PRC scholarship, mainland historians now recognize in openly published sources some key facts never before discussed in public in the PRC: the North started the war; Mao and Kim met to discuss the invasion in advance; and Mao and Stalin exchanged telegrams that clearly placed the last chance at stopping the war on Mao's shoulders by giving him veto power over Kim's final decision. The same histories tend to insist, however, that Mao had very little or no room for maneuver in this situation, pinning the blame for the war on Stalin and Kim.[96] Weathersby agrees, stating that "Mao Zedong had little room to voice objections to the fait accompli presented by the Koreans."[97] But, however difficult the situation might have been for Mao, Stalin offered him veto power over a decision that would be momentous in the history of Korea, the PRC, and the Cold War more generally. Mao decided to back Kim's plan on May 15, 1950, in what was arguably the most important date in PRC diplomatic history. What is still not said (or not allowed to be said) in China is that Mao made a terrible blunder in deciding to approve the invasion, thereby starting a war that ended up where it started (at the 38th parallel) and that cost his country the long-term separation of Taiwan, hundreds of thousands of casualties, increased PRC dependence on Soviet aid, and decades of exaggerated hostility with the United States.[98]

---

[94] "Weishen Guanyi Jin Richeng yu Mao Zedong Huitan Qingkuang de Dianbao" [Roschin's Telegram Regarding the Situation (surrounding) Kim Il-sung's Talks with Mao Zedong], May 13, 1950, in Shen, ed., *Chaoxian Zhanzheng*, 1:383; and Shen, *Mao Zedong, Si Dalin yu Chao Zhan*, 219.

[95] "Shi Dalin Guanyu Tongyi Chaoxian Tongzhi Jianyi Zhi Mao Zedong Dian" [Stalin's Telegram to Mao Zedong Regarding His Agreement with the Korean Comrades' Proposal], May 14, 1950, in Shen, ed., *Chaoxian Zhanzheng*, 1:384.

[96] Xu, *Mao Zedong yu KangMei YuanChao Zhanzheng*, 45–46; Song, *Kang Mei Yuanchao Zai Huishou*, 43. Shen Zhihua calls the Soviet position a "fait accompli" (*jicheng xianshi*) in *Mao Zedong, Si Dalin yu Chao Zhan*, 208.

[97] Weathersby, "Should We Fear This?" 13.

[98] In a meeting in 2003 in China, one diplomatic historian confided in me that this view is not uncommon in the PRC scholarly community, but there is simply insufficient academic freedom to report that Mao miscalculated the costs and benefits of Kim's invasion of South Korea. In particular he said that Mao and Stalin both underestimated the likelihood of a U.S. intervention.

In my opinion Mao deserves a much larger share of the responsibility for the war than Weathersby and the PRC scholars allow. That said, when one considers the nature of the communist alliance system in East Asia and Mao's reliance on the Sino-Soviet alliance for support to complete his own national goals, Mao was indeed in a tough spot. The PRC was a new ally of the Soviet Union, and, as discussed earlier, Stalin was suspicious of the CCP's revolutionary credentials. Mao had reason to know this. In January 1950 Stalin apparently informed Mao of internal reports by Stalin's main China advisor, I. F. Kovalev. Kovalev had fed Stalin's suspicions about Mao's loyalty to the international movement under Stalin's leadership and hinted that Beijing was eager for a separate peace with the United States.[99] As previously discussed, while in Moscow in January 1950, Mao had made several orders to his colleagues in Beijing that seemed consistent with a desire to impress Stalin with his internationalist fervor. We saw a similar behavioral pattern in spring 1949. In May of that year, after Mao's representative Huang Hua had been in contact with the U.S. ambassador, Leighton Stuart, Mao made a point of bringing Kovalev into a CCP Politburo meeting at which he would express his party's lack of interest in diplomatic relations with the United States. He did this despite Stalin's own April stamp of approval of eventual CCP diplomatic relations with the United States and the CCP's continued outreach to Stuart through late June 1949.[100] The historians Sergey Radchenko and David Wolff conclude that "Mao clearly played a double game, bringing Stalin's attention to the apparent U.S. willingness to recognize his regime, all the while claiming lack of interest in such an outcome to prove his revolutionary credentials to the Soviet leader."[101] If their analysis is correct, it seems quite plausible that a similar logic of impressing Stalin might have contributed to Mao's decision to back Kim's invasion in May 1950.

[99] Goncharov et al., *Uncertain Partners*, 97; and Goncharov, "Stalin's Dialogue with Mao Zedong," 45–76.

[100] "Cable, Kovalev to Filippov [Stalin], April 13, 1949," in *Cold War International History Project Bulletin (CWIHPB)* no. 16 (fall 2007/winter 2008): 158–61. For Stalin's reply, which calls on the CCP to accept diplomatic recognition from the United States as long as it forgoes future assistance of the KMT (the position adopted by Huang Hua in subsequent meeting), see "Cable, Filippov [Stalin] to Mao [via Kovalev], April 19, 1949," in *CWIHPB* no. 16 (fall 2007/winter 2008): 161. For Stalin's general approach to recognition in spring 1949, see Xia, *Negotiating with the Enemy*, 22–23. For a review of the "lost chance" arguments about U.S.-PRC relations in this period, see Warren I. Cohen, et al. "Symposium: Rethinking the Lost Chance in China," *Diplomatic History* 21 (winter 1997): 71–116. For a classic argument dismissing the idea of a "lost chance," see Steven M. Goldstein, "Sino-American Relations, 1948–50: Lost Chance or No Chance," in Yuan Ming and Harry Harding, eds., *Sino-American Relations 1945–55: A Joint Assessment of a Critical Decade* (Wilmington, Del: Scholarly Resources, 1989).

[101] Sergey Radchenko and David Wolff, "To the Summit by Proxy Summits: New Evidence from Soviet and Chinese Archives on Mao's Long March to Moscow, 1949," in *Cold War International History Project Bulletin (CWIHPB)* no. 16 (fall 2007/winter 2008): 110. Along the same lines, Mao had been very critical of Tito with the Soviets. See "Memorandum of Conversation between Anastas Mikoyan and Mao Zedong, February 3, 1949," in *CWIHPB* no. 16, (fall 2007/winter 2008): 142.

Mao could not afford to be seen as weak on revolution and anything but a seconding of Stalin's conditional approval of Kim's invasion plan would have made him appear so. Mao had also recently been on the receiving end of advice to put off national unification and to be cautious (in April 1949 at the Yangzi), and he did not want to be viewed internationally as short on internationalism or international fervor. He required not only economic and military assistance and training from the Soviet Union for his new country, but he had set for himself a difficult unification challenge of his own—taking Taiwan by force—which would require an extensive amount of help, particularly in building a navy and air force to fight across the Taiwan Strait, a process that the CCP had started in earnest in spring 1950 with Soviet help. Moscow was the only place to turn for such assistance at that time.[102] Moreover, Stalin had been decidedly reluctant to support the concept of a PRC assault on Taiwan in 1949–50.[103]

Available evidence suggests that Mao was more nervous than Stalin about the prospect that an invasion of South Korea could go poorly for Kim Il-sung's forces. It is unclear why Mao would raise these points of concern in discussions with Kim and his entourage if Mao had planned to go ahead with Kim's plans regardless of the perceived risks. In other words, Mao was hardly simply following orders from Moscow but was making his own calculation about the merits of Kim's plan.

There is little doubt that there was a systematic underestimation of U.S. resolve and commitment to South Korea in all three communist capitals. No factor mattered more than this one in the disaster. But Mao did not share Kim's drunken optimism about the war, at least not to the same degree. Instead, Mao apparently told Kim that the DPRK needed to prepare carefully. Consistent with Mao's inflated concerns about Japanese military activism, Mao warned that a major Japanese contingent of tens of thousands of Japanese soldiers might enter the war over time.[104] Mao's reported estimation for Korea was consistent with

---

[102] Shen, *Mao Zedong, Si Dalin yu Chao Zhan*, . 220; Goncharov et al., *Uncertain Partners*, 79.

[103] Radchenko and Wolff, "To the Summit by Proxy Summits," 111–12.

[104] Xu, *Mao Zedong yu KangMei YuanChao Zhanzheng*, 46, 51. In a book that seeks to bolster Mao's reputation for strategic acuity, Xu Yan admits that while CCP leaders had not totally excluded the possibility that the United States would reenter Asian mainland wars, they were "somewhat surprised" (*duoshao gandao turan*) by Truman's June 27, 1950, announcement of intervention in Korea and Taiwan as they had thought that the likelihood of direct U.S. intervention was decreasing with the rising power of the "revolution." Also see Song, *KangMei YuanChao Zai Huishou*, 43. Moreover in early October, after the initial invasion failed and UN forces were turning the tide, Mao criticized Kim to Stalin for not listening to Mao's warnings in spring about the danger of "reactionary" militaries entering Korea. The use of "reactionary" instead of American, suggests that Mao knew better than to claim to Stalin that he feared quick, and large-scale U.S. intervention in spring 1950. Mao, instead, seemed to overestimate the potential role of Japan and underestimate the role of the United States, a strange outcome of perceived lack of resolve or strength in the United States and perceived military activism in Japan following the reverse course and the U.S. decision to strengthen Japan. See Xu, *Mao Zedong yu KangMei YuanChao Zhanzheng*, 46.

high-level military analyses in December 1949 of a future conflict across the Taiwan Strait: Chinese General Su Yu believed that the United States would stay out but that it might mobilize Japanese forces to help Chiang Kai-shek in a cross-Strait conflict.[105]

In discussions with Mao, Kim called such intervention by the United States and/or Japan "improbable" but acknowledged that the United States might send a few Japanese divisions, which he was sure his forces could defeat. Mao replied that if Japan could extend the war, then the United States could later enter the war directly. Kim cleverly manipulated the lack of coordination between Beijing and Moscow by dismissing Mao's fears about the United States with reference to Stalin's own calculation that the United States would not enter into Korea because the United States had abandoned China! Kim then duplicitously told the Soviet ambassador that Mao believed Japan could not interfere and, if the United States were to interfere, Mao believed Chinese, not Soviet, troops should bear the burden![106] Mao indeed offered to prepare a Chinese military response to the prospect of a Japanese intervention in Korea followed by a U.S. intervention, but Kim, who apparently never informed him of the timing of the invasion in any case, apparently believed tight coordination with Beijing over Chinese intervention would not prove necessary.

Mao's blunder in spring 1950 occurred in large part because he was manipulated by both his weaker Korean ally and his stronger Soviet one. It also occurred in large part because Mao and Stalin misread the alliance dynamics of the enemy camp: underestimating the likelihood of robust, direct U.S. intervention and overestimating Washington's reliance on the much less formidable Japanese. Many factors were at work in producing the outbreak of the Korean War. Among the most important were the unclear signals sent by U.S. security relationships with Korea, Taiwan, and Japan. In addition political dynamics within the budding communist alliance allowed Kim Il-sung to manipulate his more powerful allies into exploiting the perceived opportunities provided by weakness in the anticommunist camp, so they could attempt to revise the status quo by force before that camp could solidify its position on the Asian mainland.

[105] See Goncharov et al. *Uncertain Partners*, 320, fn. 133. Internal strategic assessments in China from January 1950 match the Soviet archival claims that Mao worried about divisions of Japanese forces entering the fight in Korea.

[106] Weathersby, "Should We Fear This?" 12–13.

Chapter 3

ALLIANCE PROBLEMS, SIGNALING,
AND ESCALATION OF ASIAN CONFLICT

There were two intrawar deterrence failures in Korea in late summer and fall 1950, and both were related to alliance politics. Lack of coordination and mistrust in the communist camp rendered the alliance incapable of sending clear and timely signals of resolve to the United States that might have deterred the United States from crossing the 38th parallel in the crucial three weeks following MacArthur's successful Inchon landing. Once U.S. forces entered North Korea, escalation was assured in large part because previous U.S. actions in relation to its own regional security partners and future allies in recent months had made unification of the Korean peninsula by UN forces under U.S. leadership seem intolerably threatening to the long-term security of the PRC and, to a lesser degree, the Soviet Union.

The continuing lack of coordination in the communist camp made the communists militarily less effective in fall 1950, to the great benefit of the United States and other UN partners on the battlefield. But this same lack of coordination caused the communists to send weak signals of resolve regarding their willingness to protect North Korea from a full-scale invasion designed to unify the peninsula, thus undercutting communist efforts at coercive diplomacy in September and early October. What was good for the United States in terms of war-fighting in the near term was not good in terms of coercive diplomacy and was not in the long-term interest of the United States.

U.S. policies in the early weeks of the Korean War had a powerful impact on strategic thinking in Moscow and Beijing about the long-term implications of the military defeat of the North Korean communist regime and the unification of the Korean peninsula under a government friendly to the United States. U.S. policies toward Taiwan and Japan helped render such a prospect intolerable to Mao and, to a lesser degree, Stalin, and eventually undercut efforts to deter large-scale Chinese entrance into the war.

On a different front, varying national security interests and ideological differences between the Soviets and Chinese communists also help explain why the Vietnamese communists received so much external support from the latter from 1950–53. The Vietnamese were able to garner external support from China despite the distractions in Korea and the continuing relative indifference of the Soviet Union to peasant rebels in distant Southeast Asia.

ALLIANCES AND THE COMMUNIST DETERRENCE FAILURE OF FALL 1950

Why the U.S. Crossing of the 38th Parallel Was Hard to Deter

After the United States entered the war in late June 1950 and broke out of the Pusan perimeter in mid-September 1950 with MacArthur's brilliantly executed Inchon landing, President Truman and his advisors had a difficult decision: should the initial war aims in Korea to restore the Republic of Korea in the South expand to include the destruction of the aggressive North Korean regime and unification of the peninsula under a friendly, anticommunist government? In Korea itself, various factors contributed to the Truman administration's late September decision to expand the war effort and to attempt to unify the penin-sula by force, sending UN ground forces into North Korea on October 7, 1950. Washington wanted to teach a punitive lesson to the communists about the price of aggression. Less abstractly, U.S. leaders were tempted by the opportunity to eliminate the threat from the North once and for all after North Korean forces were cut in two and left in disarray by MacArthur's effective amphibious of-fensive.[1] Eliminating North Korean forces before they could reconstitute them-selves, they believed, would reduce the long-term burden of guaranteeing South Korean security on U.S. grand strategy. Soviet documents show that the North Koreans indeed planned to reorganize for future offensives if and when they were able, but that they were quite desperate by late September in the face of MacArthur's onslaught.[2] Moreover, for reasons I will discuss in further detail, U.S. elites in Washington and Tokyo did not believe that the Soviets and Chinese would or could effectively intervene to protect the wounded Northern regime, especially when they had failed to do so when U.S. forces were pinned down in southernmost Korea in summer 1950 (the Pusan Perimeter) or immediately upon news of the successful Inchon landing in September.

There was one additional fundamental factor that has received little atten-tion. The "strongpoint defense strategy" had fostered dangerously low levels of U.S. support for South Korean security prior to the war and, thereby, undercut deterrence of the initial communist invasion of the South. But there was a geo-strategic reason for Washington's relative indifference to South Korea. U.S. re-sources were more needed in other parts of the world that held greater strategic

---

[1] Stueck, *The Road to Confrontation*, 228–31; Rosemary Foot, *The Wrong War: American Policy and the Dimensions of the Korean Conflict, 1950–1953* (Ithaca: Cornell University Press, 1985), 69–70; John Lewis Gaddis, *The Long Peace: Inquiries into the History of the Cold War* (New York: Oxford University Press, 1987), 98–99; and Burton I. Kaufman, *The Korean War: Challenges in Crisis, Credibility and Command* (New York: Alfred A. Knopf, 1986), ch. 3.

[2] See "Shitekefu Guanyu Jin Richeng de Huitan Qingkuang Zhi Geweimike Dian," [Shtykov's Telegram to Gromyko Regarding a Meeting with Kim Il-sung], September 30, 1950, in Shen, ed., *Chaoxian Zhanzheng*, 2:561–62.

value. If, after intervening in the Korean War, the United States had stopped at the 38th parallel and allowed North Korean forces to reconstitute themselves north of the parallel, then the United States would have inherited exactly the posture that Acheson had feared when he excluded South Korea in his Press Club speech: costly standing U.S. defenses in an area considered of only secondary strategic importance to the United States in its global struggle with the Soviet Union. In his own recollections of his thinking on this issue in early July 1950, Dean Acheson describes the context of a letter he wrote regarding Korean unification to Policy Planning Staff Director Paul Nitze, who, along with George Kennan, was among the very few influential voices in Washington opposed to crossing of the 38th parallel in Korea:[3]

> Within the United States Government discussion of this vital policy question had been going on since the beginning of the attack. On June 29, before commitment of ground forces, I had said to the Newspaper Guild that our action in response to the UN resolutions was "solely for the purpose of restoring the Republic of Korea to its status prior to the invasion from the north and of reestablishing the peace broken by that aggression," and on July 10[,] I had written to Paul Nitze that in the immediate future "we have got to put in a force necessary to reoccupy to the 38th," subject to new problems that Russian or Chinese intervention would raise. In the longer run, if we should succeed in reoccupying the South, the question of garrisoning and supporting it would arise. This would be a hard task for us to take on, and yet it hardly seemed sensible to repel an attack and then abandon the country. I could not see the end of it. In other words, as the Virginians say, "we have bought a colt."[4]

Washington had not decided whether to make a permanent commitment to South Korean security and still feared being pulled into future wars by its Korean security partners if it did so. Rhee might attack the North, thereby potentially entrapping the United States in a wider war when it was even less prepared for one.[5] So a decision to stay south of the 38th parallel and to help provide for a strong, long-term defense against a reconstituted North Korean military was a difficult choice for Washington, but it was one that would almost certainly have prevented escalation of the war into a Sino-American conflict later in the year.[6]

---

[3] Nitze surprised many when he reported his opposition to expansion of the Korean War in his memoirs. He is generally associated with NSC-68 and the coining of the phrase "rollback" in that document. See Paul H. Nitze, *From Hiroshima to Glasnost: At the Center of Decision* (New York: Grove Weidenfeld, 1989), 106–8.

[4] Acheson, *Present at the Creation*, 450–51.

[5] In fact, on July 13, Rhee stated that his own forces would not stop at the parallel once they broke out of their positions in southernmost Korea. Acheson himself noted this problem in his recollections of the U.S. alliance dilemma. See Acheson, *Present at the Creation*, 451; also see Cha, "Powerplay."

[6] On the importance to escalation of the war of the U.S. crossing of the 38th parallel as opposed either to intervention in the South Korea or failure to create a smaller buffer in North Korea, see

Budgetary concerns also raised the perceived opportunity costs of simply defending against future communist ground attack in Korea. The initial outbreak of the Korean War mobilized the United States for the Cold War only in a limited way, with an increase in defense spending from $15 billion to $22 billion. The full NSC-68 complement of a $50 billion U.S. defense budget would be passed only after the United States crossed the 38th parallel and met stiff Chinese resistance later in the year.[7] Since Europe and other strategic "strong-points" remained more important to the United States than Korea even after that mobilization of new funds, an outcome short of total destruction of the Korean threat would leave the United States with a series of bad choices: leaving South Korea vulnerable to a replay of the Korean War; shoring it up with U.S. forces at great opportunity costs to U.S. grand strategy; or providing significant assistance to Syngman Rhee to defend his own country and hoping that he would behave moderately with that military wherewithal.

Why Deterrence Was Still Possible

Despite the success at Inchon and the other factors supporting the U.S. decision to invade North Korea, MacArthur's orders in late September to go north were contingent on the estimation that large-scale Chinese or Soviet intervention was unlikely.[8] These orders, together with more recently declassified intelligence reports, which will be analyzed in this section, strongly suggest that the United States could have been deterred from crossing the 38th parallel if a combination of earlier and more credible signals had been sent by Moscow and Beijing about the dangers of escalation if the United States chose to go beyond restoring the status quo ante in Korea.

One direct problem in communist efforts to deter the United States was the lack of direct contacts between Washington and Beijing, a circumstance that was partially a result of communist alliance dynamics described in chapter 2,

---

Christensen, *Useful Adversaries*, ch. 5. In that chapter I address the alternative hypotheses of Allen Whiting, who in his pioneering account of the war emphasized the importance to escalation of the breaching of a buffer zone between the neck of North Korea and the China border at the Yalu River, and of Chen Jian, whose excellent account of the war argues that it would have escalated to a Sino-American conflict even if U.S. forces had remained south of the 38th. For these two alternative theses, see Allen S. Whiting, *China Crosses the Yalu* (Stanford: Stanford University Press, 1960); and Chen Jian, *China's Road to the Korean War* (New York: Columbia University Press, 1996).

[7] Christensen, *Useful Adversaries*, ch. 5.

[8] Kaufman, *The Korean War* 85–86. Also see Stueck, *The Korean War*, 89–90. On September 8, 1950, before Inchon, the National Security Council had already viewed favorably the prospect of crossing the 38th parallel, "provided MacArthur's plans could be carried out without the risk of major war with the Chinese Communists or the Soviet Union." Meeting of the National Security Council, September 8, 1950, President's Secretary's Files, Harry S. Truman Library (HSTL), Independence, Missouri.

and of the U.S. refusal to abandon Chiang Kai-shek promptly and completely in 1949–50. The lack of direct contacts and personal relationships among leaders complicated coercive diplomacy in Korea, as Zhou Enlai passed a deterrent warning through the Indian ambassador to Beijing, K. M. Panikkar, an interlocutor distrusted in the United States and a channel that greatly reduced the authority of the CCP message.[9] This was particularly true because Panikkar himself, just two weeks earlier, had assured the United States that China's entrance into the war was "beyond the range of possibility" unless it was in the context of World War III.[10]

Though this factor was indeed important, there was more at work here than unsteady communications from a distrusted source. Differences in national interests and military strategies among the allies, combined with generally poor military and diplomatic coordination in the communist camp, prevented the communists from organizing early, defensive deployments by Chinese communist forces in North Korea that could have lent credibility to the Chinese threat of escalation. Although enemy lack of coordination initially assisted the United States on the battlefield, it led to poor policy decisions in Washington that extended a war unnecessarily for over two years.

U.S. misperceptions about the cohesion of the communist alliance were critically important to the failure of coercive diplomacy and the escalation of the war in fall 1950. If the Chinese had raced across the border with a large but still limited force before the United States had crossed the 38th parallel, the likelihood of a U.S. advance across the 38th would have been significantly reduced. Those misperceptions help explain why the United States underestimated the likelihood of initial Chinese intervention in October 1950 as American forces prepared to cross the 38th parallel. At that time, American intelligence knew that, in the preceding months, the Chinese had amassed hundreds of thousands of troops in Manchuria. The Americans generally did not believe that these forces had been sent there to enter Korea, however. Rather, they assumed their purpose

---

[9] Christensen, *Useful Adversaries*, ch. 5; Acheson, for instance, states that the Zhou Enlai warning communicated through the Indian ambassador did not seem an authoritative expression of Chinese policy. See *Present at the Creation*, 452. For Truman's recollections regarding distrust of Panikkar, see Harry S. Truman, *Memoirs, Vol. 2, Years of Trial and Hope* (Garden City, N.Y.: Doubleday, 1956), 362. For contemporary documentation of U.S. government official's disdain for Panikkar, see McConaughy to Rusk and Jessup, October 12, 1950, Office of Chinese Affairs, U.S. Department of State, film COO12, reel 15, National Archives. A recently declassified National Security Agency retrospective also points out that Zhou's warnings were also downplayed by some intelligence agency experts, some of whom also downplayed the importance of signs that PLA troops were amassing in Northeast China for potential entrance into Korea. See Guy R. Vanderpool, "COMINT and the PRC Intervention in the Korean War," DOCID: 340650, in National Security Archive, George Washington University, at http://www.gwu.edu/~nsarchiv/NSAEBB/NSAEBB278/index.htm, Doc 21, p. 15. I am grateful to Nancy Tucker for calling this document to my attention.

[10] Stueck, *The Road to Confrontation*, 229.

was to protect China from a U.S. ground invasion, something no one in Washington was considering at the time. Working under the assumption of a highly coordinated communist movement, some U.S. analysts concluded that, since Chinese forces did not enter the Korean War when U.S. forces were at their most vulnerable (in summer 1950 before Inchon), China was likely unwilling to risk war over Korea. After Inchon and before the U.S. and other UN forces crossed the 38th parallel, some previously quite nervous U.S. intelligence analysts similarly noted that Chinese and Soviet troops did not rush in to reverse what was clearly a shift in the balance of power in favor of the UN forces. This increased the intelligence community's confidence that China was unlikely to intervene massively in Korea even after U.S. forces had crossed the 38th parallel.[11] Since the alliance was being treated as relatively cohesive and monolithic, the lack of an earlier, coordinated response by the Chinese and Soviets signaled a lack of resolve in the movement overall. A declassified National Security Agency retrospective states that when the PRC did not react immediately after U.S. forces landed at Inchon and drove DPRK forces north of the 38th, the prospect of a large-scale Chinese intervention seemed greatly diminished, even to intelligence analysts who had previously been quite concerned about the implications of the Chinese military buildup in Northeast China during the summer. Many of those analysts began making efforts to explain away previous intelligence about those buildups and diplomatic warnings from China through third-country diplomats about the implications of UN forces crossing the 38th parallel.[12] Once the United States had dismissed the warnings from Beijing and UN forces had crossed the parallel, it was too late to stop escalation of the war.

If Chinese forces had entered North Korea in large numbers before the U.S. crossing of the 38th parallel on October 8, 1950, U.S. intelligence would likely have detected this, especially if Chinese troops moved far south of the Yalu border toward the 38th parallel. Under those circumstances, leaders in Washington would have known that an escalation of the Korean War into a significant Sino-American conflict would have been set off by U.S. forces crossing the 38th. This situation would have lent credibility to Chinese deterrent threats in early October to fight U.S. forces if they crossed the 38th. As historian William Stueck describes the view in Washington in September and October of that year, "[I]f either of the top Communist powers were going to intervene directly in

---

[11] On these points, see Christopher P. Twomey, "The Military Lens," Ph.D. diss., MIT, Department of Political Science, October 2004, chs. 3–4; and Stueck, *The Korean War*, 90. See Guy R. Vanderpool, "COMINT and the PRC Intervention in the Korean War," 15. For an account of how perceptions and politics led U.S. officials to treat the alliance as relatively monolithic at this time, see Selverstone, *Constructing the Monolith*, esp. ch. 8.

[12] See Guy R. Vanderpool, "COMINT and the PRC Intervention in the Korean War," 15.

North Korea, they would do so well before the United Nations forces reached the thirty-eighth parallel. ... Thus, when no foreign soldiers entered North Korea immediately following the Inchon landing—and the Soviets assumed a relatively conciliatory posture in New York—the way seemed clear for a United Nations military offensive to unify the peninsula."[13]

At their mid-October meeting at Wake Island, President Truman and Mac-Arthur discussed the prospect of large-scale Chinese entrance into the war if the general were to move his forces, which had already crossed the parallel, to the northernmost sections of North Korea. The president was apparently very concerned about such a large-scale escalation and wanted assurances from the general and from his intelligence agencies that such a development was unlikely to happen. As in early October, the scenario was again dismissed based on the lack of any militarily meaningful reaction by the Chinese to earlier U.S. deployments and actions in South Korea. While the intelligence community did consider seriously the possibility of Chinese entrance into the war in late October (just as Chinese units engaged U.S. forces in northernmost Korea), they still falsely imagined such prospective engagements would be part of a limited defensive strategy to maintain a friendly buffer between the Chinese border and UN forces in North Korea, not as part of a large-scale escalation of war.[14]

Even after UN forces were engaged by Chinese forces in late October and early November, the same exaggeration of communist alliance cohesion in the early phases of the war allowed CIA analysts to comfort themselves that Chinese military goals in Korea were likely limited. They surmised that the Chinese and Soviets were probably dedicated to saving some North Korean rump state and creating a deadlock with UN forces in northernmost Korea, as opposed to preparing to drive the Americans entirely off the peninsula. A recently declassified National Intelligence Estimate from November 6, 1950, is worth quoting at length in this regard:

> It is significant that the Chinese Communists refrained from committing troops at two earlier critical phases of the Korean war, namely when the UN held no more than a precarious toehold in the Pusan perimeter and later when the UN landings were made at Inchon. The failure to act on those occasions appears to indicate that Peiping was unwilling to accept a serious risk of war, prior to the U.S. crossing of the 38th parallel. Since the crossing of the Parallel, Chinese Communist propaganda has increasingly identified the Peiping cause with the cause of the North Koreans. The immediate objective of the Chinese Communists' intervention in Korea appears to have

---

[13] Stueck, *The Road to Confrontation*, 228–29. On a similar point, see Stueck, *The Korean War*, 90–91.

[14] Vanderpool, "COMINT and the PRC Intervention in the Korean War," 15–16.

been to halt the advance of UN forces. Chinese Communist military operations to date, including the nature of the forces employed, suggest an interim military operation with limited objectives.[15]

If we were to accept the CIA analysts' false assumptions regarding communist alliance cohesion, this analysis would make perfect sense. A China with more expansive military goals in a highly cohesive and hierarchically ordered alliance with the Soviets and North Koreans would have entered the war earlier.[16] If the analysts' assumptions about communist alliance cohesion had been true, then the PRC's military and political goals in Korea in late October and early November 1950 must, therefore, have been limited to preventing the quick and total annihilation of the North Korean military and government by UN forces. As we now know, this was very far from the case.

Even if one rejects the counterfactual reasoning that an earlier, large-scale and defensive Chinese deployment in North Korea would have precluded a U.S. crossing of the 38th parallel, we can still argue that escalation would have been more manageable from a U.S. perspective had the communist camp been more organized and had projected power earlier and more confidently into North Korea before the United States had crossed the parallel. Even if U.S. intelligence had failed to detect a large-scale Chinese entrance into Korea in late September or early October, U.S. military leaders at least would have known what they were up against earlier during their advance north, and would have been less likely to overextend themselves in northernmost Korea and fall into Mao's trap at the Yalu later in November.

Once Beijing made the eleventh-hour decision to enter the war, the Chinese strategy in Korea was actually more aggressive than it would have been under Stalin's initial plan for Chinese forces to buttress the 38th parallel before the U.S. forces were to cross into North Korea. On October 1, after receiving a desperate appeal from Pyongyang, Stalin requested that Mao send five to six Chinese divisions on an urgent basis to help defend the 38th parallel. Given national strategic interests that diverged somewhat from those of the Soviets and the Soviets' unwillingness or inability to provide prompt air cover for Chinese troops, Mao was reluctant to expose his weak forces to U.S. attack by sending a relatively small number of them as far south as the 38th parallel where they would be overextended and exposed to superior U.S. firepower.[17] After the United States ignored Chi-

[15] National Intelligence Council, "Chinese Communist Intervention in Korea," NIE-2, November 6, 1950, in *Tracking the Dragon: National Intelligence Estimates on China During the Era of Mao, 1949–76* (Washington, D.C.: U.S. Government Printing Office, 2004), 76.

[16] Stanley, *Paths to Peace*, 133.

[17] See "Ciphered Telegram Stalin to Mao Zedong and Zhou Enlai: October 1, 1950," in Alexandre Mansourov, "Stalin, Mao, Kim, and China's Decision to Enter the Korean War, September 16–October 15, 1950," *Cold War International History Project Bulletin* (*CWIHPB*) nos. 6–7 (winter

nese warnings in early October 1950 and crossed the 38th parallel, the Chinese were committed to war, but they were not committed to a defensive and limited war simply to assist in the reconsolidation of North Korean military north of the 38th, as the Soviets would have had it in early October. In his October 7 communication with Stalin via the Soviet Embassy in Beijing (sent several hours before U.S. forces began crossing the parallel), Mao told Stalin that it would be better if China initially sent nine divisions, rather than five to six, and allowed U.S. forces to come further north where it would be easier for Chinese forces to take them on in combat.[18] This strategy, born of weakness and the fear of protracted war, of course ran directly against the goal of deterring the United States from sending its forces across the 38th parallel in the first place and reduced the likelihood that Sino-American conflict on the peninsula would remain limited after escalation. Despite the number of forces initially sent to Korea and the desire to fight a decisive war, rather than an attritional one, Mao's initial plan in North Korea involved building a defensive line in northernmost Korea and building up forces for a massive offensive against the Americans some six months later, in spring 1951, by which time the Chinese would have received weapons and training, particularly fighter planes and fighter training, from the Soviets.[19]

Although it seems unlikely, this planned respite between the fall and the spring offensives might have provided new opportunities for coercive diplomacy on both sides, as Allen Whiting originally suggested in his path-breaking book.[20]

1995–96), Appendix document no. 10. For the Chinese-language translation of this Russian-language document, see Shen, ed., *Chaoxian Zhanzheng*, 2:571. For analysis of Stalin's strategy, see Mansourov, "Stalin, Mao, Kim," 15; and Xu, *Mao Zedong yu KangMei YuanChao Zhanzheng*, 82.

[18] See "Weishen Guanyu Mao Zedong Dui Chubing de Taidu Wenti Zhi Shi Dalin Dian" [Roschin's Telegram to Stalin Regarding the Issue of Mao Zedong's Attitude Regarding Dispatching Troops], October 7, 1950, in Shen, ed., *Chaoxian Zhanzheng*, 2: 588–90.

[19] For discussions of Mao's plans for an offensive in spring 1951 if U.S. forces stopped at the narrow neck of Korea rather than driving through to the Yalu as they did in late October and again in late November, see Christensen, *Useful Adversaries*, ch. 5. See also "Telegram to Zhou Enlai Concerning the Principles and Deployments of the People's Volunteer Army as It Enters Korea for Combat Operations" (internally circulated), in *Jianguo Yilai Mao Zedong Wengao*, vol. 1 (October 1949–December 1950), 560–61. I translated this document in its entirety in Appendix B of *Useful Adversaries*. A more recently published internally circulated text on military command seems to clear away any remaining controversy about this issue. It states, "Before the First Campaign of the War to Resist America and Aid Korea, the Military Commission [*Jun Wei*] planned first to allow the People's Volunteers to organize defenses in northernmost Korea, [and] after one half year, then implement the counteroffensive." The text goes on to say that the rapidity and extent of the UN advance in late October made Beijing adjust its strategy to a more mobile and aggressive one. Zhao Yanliang, "Zuo Zhan Yuanze yu Zhihui Yishu" [Principles of Warfighting and the Art of Command] in *Jundui Zhihui Lilun Jijin: Quan Jun Shou Jie Jundui Zhihui Lilun Yantaohui Lunwenji* [Outstanding Examples of Armed Forces Command Theory: Collection of Theses from the First Army-Wide Deliberation Sessions on Armed Forces Command Theory] (Beijing: National Defense University, October 1992) (military circulation only), 542.

[20] Whiting, *China Crosses the Yalu*.

But even if we were to accept that Mao's offensive plans might have changed in the interim if given the opportunity, the tardiness of the PLA forces' arrival in the theater quashed even this dim hope for coercive diplomacy. By the time Chinese forces had entered Korea in large numbers, U.S. forces were moving quickly up the peninsula providing few options for standing defenses. In November Mao decided instead to lure the U.S. forces further north and attack them with a full-scale counteroffensive involving over 260,000 Chinese troops (twelve divisions) later that month.[21]

While counterfactuals can never be proven, this evidence suggests the strong possibility that if the communist alliance's efforts had been coordinated more smoothly and Mao had promptly entered the war following Inchon and before the U.S. crossing of the 38th parallel, as Stalin had envisioned, escalation would have been less likely as there would have been much more concrete evidence of Chinese resolve in fighting for the North. If the United States had failed to note the Chinese entrance or attacked across the 38th in any case, escalation still would have likely been less severe from a U.S. perspective than what UN forces under U.S. leadership suffered in late November 1950: a massive Chinese surprise attack on overstretched and divided UN forces that led to the longest overseas retreat of U.S. forces in American history. In this very real sense, the communist alliance, from the perspective of U.S. security strategy and coercive diplomacy in Korea, was worse than a monolith.

WHY DID THE COMMUNISTS RESPOND SO SLOWLY TO INCHON?

One of the key factors that prevented earlier intervention and a clearer set of signals from the Asian communists in the late summer and early fall of 1950 was a lack of coordination and planning, combined with a fairly high degree of distrust among the three key parties: the North Korean communists, the Soviets, and the CCP. The lack of a timely North Korean request for foreign intervention after Inchon and the lack of advanced coordination between the Chinese and Soviets about the conditions of Soviet support and air cover for Chinese troops entering the war precluded the type of early foreign communist intervention in North Korea that would have signaled Moscow's and/or Beijing's resolve to the Americans before they crossed the 38th parallel.

The Soviets, for their part, had envisioned such an early defensive role for the Chinese communist forces in North Korea, and on October 1, 1950, they

---

[21] For a discussion of Mao's strategy of luring the U.S. deep, see Christensen, *Useful Adversaries*, ch. 5. For the role of divergences in Chinese and U.S. doctrines in deterrence failure in the Korean War, see Twomey, "The Military Lens."

requested that China send such early defensive deployments toward the 38th parallel immediately.[22] This Soviet request to Beijing itself came two full weeks after MacArthur's Inchon landing and only one week before the United States took the fateful step of crossing the 38th parallel. For reasons discussed in detail over the next pages, the North Koreans requested direct Soviet and Chinese entrance in the war via the Soviet Union only in the early hours of October 1. After receiving the North Korean plea for help, Stalin immediately requested Chinese entry into the war (one hour later). Only then did China begin its own lengthy decision process that included an initial expression of reluctance to enter the war; followed by a much more positive response to Stalin on October 7; a renewed period of indecision on October 12–13; and finally a reaffirmation of the earlier positive decision on October 13. This Chinese decision process alone took roughly two weeks, meaning Mao made the final decision for war only on October 13, five days after Beijing's coercive diplomacy had failed (with U.S. and other UN forces crossing the 38th parallel on October 7–8) and five days before those forces would seize Pyongyang. In fact, Chinese troops would begin crossing the Yalu late on the night of October 18–19, the same day that UN forces seized the North Korean capital. Since Beijing made its final deterrent warning to the United States via the Indian ambassador to China at midnight on October 2, those signals of communist weakness could not have come at a worse time for the purposes of successful coercive diplomacy.[23] The weak coordination between Pyongyang and its allies, the lack of sufficiently coordinated pre-Inchon preparation by the Soviets and the Chinese, and the differences between Stalin's defensive strategy for Chinese troops to enter early and defend the 38th, and Mao's offensive strategy of luring the United States deep and then launching a counteroffensive, all combined to undermine successful coercive diplomacy by the communist camp in its efforts to prevent the U.S. from attempting armed unification of the peninsula under UN control.

For the Americans, then, if enemy actions had been more coordinated and more clearly under the control of the main enemy of the United States, Stalin's Soviet Union, in June through October 1950, the calamitous attacks on U.S. forces in northernmost Korea in November 1950 might have been avoided. Of course, the lack of coordination, the stubborn optimism of Kim Il-sung, Stalin's distrust of Mao, and the related delay in Pyongyang's direct request for Chinese

---

[22] For an English translation of this document, see "Ciphered Telegram Stalin to Mao Zedong and Zhou Enlai, October 1, 1950," in Mansourov, "Stalin, Mao, Kim," Appendix Document 10. For a Chinese translation, see "Shi Dalin Guanyu Jianyi Zhongguo Pai Budui Yuanzhu Chaoxian Wenti Zhi Weishen Dian" [Stalin's Telegram to Roschin Regarding the Issue of Suggesting China Send Forces to Assist Korea], October 1, 1950, in Shen, ed., *Chaoxian Zhanzheng*, 2:571.

[23] On this last deterrent effort and the response from the United States, see Christensen, *Useful Adveraries*, ch. 5.

intervention, and so forth, assisted the United States greatly in war-fighting during those months. But to the degree that all participants in the war would have been better off with a ceasefire and a divided Korean peninsula in October 1950 than in July 1953, no one benefited from the failure of communist coercive diplomacy that came along with that communist military weakness. So, in this important sense, the United States suffered because the enemy communist alliance was still disorganized and poorly prepared in September–October 1950. Because the Americans assumed the alliance was more monolithic and well coordinated at the time, they misread a delayed communist response to the Inchon landing as a willingness to accept the defeat of Kim Il-sung's regime by military means without a massive Chinese response.

Asian Communist Alliances and PRC Strategy in Korea

When faced with weakness among anticommunist forces, as he was in early 1950, Mao supported communist expansionism in order to fulfill his internationalist agenda, protect his movement's reputation in the international communist arena, and create larger strategic buffers on his nation's periphery. But when faced with resolute opposition from the United States and others, Mao lashed out reactively, fearing for his nation's security in a way that a more distant Soviet leadership did not. After North Korea invaded the South and the United States responded with military intervention in Korea and the Taiwan Strait, Mao's nationalist and defensive side became more important than his internationalist Leninist motivations. At an August Chinese Communist Party meeting to discuss war preparations, Korea was not just "an issue related to a fraternal country, but viewed as integrally linked with the interest of our nation's northeast (Manchuria)."[24] In early October both Soviet and Chinese leaders would argue for intervention with some reluctant Chinese colleagues in the CCP Politburo by appealing to the national security risks to the PRC, including the costs of an extended Cold War on the Chinese border with Korea and, worse still, the prospect of future attacks on China from Korea and Taiwan, if the United States were able to unify the peninsula under a pro-U.S. regime. Fears of future U.S. attacks on China pushed Mao toward massive intervention in the Korean War.[25]

As early as the first week of July, the Soviets and Chinese were discussing Chinese entrance into the war in anticipation of a possible U.S. crossing of the 38th parallel (which would not occur until October). Stalin seemed impatient for Mao to amass forces on the Sino-Korean border and to coordinate with the

---

[24] August 26, 1950, entry in *Zhou Enlai Nianpu* [Zhou Enlai's Chronicle] (Beijing: Zhongyang Wenxian Chubanshe, May 1997), 1:69.

[25] Christensen, *Useful Adversaries*, ch. 5.

North Korean leadership a plan to save the North Korean military from annihilation if the U.S. intervention in Korea were to prove effective and an invasion of North Korea seemed imminent. The Soviets clearly wanted Chinese forces to help defend against that outcome by helping bolster defenses in the North. For this reason, in the weeks and months following Kim's failed gambit to unify the peninsula under his leadership, the Soviets urged China to prepare between five and nine divisions at the border for rapid assistance to North Korea if and when the United States threatened to cross the 38th parallel.[26] Leaders in both Beijing and Moscow feared that the United States might attempt unification of the peninsula under UN control if insufficient resistance was offered by North Korea and its allies. Beijing in particular had warned the North Koreans about overextension of its forces and overconfidence regarding its operations in the South. Before Inchon, Zhou Enlai had urged the North Koreans to consider the possibility of strategic withdrawal in the event of such a UN counterattack, but Kim Il-sung reportedly replied that "he ha[d] never considered retreat."[27] Zhou's advice followed a pattern of Chinese warnings to North Korea in the weeks prior to the Inchon landings.[28]

[26] See Stalin to Zhou Enlai, July 5, 1950 in Weathersby, "New Russian Documents: on the Korean War: Introduction and Translations," *Cold War International History Project Bulletin* (CWIHPB) 6, no. 7 (1996): 43; "Ciphered Telegram Stalin to Mao Zedong and Zhou Enlai, October 1, 1950," in Mansourov, "Stalin, Mao, Kim," Appendix Document 10. On July 8, 1950, Stalin seemed impatient for China to send representatives to North Korea to coordinate their activities. See "Ciphered Telegram Stalin to Soviet Ambassador Roschin in PRC transmitting message to Mao Zedong," July 8, 1950 in Weathersby, "New Russian Documents," Appendix, document no. 21. Also see Liang Zhensan, "Chaoxian Zhanzheng Qijian ZhongChao Gaoceng de Maodun, Fenqi ji qi Jiejue: Lengzhan Zhong Shehuizhuyi Zhenying Nei Guojia Guanxi Yanjiu Anli zhi Yi" [Contradictions, Conflicts, and Their Resolution among Chinese and Korean Elites During the Korean War Period: A Case of International Relations Within the Socialist Camp During the Cold War], The Central Research Center of the Modern History Center, Taipei, Paper No. 40, June 2003, pp. 60–61. On July 13, 1950, Stalin complained that he still did not know whether China had begun the process of deploying nine divisions on the Korean border. See "Ciphered Telegram Stalin to Zhou Enlai or Mao Zedong (via Roschin)," July 13, 1950 in Weathersby, "New Russian Documents," Appendix, Document 22; also see Shen Zhihua, "Sulian Kongjun Chudong: Chaozhan Chuqi Zhong Su Chao Tongmeng de Neizai Guanxi" [The Soviet Air Force Goes into Action: Relations within the Chinese-Soviet-Korean Alliance in the Early Stages of the Korean War], paper presented at the International Symposium on "The Cold War in East Asia," Hokkaido University, Japan, June 24–27, 2008, pp. 2–3. I am grateful to Andrew Kennedy for calling this paper to my attention.

[27] Liang, "Chaoxian Zhanzheng Qijian," 63. As Liang puts it, "Under conditions at that time, if China's forces had immediately moved out [into Korea], regardless of whether it was to provide defense in the rear against a U.S. amphibious landing, or on the front-lines in Inchon to assist with the offensive, this could have had a clear influence on the course of the war. However, owing to the huge Sino-Korean disputes regarding analysis of the war situation and strategic preparations, even if there were no other reasons (such as the remaining scruples that Kim Il-sung had about China's dispatch of troops) the Korean side would not be able to consider inviting China to send troops to help."

[28] Xu, *Mao Zedong Yu KangMei YuanChao Zhanzheng*, 67.

After MacArthur's September 15 Inchon landing, the Soviets themselves urged a fast-paced retreat northward of North Korean forces in South Korea to protect them from being cut off from their brethren further north. These orders were not implemented because of some combination of continued North Korean stubbornness and over-optimism, the failure of Soviet advisors on the ground to implement these orders from Moscow, and the poor communications between North Korean political leaders and their military counterparts in the field. As a result, several divisions of Korean People's Army (KPA) forces in southernmost Korea were fully cut off from the remaining forces in and around Seoul and to the north of the South Korean capital.[29]

Kim Finally Cries Uncle, Stalin Orders Out for Intervention, and Beijing Balks

One hour after receiving a desperate plea for direct Soviet and Chinese intervention from Kim Il-sung in the early morning hours of October 1, Stalin urgently suggested to Mao that he dispatch five to six divisions of troops "at once" to the 38th parallel to help Kim regroup. The goal was to defend against an impending U.S. invasion of the North.[30] This request was made from Stalin's Black Sea dacha one week before U.S. forces would cross the parallel and set in motion the escalation of the war. Stalin was asking for early, immediate, and defensive deployment of Chinese forces deep into Korea on October 1. But the lack of prior coordina-

---

[29] On Chinese concerns as early as August regarding North Korean overextension and the risk of an enveloping amphibious strike by the U.S., see Liang, "Chaoxian Zhanzheng Qijian," 62. For Stalin's complaints that his orders to evacuate North Korean forces from southernmost Korea after Inchon were not carried out by his advisors in Korea, see "Shi Dalin Guanyu Dui Chaoxian Jushi de Chuli Yijian Zhi Shitekefu He Mataweierfu Dian" [Stalin's Telegram to Shykov and Mataaev [Zakharov] Regarding Suggestions on How to Handle the Situation in Korea], October 1, 1950, in Shen, ed., *Chaoxian Zhanzheng*, 2:573–74. It is not clear if the Soviet advisors were insubordinate, as Stalin claimed, or whether they were simply confused by the optimistic intelligence they were receiving from the North Koreans. In a message sent September 29, 1950, to Stalin asking whether Moscow thought the enemy would cross the 38th parallel, Kim Il-sung confesses that until that point, the North Koreans had thought they could defeat the enemy alone. Shen's Chinese translation of this document differs in an important way from Alexandre Mansourov's English translation. See Mansourov's "Stalin, Mao, Kim," Appendix, Document 5 and "Shitekefu yu Jin Richeng de Huitan Qingkuang Zhi Geweimike Dian," [Shtykov's Telegram to Gromyko Regarding the Situation at a Meeting with Kim Il-sung], September 30, 1950, pp. 561–62. In Mansourov's translation it sounds as if Kim is "reiterating" and sticking by his original plans despite changes in the situation on the ground; in Shen's translation Kim is reporting his original plans (*shuochu ta yuanlai de xiangfa*) and the fact that they are becoming quickly obsolescent as the UN forces threaten to cross the 38th in the near term.

[30] For an English translation of this document, see "Ciphered Telegram Stalin to Mao Zedong and Zhou Enlai, October 1, 1950," in Mansourov, "Stalin, Mao, Kim," Appendix Document 10. For a Chinese translation, see "Shi Dalin Guanyu Jianyi Zhongguo Pai Budui Yuanzhu Chaoxian Wenti Zhi Weishen Dian," in Shen, ed., *Chaoxian Zhanzheng*, 2:571.

tion between the Soviets, Chinese, and the North Koreans in advance of that date precluded such a rapid Chinese response. In the broadest sense, the CCP's military preparations had been generally consistent with Stalin's wishes in the summer. The Chinese leaders had sent a large number of forces to Manchuria from July through early October 1950 and continued to consider options relating to Chinese forces crossing the Yalu. In fact, on September 21, Liu Shaoqi had reported a high degree of morale among the Chinese forces, having observed the PRC's general readiness to assist in guaranteeing North Korea's survival.[31] But in very important and consequential ways, the poor coordination among the three capitals prevented a timely North Korean request for Chinese assistance after Inchon and did not provide the proper levels of Soviet support and assurance to the Chinese to allow them to enter immediately after both Stalin and Kim Il-sung had requested Chinese intervention.

For his own part, once Mao learned of the seriousness of the North Korean military situation and received, on October 1, a formal request from Pyongyang for intervention, he himself seemed quite committed to entering the war with massive numbers of Chinese forces, as is evident in an October 2 telegram that he drafted to Stalin but never sent.[32] But Mao felt obliged to meet with his comrades in the Politburo in Beijing before making such an important decision. The majority of them were initially opposed to the war because of the risks of escalation and the military, economic, and political costs that the newly founded PRC would have to endure in a war with the world's strongest military. So, in the October 2 telegram that Mao actually sent to Stalin via the Soviet ambassador to China, the Chairman discussed how "many comrades" on the Central Committee were worried about the international and domestic costs of entering the war and stated that China would not enter the war at this time. Mao emphasized that Chinese forces were ill prepared militarily for battle with the United States, that war could harm the economy, and that it could cause domestic dissatisfaction against his new government. Mao made clear that a "final decision had not been taken." He said that there would be meetings on the question and that he would send Zhou Enlai and Lin Biao to the Soviet Union to consult with Stalin on the matter. Despite Mao's final statement that the matter was still open to discussion, Soviet Ambassador Roschin attached a somewhat overwrought analysis of Mao's telegram calling this a full abandonment of China's earlier "basic" policy toward Korea. Roschin speculated about what threats from the United States had been passed on to China via India to spook the Chinese into backing down.[33]

---

[31] Liang, "Chaoxian Zhanzheng Qijian," 64.

[32] For a discussion and translation of this telegram, see Christensen, *Useful Adversaries*, ch. 5 and Appendix B.

[33] For the English translation from the original Russian, see "Ciphered Telegram From Roschin in Beijing to Stalin, 3 October 1950, Conveying 2 October 1950 Message from Mao to Stalin," in

The Soviets apparently had been naïve to think that the Chinese would be ready to enter the war and project power as far south as the 38th parallel on a moment's notice and therefore they overreacted to the initial Chinese telegram of October 2. One problem China faced in quickly responding to a request such as Stalin's was that the North Koreans, for their own nationalist and psychological reasons, had tried consistently to limit communications and consultation with the Chinese communist representatives in Korea. This precluded proper military and political preparations for entering the war when the North Korean forces finally collapsed. In his May 1950 meetings with Mao, Kim was cocky about his prospects for success and thereby dismissed as "unnecessary" Mao's offer to prepare forces on his border in case his invasion plans were to fail and the North Koreans were to fall into trouble. Then Kim failed even to inform Mao of the timing of his invasion, of which Mao learned from the international press.[34] After the war started, the Chinese communists and Korean communists were slow to establish official contacts, a source of real frustration for Stalin.[35] Once those contacts were established, the CCP complained that very little military intelligence was shared with Chinese representatives in Pyongyang. Chinese representatives in Korea such as Chai Chengwen and Ni Zhiliang enjoyed extremely limited access to top leaders in Pyongyang. According to a recently published secondary account, the former complained that military intelligence was basically a "forbidden zone" for the Chinese. For example, in the first half of September, the Chinese (PLA) General Staff proposed sending a high-level entourage to North Korea, but the North Koreans refused.[36]

Even in the days immediately after MacArthur's Inchon landing, when the KPA's situation was increasingly desperate and UN forces began moving on Seoul, Zhou Enlai complained to the Soviets that Chinese advisors were in the dark about the situation on the battlefield.[37] A careful reading of Russian docu-

---

Mansourov, "Stalin, Mao, Kim," Appendix, document 12. For the Chinese-language version, see Shen, ed., *Chaoxian Zhanzheng*, 2:576. Such threats were indeed made by the United States via India in July and August. See Christensen, *Useful Adversaries*, 165.

[34] Xu, *Mao Zedong yu KangMei YuanChao Zhanzheng*; and Liang Zhensan, "Chaoxian Zhanzheng Qijian," 59.

[35] On July 8, 1950, Stalin seemed impatient for China to send representatives to North Korea to coordinate their activities. See "Ciphered Telegram Stalin to Soviet Ambassador Roschin in PRC transmitting message to Mao Zedong," July 8, 1950, in Weathersby, "New Russian Documents," Appendix, document no. 21.

[36] For a fascinating new article on the lack of coordination between North Korean leaders and Chinese leaders in the months leading to China's crossing of the Yalu, see Liang, "Chaoxian Zhanzheng Qijian," 59–65.

[37] Ibid., 63–65. Mansourov, "Stalin, Mao, Kim," 6. Xu Yan's account of this period does not accord with these documents or Liang's analysis. He claims that there were a lot of contacts between Chinese ambassador to North Korea, Ni Zhiliang, and the Korean leadership. See Xu, *Mao Zedong Yu KangMei YuanChao Zhanzheng*, 70–71; Shen Zhihua, "China and the Soviet Union Dispatch Troops to Aid Korea: The Establishment of the Chinese-Soviet-Korean Alliance in the Early Stages

ments reveals that the Soviets apparently believed that Zhou's complaint about Pyongyang's intentional exclusion of Beijing from the decision process was quite justified. In an initial draft of a September 20 telegram to China on the subject, Gromyko criticized as "entirely mistaken" the North Korean leadership's reluctance to "conscientiously inform the Chinese Comrades of the military situation and all the decisions the Korean commanders and political leadership made regarding issues arising in the course of military operations." In this initial draft Gromyko wrote that he believed that Kim Il-sung himself "must correct this point." Gromyko did not, however, transmit this draft to Beijing. The telegram that was actually sent was quite different and was almost certainly geared toward manipulation of the alliance to minimize China's anger at Pyongyang and to maximize the chance that Beijing would still come to its defense if and when needed. In the revised telegram the situation is described more neutrally as "not right" (or "abnormal") but then is quickly explained away by reference to the "weak links" that the central command in North Korea has with its front lines. Gromyko says this situation arose because of technical difficulties not because the "Korean Comrades are unwilling" to share information with the Chinese.[38]

Moscow's first interpretation of the motivations behind Pyongyang's delay in collaborating with Beijing fits much better than the second with what we know about the overall situation. After all, Pyongyang only formally decided to request Soviet and Chinese entrance into Korea at the very end of September and at first directly contacted only Moscow. Pyongyang's initial telegram to Stalin requesting Soviet or Chinese "volunteers" to enter the war was received in the early morning of October 1 at Stalin's Black Sea dacha.[39] Stalin, ever conscious of alliance politics and manipulations, then did two things that were consistent with his strategy of maintaining his position in the communist alliance system. First, he sent a highly polite and respectful telegram to Mao, neglecting to mention Kim's request for direct Soviet intervention. Instead, Stalin suggested that Beijing send five to six divisions of Chinese volunteer forces "at once" to help North Korea defend against a U.S. crossing of the 38th parallel. Stalin also

---

of the Korean War," unpublished working paper, 2008, translated by Yang Jingxia and Douglas A. Stiffler, pp. 11–12. I am grateful to Andrew Kennedy for calling this last work to my attention.

[38] "Geweimike Guanyu Dui Zhou Enlai de Dafu Zhi Weishen Dian" [Gromyko's Telegram to Roschin Regarding a Reply to Zhou Enlai], Sept. 20, 1950, in Shen, ed. *Chaoxian Zhanzheng*, 2:542–44. Shen offers full texts of both the telegram as sent and the original draft. Mansourov, "Stalin, Mao, Kim," 7, discusses only the final draft and describes it as a telegram from Stalin, rather than Gromyko. Moreover he seems to take at face value the explanations of the "abnormal" conditions in which Kim Il-sung is unable to inform his Chinese comrades of his activities. In the Chinese version of the original draft, the situation is described as something Kim Il-sung can fix and as *wanquan cuowu de* (totally mistaken), as opposed to the final version's *bu zhengque de* (not right), translated as "abnormal" by Mansourov.

[39] For Kim's initial request for direct Soviet and Chinese intervention, see Mansourov, "Stalin, Mao, Kim," Appendix, document no. 10.

promised that the Chinese would have command authority over those forces. He pledged that he had not discussed his request to Beijing with Pyongyang and that he had no plans to discuss this matter with Kim Il-sung. He thereby respected the confidentiality of Mao's decision process and appeared not to pressure Mao into accepting his advice.[40] Stalin then contacted Pyongyang on the same day, telling Kim that "they [elites in Moscow] believe that the most acceptable form of assistance is to dispatch [Chinese] volunteer forces. But regarding this question we must first consult with the Chinese comrades."[41] In essence, Stalin was both gently refusing Kim's request that he send Soviet troops directly into Korea, and suggesting that he would arrange for Chinese volunteers to enter Korea. Therefore, as in spring 1950, Stalin would get a good deal of the credit if China supported Pyongyang but he could pass on the blame if China did not. He was, of course, also breaking his pledge to Mao that he would not discuss his request to the Chinese with Pyongyang.

It was only after receiving this message from Stalin that Kim Il-sung would, for the first time, formally and directly approach the Chinese ambassador for direct Chinese intervention. He reportedly did so on the night of October 1, 1950.[42] It was then only after these requests were received and deciphered in China that the formal decision process for war began in Beijing. To be sure that the message was delivered, on October 3, 1950, an emissary from North Korea also carried a letter from Kim Il-sung to Beijing (dated October 1, 1950) asking for direct Chinese intervention. The letter warned that if the United States continued unopposed, it would be able to complete its mission of turning Korea into a colony and a military base.[43]

A combination of mistrust within the alliance, particularly between the North Koreans and the Chinese and between the Soviets and the Chinese, prevented a smoother and more coordinated response to the UN victories in mid-September. In a series of excellent writings on this period based on recently declassified

[40] For an English translation of this document, see "Ciphered Telegram Stalin to Mao Zedong and Zhou Enlai, October 1, 1950, in Mansourov, "Stalin, Kim, Mao," Appendix, document no. 10. For the Chinese-language translation, see Shen, ed., *Chaoxian Zhanzheng*, 2:571. For analysis of Stalin's strategy in not appearing to pressure Mao, see Mansourov, "Stalin, Mao, Kim," 15; and Xu, *Mao Zedong Yu KangMei YuanChao Zhanzheng*, 82.

[41] "Shi Dalin Guanyu Dui Chaoxian Jushi de Chuli Yijian Zhi Shitekefu he Mateweierfu" [Stalin's Telegram to Shtykov and Matveev (Zakharov) Regarding Suggestions on How to Handle the Korean Situation], October 1, 1950, in Shen, ed., *Chaoxian Zhanzheng*, 2:573–74.

[42] Liang, "Chaoxian Zhanzheng Qijian," 64–65. Liang argues that there were two reasons why Sino-Korean relations were so poorly coordinated. First, Kim was overly confident for too long that he would win his war without significant Chinese help. Second, Kim, for historical reasons relating to Chinese domination of Korea, mistrusted the Chinese and hoped to limit Chinese participation in his unification efforts, preferring to rely whenever possible on the Soviets, rather than the Chinese. Also see Xu, *Mao Zedong Yu KangMei YuanChao Zhanzheng*, 82.

[43] Xu, *Mao Zedong yu KangMei YuanChao Zhanzheng*, 89–90.

Soviet archives, Chinese diplomatic historian Shen Zhihua chronicles how Stalin's caution about being dragged into World War III by his Asian allies and his jealous distrust of the Chinese communists' intentions in Korea prevented him from responding earlier to North Korean suggestions that Chinese intervention in the war would be useful to Pyongyang. Following the initial U.S. intervention in Korea and in accord with Stalin's request for China to assemble forces on the border to assist in North Korean defense in the event of a U.S. crossing of the 38th parallel, Mao had promised Kim in July that, if he ran into problems in the war, the PRC would prepare the necessary forces to dispatch to Korea. Zhou Enlai similarly told the Soviets that China would assemble and ready forces in Manchuria for such a contingency, but also requested Soviet air cover if and when they were to be deployed to Korea. It was very clear that Stalin offered to provide such support to the Chinese if and when they decided to enter the war but that he envisioned PRC entry only if UN forces clearly threatened to move ground troops into North Korea itself. He seemed quite concerned that any Chinese military action in Korea before PRC intervention was entirely necessary to save North Korea would only make escalation of the war, perhaps to a global war, more likely. Stalin wanted to maintain control over the situation by preventing the Chinese and North Koreans from taking action without his approval.[44]

One major problem precluding a more timely Chinese response was the strained lines of communication and decision-making within the alliance. The North Koreans, of course, wanted the Chinese to be standing at the ready to defend the North if they were needed, but wanted to avoid excessive Chinese influence in Korea, so they turned to the Soviets for guidance on all important matters, including Chinese intervention. In July the North Koreans sent reports to Stalin asking for his opinion about the prospect of eventual Chinese intervention, but also sent relatively rosy assessments about the prospects of victory for his forces against UN troops in the South. As the cautious, more globally conscious leader of the communist movement, such reports from Pyongyang only increased Stalin's desire to rein in what he saw as the potential for North Korean and Chinese adventurism in Korea that could lead to unnecessary escalation and, potentially, a loss of Soviet prestige to the Chinese communists in Korea.[45] Shen writes, "[T]he eagerness Mao Zedong showed to send troops to Korea most likely aroused Stalin's suspicions: one consequence of China sending troops would be to expand her status and influence in Korea. From a long-term point of view, this would not be to the Soviets' advantage."[46]

[44] Shen, "China and the Soviet Union Dispatch Troops to Aid Korea," and Shen, "Sulian Kongjun Chudong."

[45] Shen, "Sulian Kongjun Chudong," 3–5; and Shen, "China and the Soviet Union Dispatch Troops to Aid Korea," 8.

[46] Shen, "China and the Soviet Union Dispatch Troops to Aid Korea," 8.

For their part, the North Koreans were suspicious of the Chinese and were certainly unwilling to risk alienating Stalin by upgrading their military coordination with the Chinese communists without clear direction from Moscow. According to Shen, in August Mao reached out to the Soviet philosopher Pavel Yudin as a channel to Stalin about the dangers facing the Korean communists over time and the need for early Chinese intervention if and when the war turned against North Korea. But Stalin apparently did not offer any supportive signals to the Chinese in response to this outreach.[47] In July and August, the North Koreans continually raised the prospect of Chinese intervention with the Soviets, and, in late August, began to send more pessimistic prognoses about frontline conditions and the prospect of a UN amphibious invasion behind their forward positions in the South. Stalin responded in a reassuring but rather disingenuous fashion, assessing the likelihood of North Korean victory as great despite recent setbacks, and promised Soviet material support to North Korean forces. He failed to address Kim's questions regarding potential Chinese intervention. In fact, throughout July and August, Stalin would simply refuse to respond to queries from his representatives in North Korea about Kim Il-sung's questions regarding the prospect of Chinese intervention.[48] Shen argues that after several iterations of this kind, Kim would come to understand that Stalin would not even consider Chinese intervention until a critical moment had arrived in the war and North Korea found itself in dire straits.[49]

In an article on the PRC-DPRK relationship during the war published in Taiwan, the apparently pseudonymous author, Liang Zhensan, argues that more prompt Chinese military backing of the North Koreans after Inchon was precluded by two factors: differences in strategic assessments between the overly self-confident North Korea and the more cautious Chinese; and North Korea's misgivings about Chinese involvement in their affairs, which made them rely on and communicate with the Soviets exclusively until the situation was very dire. Liang suggests that if Kim had informed China earlier of the depth of North Korea's military problems and directly asked for assistance before October 1, 1950, such assistance might have been forthcoming at a time when it could have made a decisive difference in the war.[50] He writes, "China certainly dispatched forces

[47] Shen, "Sulian Kongjun Chudong," 4; and Shen, "China and the Soviet Union Dispatch Troops to Aid Korea," 8.

[48] Shen, "Sulian Kongjun Chudong," 3; and Shen, "China and the Soviet Union Dispatch Troops to Aid Korea," 8–9, 11.

[49] Shen, "Sulian Kongjun Chudong," 4.

[50] Liang, "Chaoxian Zhanzheng Qijian." Liang's article focuses mostly on the military implications of Kim's late request. But the greater impact was arguably on coercive diplomacy. William Stueck points out that the late nature of the Chinese entrance into the war, due in part to this late request from North Korea, gave the Americans false confidence after Inchon that China would stay out of the war. See Stueck, *The Korean War*, 98.

into Korea to 'Oppose the United States, Help Korea, and Defend the Nation,' in order to fulfill revolutionary goals, and carry the responsibilities of internationalism. But the North Korean leadership's innermost thought (*neixin*) was to resist China's direct provision of military assistance. In analyzing the reasons why, there are generally two aspects [to consider]: first, there is Kim Il-sung's excessive self-confidence (*guoyu zixin*), his excessively optimistic estimation of the situation, and his belief that he could rely on his own power to solve the entire problem; second, is the special historical Sino-Korean relationship that existed; China's ability to have suzerainty (*zongzhuquan*) over Korea as well as the Chinese Communists' influence over the Korean Workers Party caused the Korean leaders to have some misgivings and concerns (*danxin he youlü*) toward China's dispatching of troops."[51]

Shen Zhihua argues that once China learned of the successful Inchon landing, Zhou Enlai reached out to DPRK's ambassador to China, Yi Chu-yon, on September 19 and asked what the North Koreans might be requesting from China in terms of support. The North then turned to the Soviet representative in Pyongyang for a response to the Chinese request, emphasizing the potential utility of Chinese troops further north in Korea and, as before, received no reply from Stalin. According to Shen, at a September 21, 1950, North Korean Workers' Party Politburo meeting, Kim Il-sung would pour cold water on the idea of responding to the Chinese offers positively without first getting Stalin's approval. In rejecting such a direct outreach to Beijing, Kim cited the following factors: the critical material support that the North had received from the Soviet Union; the likely jealousy Stalin would feel if Pyongyang now turned directly to Beijing for help without approval from Moscow; the likelihood that efforts to rebuild the DPRK army might be sufficient to hold the line against the UN forces; and Kim's certainty that his two larger allies would save him in the end if the situation deteriorated further. Shen sums up the situation by saying, "[T]he result of the meeting was that no resolution [on a request for Chinese entrance] was passed. Now Kim Il-sung was acting completely by reading Stalin's mind, while Stalin was trying his best not to let China enter the war until the final moment; i.e. the moment the enemy crossed the 38th parallel."[52]

Of course, since Stalin's direct request to the Chinese on behalf of the North (and the North's subsequent direct request to the Chinese) would not come until October 1, this led to a critical delay in the alliance responding to the new strategic situation. A prompt response might have signaled to the United States their resolve in preventing the destruction of the DPRK. As discussed earlier, the

---

[51] Liang, "Chaoxian Zhanzheng Qijian," 65.

[52] Shen, "China and the Soviet Union Dispatch Troops to Aid Korea," 13; and Shen, "Sulian Kongjun Chudong," 7.

buildup of forces in the Northeast, which Zhou Enlai had hoped might deter UN forces from crossing the 38th parallel, had indeed been watched closely by U.S. intelligence, but the failure of the alliance to act in the weeks immediately following Inchon led intelligence analysts in the United States to start discounting the information that they had collected about the buildup and to reduce greatly the probability assigned to a large-scale Chinese intervention in Korea.[53]

There were still other critically important alliance-related hurdles that delayed Chinese entrance into the war until after the first U.S. forces crossed the parallel on Oct. 7. The lack of tight Sino-Soviet coordination and preparation before Inchon regarding Sino-Soviet burden-sharing in a wider war also delayed the Chinese decision to enter this war. As a result, Beijing made the final decision to enter the war nearly two full weeks after Stalin's and Kim's initial requests for Chinese intervention on October 1 (and those requests themselves came quite late in the U.S. decision process regarding the crossing of the 38th parallel). Once Korea had finally invited the Chinese to enter the country with military force, Beijing's decision process was complicated and lengthened further by the lack of prewar coordination between Moscow and Beijing regarding the degree and type of Soviet assistance to Chinese troops in Korea once those troops crossed the Yalu River border. In particular, the initial telegram Mao sent to Stalin on October 2, 1950, foregoing intervention for the time-being, discusses how poorly equipped the Chinese forces were in fall 1950. Mao seemed concerned that if such troops raced down as far south as the 38th parallel, as Stalin wished, they could draw the United States north and be forced into retreat on the enemy's terms.[54] Although they had collaborated for months in the long-term development of indigenous Chinese military capabilities, Beijing and Moscow had failed to work out in advance the details of how, in the nearer term, Soviet military assistance would be supplied to the Chinese forces as they prepared to enter the field of battle in Korea.[55] In fact, the entire purpose of Zhou

[53] For Zhou Enlai's estimation about the deterrent effect of amassing forces in the Northeast, as expressed to the Soviets in discussions on September 18, 1950, see Shen, "Sulian Kongjun Chudong," 6. For the change in U.S. estimates about the meaning of that buildup in the weeks following Inchon, see Vanderpool, "COMINT and the PRC Intervention in the Korean War," 15.

[54] For the English translation from the original Russian, see "Ciphered Telegram from Roschin in Beijing to Stalin, 3 October 1950, Conveying 2 October 1950 Message from Mao to Stalin," in Mansourov, "Stalin, Mao, Kim," Appendix, document no. 12. For the Chinese-language version, see Shen, ed., *Chaoxian Zhanzheng*, 2:576.

[55] This was a fact that Mao cited in his initial, somewhat conditional decision to enter the war on October 7 and in his final decision of October 13 when explaining his comrades' delay in reaching their final decision to enter the war. On October 13, 1950, Mao telegrammed to Stalin via Ambassador Roshin the Politburo's final decision to enter the war after a sixty-hour period of reversal during October 10–13 following concerns about the nature and depth of Soviet assistance to Chinese troops entering Korea. He wrote, "Our comrades' earlier inability to be resolute was because they were still not clear about the issue of the international situation, the issue of Soviet military support, and

Enlai's and Lin Biao's trip to the Soviet Union on October 8 would be to ensure and coordinate Soviet aid for the Chinese forces.[56] On the issue of air cover, the PRC and the USSR had been working quite closely in developing and training a PLA air force tentatively scheduled for deployment in spring 1951, but they had apparently not coordinated the details of how, before that time, Soviet planes and pilots would assist in protecting Chinese ground deployments in Korea and base areas and industrial sites in China.[57] As Mao offered conditional agreement with Stalin's request to send Chinese forces into Korea on October 7, Mao complained that Chinese forces on the border were still terribly ill equipped.[58]

---

the issue of air force cover." See "Weishen Guanyu Mao Zedong Jueding Chubing Deng Wenti Zhi Shi Dalin Dian" [Roschin's Telegram to Stalin Regarding the Issue of Mao Zedong's Decision to Dispatch Troops, etc], October 13, 1950, in Shen, ed., *Chaoxian Zhanzheng Qijian*, 597. For discussions of Mao's requests for Soviet planes and training for a jet-fighter force capable of supporting Chinese offensives in spring 1951, see "Weishen Zhuanfa Mao Zedong Guanyu Dui Zhongguo Tigong Kongjun Yanhu Deng Wenti Zhi Shi Dalin Dian" [Roschin Passing on MaoZedong's Telegram to Stalin Regarding the Issue of Providing China Air Cover, etc.], July 22, 1950, in Shen, ed., *Chao Xian Zhanzheng*, 483.

[56] Mao first raised the idea of sending Zhou and Lin on such a mission in the October 2, 1950, telegram to Stalin. Zhou and Lin would travel after Mao's initial decision to enter the war on October 8, 1950. Because of uncertainties related to the timing and degree of Soviet assistance for his forces, Mao reconsidered that October 8 decision during October 10–12. See "Weishen Guanyu Mao Zedong Dui Chubing de Taidu Wenti Zhi Shi Dalin Dian" [Roschin's Telegram to Stalin Regarding the Issue of Mao Zedong's Attitude Regarding Dispatching Troops], October 7, 1950, in Shen, ed., *Chaoxian Zhanzheng*, 2:588–90; and "Weishen Guanyu Mao Zedong Jueding Chubing Deng Wenti Zhi Shi Dalin Dian" [Roschin's Telegram to Stalin Regarding the Issue of Mao Zedong's Decision to Dispatch Troops, etc.], October 13, 1950, in Shen, ed., *Chaoxian Zhanzheng*, 2:597. For an English translation of this October 13 document, see "Ciphered Telegram, Roschin to Stalin, 14 October 1959, re: Meeting with Mao Zedong," in Mansourov, "Stalin, Mao, Kim," Appendix, document no. 19.

[57] For documents outlining Mao's and Stalin's ambitious plans for development of a PLAAF with Soviet jets and pilot training, see, for example, "Weishen Zhuanfa Maozedong Guanyu Dui Zhongguo Tigong Kongjun Yanhu Deng Wenti Zhi Shi Dalin" [Roschin Passing on Mao Zedong's Telegram to Stalin Regarding the Issue of Providing Air Force Cover, etc.], July 22, 1950, in Shen, ed., *Chaoxian Zhanzheng*, 2:483. Stalin agreed to assist in accordance with the plan on July 25. See "Weixinsiji Guanyu Tongyi Xunlian Zhongguo Feixingyuan Wenti Zhi Weishen Dian" [Vishinsky's Telegram to Roschin Regarding Agreement on the Question of Training Chinese Pilots], July 25, 1950, in ibid, 487.

[58] See "Weishen Guanyu Mao Zedong Dui Chubing de Taidu Wenti Zhi Shi Dalin Dian" [Roschin's Telegram to Stalin Regarding the Issue of Mao Zedong's Attitude Regarding Dispatching Troops], October 7, 1950, in Shen, *Chaoxian Zhanzheng*, 2:588–90; Shen, "Sulian Kongjun Chudong," 8. There was apparently something to the complaint. One telegram from Mao to Stalin as late as November 7, just two weeks before the massive PLA counteroffensive against MacArthur's troops, is particularly telling regarding the lack of tight coordination between the Soviet and Chinese militaries in preparation for China's entrance into the war. In the telegram, Mao describes the haphazard nature of his forces' equipment, most of which had been confiscated from previous enemies in previous wars. He said the lack of uniformity in the equipment, including caliber of guns, meant sustaining forces in the field would be terribly difficult. He then offered a laundry list of weapons and ammunition that he claimed to need by January or February of 1951, months after the large-scale escalation that would occur in late November! See "Mao Zedong Guanyu Qingqiu Sulian Tigong Wuqi Zhuangbei Zhi Shi Dalin Dian" [Mao Zedong's Telegram to Stalin Regarding a Request for the Soviet Union to Supply Arms], November 7, 1950, in Shen, ed., *Chaoxian Zhanzheng*, 2:616.

One basic problem in alliance coordination in Korea was that, for reasons related to China's own national security, Mao did not want his forces to be bogged down in a war of attrition in Korea. Therefore, once he decided to enter the war, Mao searched for ways to achieve total victory on the peninsula for North Korean and Chinese forces. While there is ample evidence that Beijing entered the war out of defensive motivations related to the UN crossing of the 38th parallel and the threat that would pose for his new country over the long run, China's strategy in Korea was offensive and required the ability to deliver a knockout blow to U.S. forces. From all appearances, and as we will see in the next chapter, the Soviets seemed much more dedicated to the prevention of a U.S. rollback in North Korea than to total victory and seemed more comfortable with the prospect of a longer war that sapped U.S. global strength in Korea and kept both China and North Korea highly dependent on Soviet assistance. Reflecting the differing national interests of the various members of the alliance, Stalin's strategy allowed for fighting to achieve these goals, but only if that fighting was being done by Chinese and North Korean forces, not his own.

U.S. ALLIANCE POLICY AND FAILED COERCIVE DIPLOMACY IN NORTHERN KOREA

In his account of the Soviet strategy toward Beijing and Pyongyang in Autumn 1950, political scientist Alexandre Mansourov concludes that "had the United States been less ambivalent, more consistent, and more persuasive on the diplomatic front in stating to Moscow and Beijing the goals of its Korean campaign—for example, that it had no desire to attack mainland China or threaten the territory of the Soviet Far East—the Soviet and Chinese governments could well have decided to let Kim Il-sung's regime go under." Mansourov blames MacArthur's brusque statements and demands, in particular, as the key factor that undercut U.S. coercive diplomacy.[59] I agree with Mansourov's general notion that the Soviets could have lived with a unified Korea under UN control more easily than could Mao, and most obviously, than could Kim Il-sung. But the historical data now available strongly suggest that it would have been much harder for the United States to reassure China than Mansourov suggests, regardless of the diplomatic niceties. In this sense, the divided nature of the communist alliance made it more difficult to engage in coercive diplomacy.

The main reason that China reacted so strongly to the U.S. crossing of the parallel is that a series of decisions by the United States regarding its future allies in Japan, Korea, and Taiwan practically rendered meaningless any of its

[59] Mansourov, "Stalin, Mao, Kim," 29.

diplomatic efforts to reassure China about the prospects of a unified Korea allied with the United States. While the threat of U.S. escalation if China were to enter the war— passed in part through a "calculated indiscretion" by the U.S. ambassador to India, Loy Henderson—were not discounted in Beijing, assurances that the PRC would be secure if UN forces were permitted to unify the peninsula under a pro-U.S. regime were simply not credible in Beijing. China's suspicions were largely rooted in U.S. policy toward its future allies in Taiwan and Japan.[60]

In the early weeks of the Korean War, the United States adopted three policies related to its East Asian security partners and future allies that greatly affected the prospect of limiting the escalation of the Korean War and preventing a U.S.-PRC war there. First, the Truman administration decided on June 27, 1950, to block the Taiwan Strait with elements of the U.S. Navy's Seventh Fleet. Washington did so, ostensibly, to prevent attack from either side of the Strait. It was at this time that the U.S. government shifted from recognizing Taiwan as part of Chinese territory to stating that the island of Taiwan's sovereignty was "undetermined," a stance that remains to this date.[61] In Beijing, unfortunately, the U.S. intervention was seen as an unambiguous reversal of previous pledges by the Truman administration not to involve itself in the Chinese Civil War. Second, the Korean War led Washington to redouble its efforts, ultimately largely unsuccessful, to convince Japanese authorities to contribute more to the future U.S.-Japan alliance by building up its own military power. For Moscow, Beijing, and Pyongyang, these efforts signaled a frightening prospect: a rearmed and assertive Japan in the future. Third, the United States decided for a range of military, political, and budgetary reasons to cross the 38th parallel and attempt to unify the Korean peninsula under friendly rule by means of military force. These three policies in combination convinced a very reluctant Beijing to send

---

[60] For the July 1950 threat via India to China that, if Beijing attacked in either Korea or the Taiwan Strait, "the United States will consider itself at war, and not only deal with such Chinese forces as may be met in the field but also strike at the bases of Chinese power," see Davies Draft, July 11, 1950, and Herbert Feis's note on Memorandum by John Davies, entitled "Calculated Indiscretion to be Committed by Ambassador Henderson," July 12, 1950, PPS Records, box 14, file: China 1950–51, National Archives. When the Chinese communists initially balked at the Soviet request to send in troops in early October 1950, the Russian ambassador in Beijing opined in a telegram to Stalin that the Chinese must have been deterred by U.S. threats passed through India. For the English translation from the original Russian, see "Ciphered Telegram from Roschin in Beijing to Stalin, 3 October 1950, Conveying 2 October 1950 Message from Mao to Stalin," in Mansourov, "Stalin, Mao, Kim," Appendix, Document 12. For the Chinese-language version, see Shen, ed., *Chaoxian Zhanzheng*, 2:576. See Christensen, *Useful Adversaries*, ch. 5, for how Mao understood but accepted the risk of such escalation when he dispatched Chinese troops to Korea.

[61] See Romberg, *Rein In at the Brink of the Precipice*, ch.1; also see Nancy Bernkopf Tucker, *Strait Talk: United States-Taiwan Relations and the Crisis With China* (Cambridge: Harvard University Press, 2009), 13.

large numbers of forces into Korea and to adopt aggressive, relatively expansive military strategies once in the country. Moreover, Beijing chose to do so despite concerns about the timing and scope of Soviet assistance to its efforts.

The direct U.S. intervention in the Taiwan Strait was viewed as a shock and as an invasion of Chinese territory by Mao and his colleagues. Even before Truman's January 5, 1950, speech on nonintervention in the Chinese Civil War and the Acheson Press Club speech one week later, Chinese military strategists did not expect direct U.S. intervention in the conflict across the Taiwan Strait. As fit their assessments for Korea, they anticipated that the United States might send Japanese "volunteers" to help the "Chiang clique" but would forego direct intervention.[62] After the Truman and Acheson speeches on the American side and the signing of the PRC-USSR Treaty of Mutual Defense on the communist side, the prospect of direct U.S. intervention in cross-Strait relations must have seemed smaller still. The Truman administration's reversal in both Taiwan and Korea in June affected Beijing's perceptions of U.S. Korean War intervention in three key ways. First, future assurances of U.S. restraint either in Korea or at the Yalu border with China carried little weight in Beijing. If Washington could reverse its pledges on Taiwan policy so easily, why could it not reverse its stated policy in Korea as well? Second, Taiwan provided a second front from which the United States and Japan could attack China at any time of Washington's choosing. The United States would control both Northern Korea, adjacent to China's industrial center in Manchuria, and Taiwan, in striking distance of China's second industrial area of Shanghai. Third, U.S. direct support for the KMT on Taiwan meant that the United States would be protecting and encouraging Mao's domestic anticommunist enemies at a time when communism was being rolled back in neighboring Korea. The prospect of a reinvigorated domestic opponent significantly raised the costs of settling for a permanent standoff along the Yalu border with large numbers of PLA forces. After all the PLA was a Party army, not a national army, and its forces to this day are dual purpose forces designed to fend off domestic enemies of the CCP as well as enemies abroad.[63] China's

[62] For example, see the December 1949 analysis of General Su Yu in preparation for an invasion of Taiwan cited in Goncharov et al., *Uncertain Partners*, 320, n. 19. For discussion of this report see Xu Yan, *Di Yi Ci Jiaoliang: KangMei YuanChao Zhanzheng Lishi Huigu yu Fansi* [The First Trial of Strength: Reflection and Retrospective on the History of the Resist America, Aid Korea War] (Beijing: Chinese Broadcast Television Press, 1990), 11–12.

[63] For the importance of the U.S. reversal on Taiwan in the calculations of Mao and Peng Dehuai regarding Korea and the fear of a long-term standoff at the Yalu border, see Yufan Hao and Zhai Zhihai, "China's Decision to Enter the Korean War: History Revisited," *The China Quarterly* 121 (March 1990): 101–8; Ye Yumeng, *Chubing Chaoxian* [The Korean War], (Beijing: Beijing Shiyue Wenyi Chubanshe, 1990), 93. In his rallying of support for Mao's plan to enter the war, Peng Dehuai emphasized the fear of a future two-front war from Korea and Taiwan. See Chai Chengwen and Zhao Yongtian, *Banmendian Tanpan* [Negotiations at Panmunjom] (Beijing: Liberation Army Press, 1989), 81–82; Ye, *Chubing Chaoxian*, 3–6, 93; Wang Suhong and Wang Yubin, *Kongzhan*

concerns on this score could only have been aggravated by an unauthorized and highly public trip by General MacArthur to Taiwan in late July with three squadrons of fighter jets, during which he publicly criticized the "threadbare argument by those who advocate appeasement and defeatism in the Pacific that if we defend Formosa we alienate Continental Asia."[64]

The Truman administration was primarily interested in the reputational costs of appearing weak in the face of an armored invasion by North Korea that violated UN resolutions on Korea. (This is, perhaps, the first major post–World War II invocation of the Munich analogy.) The U.S. leadership viewed the war in Korea as most directly related to the future security of Japan. The Korean War, therefore, amplified voices in the administration calling for a rearmed Japan. For example, on September 15, 1950, Dulles would argue to the State Department that there should be no limits set on Japanese military power.[65] Dulles would gain little satisfaction on this score in Tokyo, as discussed in the last chapter, but there is little doubt that Dulles's desires for Japan were consistent with Stalin's and Mao's fears about the future strategic environment in East Asia. As we will see, those concerns in the communist capitals about the future of Japan played prominently in Chinese and Soviet deliberations in October regarding the meaning of U.S. occupation of northern Korea and the need for Chinese entrance into the Korean War to prevent such an outcome.

Windows and War: Mao Pulls and Stalin Pushes the CCP Politburo into Korea

Especially given Beijing's lack of detailed coordination with Pyongyang and Moscow, it is hardly surprising that Mao and his colleagues did not simply accept Stalin's request to immediately move ill-equipped Chinese forces without air cover far from the Chinese border to defend the 38th parallel against the Americans. Such a momentous decision for such a young country would require deliberations. The initial telegram Mao sent to Stalin on October 2 reflected opposition to China's entering the war among many members of the Central Committee, whose views Mao learned of at Politburo meetings on the evening of October 1 and October 2.[66] Many line officers and soldiers were also apparently

---

Zai Chaoxian [Aerial Combat in South Korea] (Beijing: Liberation Army Press, 1992), 107; Hong Xuezhi, *KangMei YuanChao Zhanzheng Huiyi* [Memories from the Korean War] (Beijing: Liberation Army Press, 1991), 1. Also see Xia, *Negotiating with the Enemy*, 44.

[64] "General MacArthur's Message on Formosa," Aug., 17, 1950, Acheson Papers, box 65, Harry S. Truman Library (HSTL). For clear evidence that the trip was actually opposed by President Truman, see Meeting with the President, Policy Planning Staff Records, box 14, file: China, National Archives.

[65] See Momoi, "Basic Trends in Japanese Security Policies," 343–45.

[66] Xu, *Mao Zedong Yu KangMei YuanChao Zhanzheng*, 86; also see Shen, *Mao Zedong, Si Dalin, yu Chaoxian Zhanzheng*, 228. For the most detailed description in English of these contentious

quite nervous about taking on the Americans in the leadup to war.[67] The telegram that Mao himself had drafted on October 2, 1950, but never sent to Stalin, suggests that the Chairman himself was apparently committed to go to war very early in the process, basically as soon as he had received a formal request for assistance from Kim Il-sung via the PRC embassy in Pyongyang the previous evening.[68] He and like-minded leaders like General Peng Dehuai needed to convince their colleagues of the wisdom of this policy and the dangers of the alternatives. According to various analyses, Mao was on board early because he saw both internationalist and nationalist reasons to fight the Americans in North Korea.[69] According to Xu Yan, a leading Chinese military and diplomatic historian at the PLA's National Defense University, an expanded session of the Chinese Politburo was hurriedly convened on October 4 to discuss the Korea situation following the requests for Chinese intervention by Stalin and Kim.[70]

Xu reports that no minutes were kept of the relevant meetings on the 4th and 5th. So his version of events is based on the recollections of participants to which, given his research position at the National Defense University, Xu should have full access. When they initially met on October 4, the majority of Politburo members were opposed to entering the war. Drawing upon General Peng's account of the meetings ten days after they convened, Xu compiles the following list of reasons offered by those opposed to entering Korea: "1) the wounds of [previous] wars have not yet healed; 2) the land reform process is not yet completed; 3) domestic bandits and special agents have not yet been thoroughly eliminated; 4) military equipment and training is not yet ample; 5) a portion of the military are sick and tired of war. To sum up, preparations are insufficient."[71]

Those who supported the war in the Politburo argued that while China's "preparations are insufficient, the imperialist clique's preparations are also insufficient."[72] They believed that the strategy of the United States and its part-

---

meetings in early October in Beijing, see the fascinating study by Andrew Kennedy, "The Origins of Audacity," unpublished manuscript, 2009, ch. 4.

[67] Xu, *Mao Zedong Yu KangMei YuanChao Zhanzheng*, 59.

[68] Ibid., 83–86. For a full translated copy of the unsent October 2 telegram, see Christensen, *Useful Adversaries*, 271–75.

[69] Liang, "ChaoXian Zhanzheng Qijian," 65. One diplomatic historian in China confided in me privately in 2004 that there is increasing evidence that Mao did not fully buy into the national security arguments for entering the war that he offered to his colleagues, but rather simply saw the Korean situation as a chance to fight the Americans in the name of internationalism. This historian made this point in agreement with my analysis that Mao in 1950 was an unusually aggressive leader, even by the standards of the Chinese Communist Party at the time.

[70] Xu, *Mao Zedong yu KangMei YuanChao Zhanzheng*, 90.

[71] Ibid., 91; a similar account can be found in Zhang Xi, "Peng Dehuai Shouming Shuaishi KangMei YuanChao de Qianqian Houhou" [The Events Preceding and Following Peng Dehuai's Commissioning as Commander of the Oppose America, Assist Korea War], in *Zhonghong Dangshi Ziliao* [Chinese Communist Party History Material] 31 (1989):132.

[72] Ibid., 91.

ners was to exploit internal difficulties in other countries in order to seize them and expand their influence. This alleged U.S.-led effort needed to be stopped sooner, rather than later. Mao himself weighed in saying it would be hard to watch while a neighboring country suffered great difficulties. Moreover, he stated that the Chinese had a chance to achieve a great victory over imperialism in Korea. According to Xu, Zhou Enlai added reasons based on national defense for sending in troops. Although Zhou left no record of his own statements at the October 4–5 meetings, twenty days after the decision he gave a relevant report to the leadership of the People's Consultative Congress. In that report Zhou Enlai estimated that, even if the enemy were not to attack China right away after conquering North Korea, the preparation and maintenance of standing defenses along the Yalu river to counter the Americans and Koreans and the relocation of Manchurian industries away from positions vulnerable to enemy air attack would be very costly for the PRC over the long run. China would never know when the enemy would choose to attack and, Zhou added ominously, "once the enemy has occupied Korea, it won't just stop there."[73] This accords with Zhou's and Mao's consultations in late September, in which China was viewed as a future target of the United States after UN forces crossed the 38th parallel.[74] Of course, as Xu notes, fighting the United States over Korea could also have caused massive damage to China and its economy, and so Mao faced a difficult dilemma.[75]

According to an internal Party history, Peng weighed in decisively in support of sending troops into Korea in the critically important meeting on the afternoon of October 5. After arriving in Beijing on October 4 and staying silent at the meetings that evening, Peng decided overnight to support the dispatch of Chinese troops and at the October 5 meeting helped Mao convince wavering comrades by saying that allowing the United States to occupy both Taiwan and all of Korea up to the Yalu river border posed too large of a security risk to the young PRC. Such deployments would allow Washington to find "some excuse" to invade China at any time of its choosing and would encourage "reactionary" domestic opponents to the new Chinese regime. So, even if the risks were great and the costs of war high, for defensive reasons China had no choice but to enter the war.[76] Author Zhang Xi reports that after Peng's presentation, Mao suggested

[73] Xu, *Mao Zedong Yu KangMei YuanChao Zhanzheng*, 92. For the original speech, see "KangMei YuanChao, Baowei Heping" [Oppose America, Aid Korea, Preserve Peace], October 24, 1950, in *Zhou Enlai Junshi Wenxuan* [A Selection of Zhou Enlai's Writings on Military Affairs] (Bejing: Renmin Chubanshe, 1997), 4:72–78; this line of reasoning was consistent with Zhou's own logic in exchanges with Mao in late September. See Xu, *Mao Zedong Yu KangMei YuanChao Zhanzheng*, 74.

[74] Ibid., 74.

[75] Ibid., 94.

[76] Zhang, "Peng Dehuai Shouming Shuaishi KangMei YuanChao de Qianqian Houhou," 136.

that he had long ago reached the same conclusion, stating that Peng had "hit the nail on the head (*yizhen jianxie*)."[77]

Two parallel and related logics won the day in Beijing: the first was the desire to jump through a window of opportunity to attack the enemy before the United States and its allies were more fully prepared for war; the second was the need to prevent a window of vulnerability from opening, by not allowing the United States and Japan to amass forces and use bases in Korea and Taiwan to attack the PRC later. In his briefing to his division commanders just before entering Korea, Peng pointed out in October 1950 that the United States and its allies were overextended, and, while the Americans were indeed mobilizing, it would take the United States until June 1951 to draft and train 500,000 additional troops. Peng said that this was the reasoning used by those in the Politburo who decided that fighting a war sooner rather than later was the right decision, despite China's own lack of full readiness.[78] As Peng explained, "Our preparations [for war] are insufficient, [but] the enemy's preparations are also [currently] insufficient, especially the American imperialist's preparations are insufficient."[79] Mao's and Peng's argument to the Politburo for action was that a combination of geography and Chinese military backwardness made ground warfare the most desirable option to employ against the Americans, who could attack by bombardment from Korea and Taiwan at any time of its choosing. Waiting to fight would also allow the Americans and it allies to bring more force to bear against economically more developed targets. So, the winning argument for war in Beijing was that fighting a war in the short term would be less painful than fighting a larger war later.[80]

The same kind of windows logic was emphasized in a telegram that Stalin sent to Mao on October 5 during the important CCP Politburo meeting. In the telegram Stalin explained why he was disappointed that China had not simply accepted his October 1 request to send five to six divisions toward the 38th parallel to help defend North Korea. Stalin emphasized China's national security concerns as paramount in his reasoning for why China should send troops into Korea, stating that he had assumed China had prepared to enter the Korean War over the previous months "in order to prevent Korea from becoming a military base from which the Americans and the future militaristic Japan can oppose

---

[77] Ibid.

[78] Peng Dehuai, "Talk to the Meeting to Mobilize Cadres of Division Commander and Above of the People's Volunteer Army," October 14, 1950, in *Peng Dehuai Junshi Wenxuan* [Selected Military Writings of Peng Dehuai] (Beijing: Central Documents Publishers, Sept. 1988), 321.

[79] Ibid.

[80] See Peng Dehuai, *Memoirs of a Chinese Marshal: The Autobiographical Notes of Peng Dehuai 1898–1974* (Beijing: Foreign Languages Press, 1984), 473–74; and Christensen, *Useful Adversaries*, ch. 5.

China. This is all closely bound up with China." He continued by suggesting that there were four key reasons for him to advise China to enter the war with five to six divisions.

1. As the Korean War has shown, the United States is not now prepared to fight a wider war.
2. The militarist forces in Japan have not yet been rehabilitated, and cannot give the United States military assistance.
3. Owing to this, the United States will be compelled to yield on the Korea question by China, which has standing behind it its Soviet ally, and [so, the United States] cannot but accept terms of settlement of Korean issue; these terms would be beneficial to Korea and would cause the enemies to have no way to transform Korea into their military base.
4. For the same reasons above, the United States would not only have to abandon Taiwan, but also reject the signing of a unilateral peace treaty with Japan, abandoning their plot to revive Japanese militarist activities and to make Japan become their springboard in the Far East.

From this I calculated that, if China were to adopt a passive, wait-and-see approach, and would not launch a serious test of strength, and [would not] once again impress people with a display of power, then China would be unable to obtain these concessions. China would not only be unable to get these concessions, but would even be unable to obtain Taiwan. The United States clings to Taiwan so it can be a military base. It does this not for Chiang Kai-shek, who has no hope of achieving victory, but for itself or for a future militaristic Japan. [81]

Stalin continued that, on the off chance that the Americans chose to escalate and cause a world war out of concern for its "prestige," then it was better to fight such a war now rather than later. Applying windows logic, he said "then let us fight now, and not after several years. By that time, a militaristic Japan will have been revived and will be a U.S. ally, and the United States and Japan will have a bridgehead on mainland Asia under conditions in which Syngman Rhee controls all of Korea." Stalin concluded his telegram by expressing a bit of bewilderment

---

[81] The translation of the documents offered here is based on two separate documents. The first is the Chinese version of the October 5 document from Mao to Stalin. The second is the translation of an October 8 telegram from Stalin to Kim that claims to include the October 5 telegram to Mao. The two seems sufficiently similar that they are likely fully identical, but because I was not sure and because the October 8 letter deleted certain sections regarding Chinese domestic politics, I used the Chinese version when differences appeared between the two translations. For the Chinese version, see "Shi Dalin Guanyu Zhongguo Chubing Wenti Zhi Mao Zedong Dian" [Stalin's Telegram to Mao Zedong Regarding the Issue of Dispatching Troops], October 5, 1950, in Shen, ed., *Chaoxian Zhanzheng*, 2:581–82. For the English version, see Mansourov, "Letter, Stalin to Kim Il-sung," October 8, 1950, in Mansourov, "Stalin, Mao, Kim," Appendix, document no. 13.

regarding Mao's discussion of war and domestic politics in the PRC contained in the October 2 telegram that the Chairman had sent to him. Mao had said that many in China yearn for peace and that, therefore, war would be domestically risky for the PRC. Seeming to call into question Mao's movement's internationalism, Stalin said he understood this to mean that capitalist party factions in China's government would be able to use domestic dissatisfaction in wartime conditions to attack the Communist Party and its leaders.[82]

Stalin's lengthy and detailed October 5 telegram provided reasons grounded in national security concerns and internationalist ideology explaining why Mao should push his party toward war. However, he placed a heavy emphasis on the national security issues for China. In laying out the argument for war now rather than war later, Stalin emphasized issues related to all three of Washington's future East Asia alliance relationships. He argued that Korea and Taiwan were linked together and that the United States had chosen to protect Taiwan and to unify Korea as future launching pads for attack against China. The protection of Chiang Kai-shek on Taiwan helped undercut assurances made by Washington that its intention of unifying the Korean peninsula was not aimed at the Chinese communists or any other third party. With regard to Japan, Stalin ascribed to the United States the intention of rebuilding Japanese military strength but also pointed out that Washington had not yet succeeded in doing so.

So, in Stalin's mind, the prospect that the United States could sign a peace treaty with Japan unilaterally and then encourage a more robust Japanese military posture undercut both sides of successful coercive diplomacy: the credibility of near-term threats and the credibility of long-term assurances that the threats are conditional on the target's behavior. U.S. policy on Japan helped create a double failure of coercive diplomacy. The difficulty that Dulles and other U.S. officials had in convincing Japan to play a more active military role helped undercut the credibility of the United States' deterrent threat to escalate effectively in the near term if China were to enter the war in Korea. At the same time, the longer-term prospect of a remilitarized Japan undercut Washington's assurances to China that a unified anticommunist Korea could be viewed as an acceptable and nonthreatening outcome for China and the Soviet Union. Finally, Stalin tied all the elements of his threat assessment together by saying that Korea and Taiwan were a threat because of the future opportunities they would supply not only to an aggressive United States, but also to a militarily revived Japan.

---

[82] Ibid. For the Chinese version, see "Shi Dalin Guanyu Zhongguo Chubing Wenti Zhi Mao Zedong Dian, (Stalin's Telegram to Mao Zedong Regarding the Issue of Dispatching Troops," October 5, 1950 in Shen, ed., *Chaoxian Zhanzheng*, 2:581–82. For the English version, see Mansourov, "Letter, Stalin to Kim Il-sung," October 8, 1950, in Mansourov, "Stalin, Mao, Kim," Appendix, document no. 13.

Infamous for his diplomatic duplicity, Stalin was almost certainly not fully sincere in these strategic assessments. We can be fairly certain that the confidence Stalin expressed to Mao about a superpower conflict in 1950 was just bravado. As Kathryn Weathersby argues, no matter how concerned he might have been about the future balance of power if North Korea were to fall, Stalin did not see the USSR as having the upper hand in 1950 and he therefore seemed committed to going to great lengths to avoid war with the United States.[83] We can probably also take with a grain of salt Stalin's predictions that, if China entered the Korean War, the United States would forego a unilateral peace treaty with Japan and abandon Taiwan. In fact, as discussed in the last chapter, it seems just as likely, if not more so, that Stalin hoped that Chinese involvement in Korea would distract Chinese attention from Taiwan and also lock the United States and China into a hostile, long-term relationship, preventing Beijing from straying too far from Soviet alliance leadership.

Those caveats having been stated, Stalin's concerns about the threats that failure in Korea would pose for China and the Soviet Union were likely fully sincere. Stalin seemed to have shared for quite some time many of China's apprehensions about a revived Japanese military presence in Northeast Asia, particularly on the Asian mainland. He probably was quite concerned, albeit not to the same extent as were CCP elites, that a unified Korea and an increasingly confident United States would lead only to a more assertive and capable Japan in Asia in the future.[84]

It is not entirely clear how important Stalin's October 5 telegram to Mao was in the CCP leadership's deliberations for war. It seems quite likely that the key decisions may have already been made in Beijing by the time the cable arrived. Military historian Xu Yan argues that the Politburo had reached the same conclusion on its own and that the Stalin telegram was not decisive for that reason. Xu points out that Stalin's argument that it would be better to fight sooner rather than later was precisely in line with the arguments already made in the CCP Politburo session by war supporters such as Mao Zedong and Peng Dehuai.[85] For our purposes, what is important is that Mao and Stalin saw eye to eye on the logic of the latter's telegram, which emphasized the importance of U.S. defense relationships in the region as a reason to fight now, rather than later. Late in the

[83] Weathersby, "Should We Fear This?" 19. Agreeing with this basic viewpoint, Xu Yan argues that, if anything, Stalin excessively feared the United States at the time. Xu places this misperception in the same league as the U.S. underestimation of Chinese strength and resolve as causes of the tragedy of Korean War escalation. Xu, *Mao Zedong Yu KangMei YuanChao Zhanzheng*, 96. Xu writes that in the Soviet Politbturo sessions on October 5, Stalin emphasized that war with the United States needed to be avoided even if Korea needed to be surrendered.

[84] Weathersby, "New Findings," and "Should We Fear This?"

[85] Xu, *Mao Zedong yu KangMei YuanChao Zhanzheng*, 100.

evening of October 6, 1950, Mao told Soviet Ambassador Roschin that the CCP had decided in principle to send at least nine divisions to Korea. In this message, Mao stated that he "totally agreed with" (*wanquan tongyi*) Stalin's analysis of the international situation in the October 5 telegram.[86] In a telegram on October 8 to the North Koreans, Stalin described how he had convinced the wobbly Chinese, including Mao, about why they had to enter the war.[87] As was usually the case with Stalin's version of events, the truth of the matter was almost certainly more subtle.

We know Mao himself was resolved to enter the war in the first days of October, but he did need to convince more nervous comrades.[88] Stalin's October 5 telegram may have assisted in that process, but we are not sure exactly what role it played. As Xu recognizes, even if Mao and the CCP leaders came to the same general conclusions on their own, Stalin's thinking on the issue, as laid out in his telegram, was still quite important in Beijing. First, no matter how resolute he was about fighting in Korea, Mao would need Soviet military support for a Chinese effort against the United States, including air cover for key points (*yaodian*) in the interior.[89] So, it was important that the Chinese and Soviets were on the same strategic page. Second, doubts about the decision to enter the war lingered in Beijing long after the initial decision was made in the Politburo on October 5. For example, on October 6, at a meeting to discuss how to implement the October 5 decision, Lin Biao once again stated that he opposed sending troops to Korea.[90] Moreover, Mao's message to Stalin at midnight on October 6–7 regarding the Chinese decision to enter the war was, in some basic sense, no more final or definitive than was his October 2 telegram declining Stalin's October 1 request for China to enter the war immediately. Both statements were heavily conditioned on how much support China was to expect from Stalin's USSR.

However confident Stalin himself might have been after he received Mao's message on October 7, we now know that the CCP Politburo's decision was still conditional on the timeliness and scope of the military support that the

---

[86] See "Weishen Guanyu Mao Zedong Dui Chubing de Taidu Wenti Zhi Shi Dalin Dian" [Roschin's Telegram to Stalin Regarding the Issue of Mao Zedong's Attitude Regarding Dispatching Troops], October 7, 1950, in Shen, ed., *Chaoxian Zhanzheng*, 2: 588–90.

[87] "Shi Dalin Guanyu Zhongguo Chubing Wenti Zhi Jin Richeng de Xin" [Stalin's Letter to Kim Il-sung Regarding the Issue of China Dispatching Troops], October 8, 1950, in Shen, ed., *Chaoxian Zhanzheng*, 2: 591–92.

[88] Both Xu, *Mao Zedong yu KangMei YuanChao Zhanzheng*, and Liang, *Chaoxian Zhanzheng Qijian*, argue that Mao's own relatively high degree of resolve in supporting North Korea flowed from his commitment to communist internationalism as well as his concerns about PRC national security.

[89] Xu, *Mao Zedong yu KangMei YuanChao Zhanzheng*, 100.

[90] According to Xu, Zhou Enlai criticized Lin Biao severely for breaking ranks with a decision that had already been made by Mao and the Politburo. Xu, *Mao Zedong yu KangMei YuanChao Zhanzheng*, 100–101.

Soviets were willing to lend to their Chinese allies.[91] On October 6–7, Mao told Stalin's envoy that the PRC was preparing to send nine divisions, not just six, but that they could not be dispatched immediately and that, in the meantime, Zhou Enlai and Lin Biao should travel to the USSR to discuss Soviet aid.[92] Since Stalin was at his dacha on the Black Sea, Zhou would need to fly to Crimea, along with Lin Biao, and would not meet with Stalin until the night of October 9. Apparently dissatisfied with the initial messages that Zhou was sending back to Beijing on the question of prompt Soviet provision of air cover (an urgent telegram, it seems, was sent from Zhou to Beijing from the PRC Embassy in Moscow on the night of October 10), on October 12 Mao ordered that preparations for entering Korea be halted and that troops simply train pending further orders. He also recalled Peng Dehuai to Beijing and called for a second emergency Politburo meeting on October 13 to discuss the Korean War.[93] It was only on October 13 that Mao would reaffirm China's commitment to the war in a telegram to Stalin.

Mao's eleventh-hour hesitation and Zhou Enlai's continued arguments to Stalin for Chinese reluctance to enter the war suggest that Stalin's arguments in his October 5 telegram, repeated to Zhou Enlai during his trip to Moscow, still might have been important in steeling the CCP's resolve for the final commitment to war, even if they did not directly impact the deliberations on October 5 in Beijing.[94] Stalin's arguments, in line with those in China who supported entering the war, could then simply have made it easier for Mao and Peng to gain and maintain support among their comrades who in early October still had severe reservations about a war and whose concern could only have grown when it appeared to top Chinese observers that Stalin was backing out on his earlier promise to provide air cover promptly to Chinese forces crossing the Yalu river.

[91] This is true despite China's telegram to Stalin stating that China would enter the war and despite October 8 telegrams from Mao to both his field commanders and Kim Il-sung announcing China's plans to enter the war. For Mao's October 8 telegrams to the commanders and to Kim Il-sung, see "Guanyu Zucheng Zhongguo Renmin Zhiyuan Jun de Mingling" [Orders Concerning the Formation of a Chinese People's Volunteer Army], October 8, 1950; and "Guanyu Paiqian Zhiyuan Jun Ru Chaoxian Zuozhan Wenti Gei Jin Richeng de Dianbao" [Telegram to Kim Il-sung Regarding Sending a Volunteer Army into Korea for Combat], October 8, 1950, in *Jianguo Yilai Mao Zedong Wengao*, 1:543–45.

[92] For a subtle discussion of this oft-misinterpreted aspect of the history, see Mansourov, "Stalin, Mao, Kim," 22–24.

[93] For an authoritative secondary history, see Zhang, "Peng Dehuai Shouming Shuaishi KangMei YuanChao Qianqian Houhou," 147–51. For Mao's October 12 orders to Peng in Manchuria, see "Guanyu Shisan Bingtuan Reng Zai Yuandi Xunlian Gei Peng Dehuai Deng de Dianbao" [Telegram to Peng Dehuai et al. Regarding the Thirteenth Group Army Remaining in Their Original Locations for Training], October 12, 1950, in *Jianguo Yilai Mao Zedong Wengao*, 1:552.

[94] According to Mansourov, Stalin employed the same type of arguments in his discussions with Zhou and Lin in the Soviet Union. See Mansourov, "Stalin, Mao, Kim," pp. 22–23.

We do not have the detailed minutes of Zhou's negotiations with Stalin during his trip to the Soviet Union in the second week of October, but we can be fairly sure from his translator's memoirs and from secondary accounts that the main topics were the nature and timing of Soviet assistance and the overall willingness or unwillingness of the Chinese to fight in Korea. Always a fine negotiator, Zhou apparently opened the discussion with all the reasons for China not to enter the war.[95] Back in Beijing, Mao was apparently unsatisfied with the nature of the initial Soviet responses to Zhou's requests for rapid and significant Soviet military assistance. He therefore entered a final period of apparent indecision regarding the war and put the plans to dispatch troops to Korea on hold in an October 12 telegram to his commanders. As will be discussed in more detail, on October 12–13 Stalin mistakenly concluded that his Chinese allies had backed out of the war, and he began to make preparations to facilitate communist abandonment of North Korea. When Mao reconfirmed to Stalin on October 13 that China would indeed enter the war, the Chairman emphasized the importance of clarification of Soviet assistance levels in the decision. Mao said, "Past hesitations by our comrades occurred because questions about the international situation, questions about the Soviet assistance to us, and questions about air cover were not clear to them. At present all of these questions have been clarified."[96] Mao added an explanation for why China needed to en-

[95] Mansourov, "Stalin, Mao, Kim," 22–25, speculates that by expressing continuing reluctance to enter the war in his negotiations with Stalin for assistance, Zhou was freelancing and attempting to come between Stalin and Mao because he was actually opposed to the war. This conclusion does not fit our general understanding of Zhou's role in the CCP or his personality, nor does it match what we know about Zhou's loyal support of Mao just days earlier in the Politburo meeting in Beijing. What is much more likely is that Zhou was simply conveying Beijing's mixed views on the matter in order to elicit the maximum amount of support from Stalin. Mansourov cites approvingly the memoirs of Shi Zhe, the Russian interpreter to Mao and Zhou. See Shi Zhe, *Zai Lishi Juren Shenbian: Shi Zhe Huiyilu* [At the Side of History's Giants: Shi Zhe's Memoirs] (Beijing: Chinese Central Documents Press, 1991), 495–96. Xu Yan argues that Shi Zhe probably got the facts about Zhou's statement correct, but failed to recognize that his stonewalling was not a violation of the Politburo's earlier decision for war but a negotiating tactic designed to coax the Soviets into supplying aid. Another way to view this is that it was not a violation of the Politburo's decision, because that decision to enter the war was conditional all along on significant and early Soviet assistance. So, as Zhou and Mao later discussed with Kim Il-sung, Zhou had gone to Moscow with a choice for Stalin: China would enter the war if the USSR was willing to provide air cover. See Xu, *Mao Zedong yu KangMei YuanChao Zhanzheng*, 114–15.

[96] "Ciphered Telegram from Roschin in Beijing to Stalin," in Mansourov, "Stalin, Mao, Kim," October 14, 1950, Appendix Document 12. Shen Zhihua dates this telegram as October 13. The difference might be in the time difference between Beijing and the western Soviet Union. For the Chinese-language version, see Shen, ed., *Chaoxian Zhanzheng*, 2:597. This telegram, apparently received in the early hours of October 14 by Stalin, was preceded by a telegram to Zhou in the Soviet Union on October 13 and was followed by a more detailed telegram to Zhou Enlai later on October 14, 1950, discussing Beijing's strategy. For the earlier October 13, 1950, and the later October 14, 1950, telegrams, see "Guanyu Wo Jun Yingdang Ru Chao Canzhan Gei Zhou Enlai de Dianbao" [Telegram to Zhou Enlai Concerning [Why] Our Troops Should Enter Korea], October

ter the war that was fully consistent with Stalin's October 5 telegram, citing the huge security costs of inaction to China, the region, and the world, and the huge potential benefits of taking the fight successfully to the Americans in Korea.[97]

It seems quite possible that Stalin sincerely agreed to provide air cover in the meetings with Zhou and Lin, but that the Chinese were disappointed to discover that the air cover would not arrive immediately. The Chinese were, perhaps, initially incredulous about the technical explanations Stalin and his colleagues offered for the delay. Historian Zhang Xi speculates that, after the stunning U.S. victories following Inchon, Stalin had grown nervous about the prospect of a direct U.S.-Soviet clash and therefore rescinded on his earlier promises of prompt air cover for Chinese forces entering Korea.[98] A simpler explanation would be that there was insufficient coordination between Beijing and Moscow on this question before Mao decided to cross the Yalu. Therefore, when Chinese "People's Volunteers" began pouring over the Yalu border on the night of October 18, 1950, the Soviet planes were not yet there to protect them. The available evidence suggests that this was more of a function of poor alliance coordination than it was Soviet reluctance to supply air cover in general. The Soviets had promised air cover in principle as early as July 1950 and once they made a concrete decision to assist in mid-October, the Soviets were able to get their pilots into the fight by November 1, hardly a long lead time for decisions that were made so late in the escalation process.[99] Judging from Chinese and U.S. documents, these Soviet air activities were quite significant, even before MacArthur made his final push to the Yalu in late November. On November 15, 1950, Mao agreed with Stalin's proposal to transfer 120 Mig-15s to China. In the same telegram Mao praised Soviet pilots for having downed twenty-three U.S. planes in the previous

---

13, 1950, and "Guanyu Zhiyuan Jun Ru Chao Zuozhan de Fangzhen He Bushu Gei Zhou Enlai de Dianbao" [Telegram to Zhou Enlai Concerning the Principles and Deployments of the People's Volunteer Army as It Enters Korea for Combat], in *Jianguo Yilai Mao Zedong Wengao*, 1:556, 560–561. The documents are translated in their entirety by the author in Christensen, *Useful Adversaries*, Appendix B, 273–75.

[97] For the full telegram to Stalin, see ibid. For such an interpretation of the document emphasizing these points, see Zhang, "Peng Dehuai Shouming Shuaishi KangMei YuanChao Qianqian Houhou," 151.

[98] Zhang, "Peng Dehuai Shouming Shuaishi KangMei YuanChao Qianqian Houhou," 150. It is important to note that he offers no evidence for this conclusion, but it is likely that the speculative impression was shared by Zhou's entourage and the leadership in Beijing. Disagreeing with this conclusion is Mansourov, "Stalin, Mao, Kim."

[99] This interpretation, if true, would help explain the difference between Kathryn Weathersby's suggestion that Stalin was relatively reluctant to enter the fight using Soviet pilots, "finally" doing so by November 1, 1950, and Alexandre Mansourov's interpretation that Stalin was consistently resolute in his willingness to provide such air cover and that only the Chinese had wavered in October. See Weathersby, "New Russian Documents," and Mansourov, "Stalin, Mao, Kim."

twelve days.[100] On November 21, 1950, Acheson discussed the downing of four U.S. B-29s over northernmost Korea by Soviet planes with French Ambassador Henri Bonnet.[101]

In one sense, exactly how and when Stalin's own analysis and argumentation to the Chinese might have mattered in the Chinese decision process might not be so important for our purposes here. The most important argument that Xu makes in supporting his claim that Stalin's October 5 telegram was not influential in the CCP decision process is that the basic arguments about future U.S. alliances and the logic of a closing window of opportunity or an opening window of vulnerability laid out in Stalin's telegram was apparently already shared and articulated by Politburo members in Beijing. In fact, in his telegram it is possible that Stalin was simply cynically playing on what he correctly believed to be Chinese leaders' existing fears about the future of Taiwan, Japan, and Korea, knowing Mao would "totally agree" with the logic of his telegram. But whichever the case, we still see the consequences that early developments in the U.S. Cold War alliance system in East Asia had on Chinese and Soviet strategic thinking in what was, by all accounts, an extremely difficult decision to make. The failure of U.S. coercive diplomacy in northern Korea in 1950 was due in part to the ways in which its currently weak but possibly strong and aggressive future alliances undercut near-term credible threats and long-term credible assurances. Under those circumstances, once U.S. forces crossed the 38th parallel, nothing MacArthur could have said or done differently in the near term likely could have prevented escalation of the Korean War to a Sino-American conflict, short of an unforced military retreat, an option advocated by no one in the U.S. government.

Stalin Nearly Accepts Defeat but China Saves the DPRK

The Soviet reaction to Mao's final days of indecision in Beijing also speaks volumes about how the communist alliance was in many ways harder for the United States to coerce successfully than a unified alliance might have been. Despite all of their own fears about a resurgent Japan and a unified Korea working in cahoots in Northeast Asia, the Soviets were apparently much more willing to accept North Korean collapse as the price of avoiding a wider war with the United States than were the Chinese. On the one hand, Stalin was clearly unable simply to order China into the war against its will, as he might have been able to do if the movement were monolithic. But it is clear that Soviet national security con-

---

[100] "Mao Zedong Guanyu Tongyi Jiaqiang Kongjun de Jianyi Zhi Shi Dalin Dian" [Mao Zedong's Telegram to Stalin Regarding Agreement on the Proposal to Strengthen the Air Force], November 15, 1950, in Shen, ed,., *Chaoxian Zhanzheng*, 2:622.

[101] Minutes of Meeting of Dean Acheson with French Ambassador Bonnet, November 21, 1950, Executive Secretariat Records, lot 53D444, box 14, National Archives.

cerns regarding unification of Korea under UN control were still different than those of the Chinese.

Stalin had grown frustrated with the CCP in the first half of October. He had expected China to snap to attention and enter the war unconditionally, immediately, and on his terms after he sent his initial telegram on October 1. Stalin again came to believe that China would enter the war after receiving Mao's telegram on October 7, 1950. So, he must have been terribly disappointed a second time when Zhou came to the Soviet Union and once again seemingly backtracked from that commitment. What he did not recognize readily is that Beijing all along was willing to enter the war as long as the Soviets provided significant assistance and air cover, a common theme that ties together Mao's October 2, October 7, and October 14 messages to Stalin.

A careful analysis of Stalin's roller-coaster ride of pessimism and confidence from October 1–14 reveals two important and related alliance dynamics that guided the process of Korean War escalation: 1) the differences between Chinese and Soviet security concerns in Korea, and the relative cautiousness of the alliance leader about confronting the United States directly; and 2) Stalin's overriding goal of preserving Soviet prestige and influence within the communist alliance. Four telegrams reveal Stalin's manipulation of his allies. In the first telegram to Kim on October 2, as discussed earlier, Stalin duplicitously reveals to Kim his allegedly confidential discussions with Mao and accuses the Chinese of backing out on their earlier commitments, even though at that point Mao had not drawn any final conclusions about the Chinese entrance into the war. On October 8, Stalin sent a telegram to Kim Il-sung regarding Beijing's October 6 and October 7 communications with the Soviet leader. Although Mao had said that the PLA was not yet ready to enter the war on October 7 and that he wanted discussions in the USSR about the terms of Soviet assistance, Stalin apparently took this new position by Beijing as a permanent and full reversal of Mao's earlier reluctance to enter the war. In his telegram to Kim, Stalin did more than reveal again all his confidential discussions with Mao, Stalin portrayed himself as the hero in the story. He revealed to Kim his initial request to Mao on October 1 to send five to six divisions into Korea and simplistically and inaccurately described Mao's "refusal" to do so. Stalin then pasted into the telegram to Kim a huge section of his October 5 telegram to Mao, in which he had urged Beijing to reconsider. Stalin concluded the October 8 telegram to Kim by claiming that, after receiving Stalin's own October 5 telegram, Beijing had finally come around to his position and was now willing to send nine divisions into Korea.[102] As in

---

[102] For the English translation of this Russian-language document, see "Letter, Stalin to Kim Il-sung (via Shtykov) 8 [7] October 1950," in Mansourov, "Stalin, Mao, Kim," Appendix, document no. 13.

early 1950, Stalin seemed more interested in portraying himself as the most helpful ally to the North Koreans than in building any solidarity between the Chinese communists and the North Korean communists. Moreover, he did this even though he had consistently been less willing to take risks than the Chinese communists had been since the outbreak of the war.

Some time after Zhou Enlai arrived at Stalin's Black Sea dacha to negotiate Soviet support, Stalin once again grew very pessimistic about the prospects of Chinese intervention. According to Mansourov's account, in response to Zhou's expressed reasons for China to avoid entrance into the war, Stalin first grew angry. The Soviet leader then startled the Chinese entourage by saying that they should encourage the North Korean government to flee Korea and set up bases in Manchuria. [103] Although it would be a most unwelcome outcome in Moscow, Soviet national security, broadly considered, would not be very greatly affected by the unification of Korea under a pro-U.S. regime. Therefore, preventing it was not worth risking a wider war involving direct Soviet intervention (Chinese intervention in a more limited war, of course, was quite another matter).[104] According to the Chinese interpreter at the meetings, Stalin proposed to Zhou Enlai and Lin Biao that while U.S. occupation of all of Korea would of course be terrible, especially for China, it would be better if it came about by a planned retreat to Manchuria by Kim Il-sung's forces than their total destruction in fighting in North Korea.[105]

This new position by Stalin may have seemed an empty threat—a pure negotiating tactic—designed to startle the Chinese into accepting the burden of North Korea's defense. But it was apparently much more than that. On October 13 Stalin appears to have sent orders to his embassy in Pyongyang to prepare to evacuate Kim Il-sung's government from North Korea and to prepare to set up a government in exile. These orders caused even more alarm among the Korean comrades (who nevertheless apparently complied with the order from their "loving father" in Moscow). Obsessed with the Soviets' relative position in the movement, in this October 13 telegram Stalin once again pinned all the blame for the North Korean collapse on the CCP's unwillingness to fight in Korea despite Soviet leadership in asking them to do so.[106] Then, in two telegrams to

[103] Ibid. 23.

[104] The interpreter Shi Zhe's memoirs confirm that Stalin was concerned mainly about the prospect of direct engagement between Soviet and American forces in Korea. Shi Zhe, *Zai Lishi Juren Shenbian*, 497–98.

[105] Ibid., 496–98.

[106] Mansourov reports that on October 13, 1950, Stalin apparently telegrammed Pyongyang to order the evacuation. Mansourov, "Stalin, Mao, Kim," 26. We know such an order was sent, because Shtykov refers to it and to Kim Il-sung's and Pak Hon-yong's surprise upon receiving it in a telegram to Stalin the next day. See "Ciphered Telegram Shtykov to Stalin," October 14, 1950, in Mansourov, "Stalin, Mao, Kim," Appendix, document no. 18; for the Chinese translation, see "Shitekefu Guanyu

Kim Il-sung on October 14 Stalin would report to an elated Kim that he should postpone the evacuation plan and that the Chinese were indeed coming to assist him. Especially in the second telegram, Stalin portrays himself as resolute and consistent in his support and the Chinese as having wavered but having finally decided to enter the war, adding gratuitously for good measure, "all technical equipping of the Chinese military will be provided by the Soviet Union."[107]

When the Chinese communists appeared unwilling to enter the war, in communications with Pyongyang Stalin blamed them for their lack of strategic acumen, their lack of resolve, and their lack of internationalism, placing the North Korean defeat firmly on Beijing's doorstep. When the Chinese communists had clearly agreed to enter the war, in communications with Pyongyang Stalin emphasized how they had wavered before they were urged on by Stalin. He also underscored how massive Soviet assistance to China was largely responsible for their ability to fight. In this way, regardless of the outcome of the Chinese decision, the USSR would be seen in Pyongyang as the clear alliance leader and as the high church of communist internationalism. Moreover, regardless of the outcome of the Chinese decision process, Stalin would maintain his key equities in the communist alliance system. If China entered the war it would be distracted from Taiwan and terribly dependent on Soviet support and therefore much less likely to go its own way and break with the Soviets. If China did not enter the war and U.S. forces were allowed to move to the Yalu, China would also be distracted from an offensive in Taiwan and would also be dependent on the Soviet Union to help deter and defend against a U.S. attack on Manchuria.

WORSE THAN A MONOLITH IN INDOCHINA: MAO'S ASSISTANCE TO VIETNAM BEFORE AND DURING THE KOREAN WAR

The First Indochina War

Although the fighting there was being done by France and Great Britain, U.S. NATO allies, and not directly by the United States, alliance politics in the communist camp also made the international communist movement harder to contain in Southeast Asia—something that the United States, Great Britain, and

---

Baihui Jin Richeng de Qingkuang Zhi Shi Dalin Dian," October 14, 1950 in Shen, ed., *Chaoxian Zhanzheng*, 2:599. Kim Il-sung would on occasion obsequiously use the equivalent of the Chinese word "cifu" or "loving father" in describing Stalin's concern about and aid to the North Korean regime. See, for example, "Shitekefu Guanyu Zhuancheng Jin Richeng de Ganxie Xin Zhi Shi Dalin Dian" [Shtykov's Telegram to Stalin Passing Along Kim Il-sung's Letter of Thanks], August 31, 1950, in Shen, ed., *Chaoxian Zhanzheng*, 2:517.

[107] "Shi Dalin Guanyu Zhongguo Chubing Yuanzhu Chaoxian Wenti Zhi Jin Richeng Dian," October 14, 1950, in Shen, ed., *Chaoxian Zhanzheng*, 2:601.

France were all committed to accomplishing. This was particularly true in Indochina, where Chinese economic and military assistance to Ho Chi Minh's young government was indispensible in its successful prosecution of war against the French colonial forces in the northernmost part of Vietnam.

In January 1950, Mao formally recognized Ho Chi Minh's Democratic People's Republic of Vietnam (DRV) before the Soviets did later in the month. Although the order of recognitions was only of symbolic importance, what it symbolized was indeed important. As a global power the Soviets could not let East Asia harm interests at the core. In some ways there was a parallel between the Soviets' valuing of Europe over Southeast Asia and the U.S. strongpoint strategy that precluded more active intervention in peripheral areas, such as mainland China. In addition, as a skeptic about rural, Third World revolution, Stalin had ideological reasons to call into question the veracity of Ho's Marxist-Leninist credentials.

Stalin's main interests were in Europe, and his reluctance to recognize Ho's regime came partially out of his desire not to alienate the French public. The bulk of French citizens supported French imperialism in Southeast Asia.[108] In fact, French communist politicians themselves voted for war credits for the Indochina war either out of heart-felt conviction or fear of electoral backlash. Stalin did not want to alienate the French public at a time when France faced decisions regarding European defense that could influence the European balance of power to the detriment of the Soviets (France's role in NATO and the European Defense Community).[109]

As a regional actor, Mao cared little about France, NATO, and the question of German rearmament, but he did care a great deal about Indochina, which bordered his own nation. In addition to these differences in national interest between the Soviets and Chinese communists, Vietnamese revolution provided a fine platform for proof of Mao's own brand of rural, Third World revolution combined with postcolonial nationalism. Both geographic and ideological factors led Mao to give significantly more assistance to Ho Chi Minh in 1950 than the Soviets did. It seems fair to argue that Mao provided more aid to Ho's movement than the Soviets would have arranged for Ho had Moscow been in tight control over events in the region (if nothing else, the relative unimportance of Vietnam as compared to France for the communist alliance as a whole probably would have limited Soviet assistance).

Mao went further than simply being the first to recognize Ho Chi Minh's government. He offered Ho economic and military assistance and created Chinese

[108] Zhai Qiang, *China and the Vietnam Wars, 1950–1975* (Chapel Hill: University of North Carolina Press, 2000), 15.

[109] Ilya Gaiduk, *Confronting Vietnam: Soviet Policy Toward the Indochina Conflict* (Washington D.C., and Stanford: Woodrow Wilson Center and Stanford University, 2003), 3.

advisory groups to assist the Vietminh in planning war with the French.[110] The initial decisions for support of Vietnam apparently came in late December 1949. Although China rejected certain Vietnamese communists' requests, including one for $10 million U.S. dollars in hard currency, the Chinese Communists offered clear, early support to their Vietnamese brethren.[111] The CCP Central Committee's priority was to assess and meet the emergency military needs of the Vietnamese communists, but it is clear that in the early months of 1950 the ultimate goal was to assist in helping Ho Chi Minh "defeat the French imperialists" in a broader struggle.[112] One internally circulated CCP Party history argues that, despite the appearance of joint Sino-Soviet support for the anti-French struggle of the Vietnamese communists, in material terms only China provided support during the early 1950s. That support included 116,000 guns as well as hundreds of artillery pieces, ammunition, and so on. Moreover, this book states that Chinese provision of firearms, food, oil, and medical supplies helped harden Ho Chi Minh's attitudes toward military struggle with the French.[113] It is difficult to believe that the Soviets would have devoted such attention and expended so many resources on distant Vietnam if they had been managing the alliance's regional strategy in Southeast Asia from Moscow.

Historians Chen Jian and Qiang Zhai point out that national security concerns were almost certainly part of the reason for Mao's early support of revolution in Vietnam, but that assistance was not merely a reaction to the heating up of the Cold War and China's increased worries about its security: it also had strong ideological components.[114] After all, the initial Chinese commitment of support occurred in the period before the Korean War, when U.S. defense spending was one third of what it would be a year later, when NSC-68 was shelved by Truman as unworkable, and when the United States had recently withdrawn troops from Korea and mainland China and had apparently written off Korea and Taiwan in

---

[110] Chen, "China and the First Indo-China War, 1950–54."

[111] For documentation, see relevant telegrams by Liu Shaoqi to Lin Biao (December 20, 1949) Mao Zedong (December 24, 1949), and the Central Committee of the Vietnamese Communist Party (December 24), in *Jianguo Yilai Liu Shaoqi Wengao*, 1:165, 186–187, 189, respectively. An internal Chinese history states that in early January China supplied the Vietnamese communists with emergency aid of 1200 artillery shells, 420,000 bullets for U.S.-style 30 caliber machine guns, 910,000 bullets for British-style 30 caliber machine gun bullets, and 20 vehicles. "Jianguo Hou Wo Dui Wai Junshi Yuanzhu Qingkuang" [The Foreign Military Aid Situation After the Founding of the Nation], *Junshi Ziliao* [Military Materials] no. 4 (1989): 33; internally circulated.

[112] For the two-step program see the March 13, 1950, instructions given to Luo Guibo, the CCP Central Committee's secret emissary to Vietnam in "Zhongyang Gei Luo Guibo de Dianbao" [Central Committee Telegram to Luo Guibo], in *Jianguo Yilai Liu Shaoqi Wengao*, 1: 478–79.

[113] Guo Ming, ed., *ZhongYue Guanxi Yanbian Sishi Nian* [40 Year Evolution of Sino-Vietnamese Relations] (Guangxi, Guangxi People's Publishers, May, 1992), internally circulated, pp. 26–27, 55.

[114] Chen, "China and the First Indo-China War, 1950–54," 90; and Zhai, *China and the Vietnam War*.

Truman's January 5 speech and Acheson's defense perimeter speech of January 12, 1950. So, especially before the Korean War, a desire to improve China's ideological stature in the international communist movement seems to have been a more important motivation behind the CCP's assistance to the Vietnamese communists. Emphasizing the importance of international ideological status in Chinese strategy at the time was Vice Chairman Liu Shaoqi. In an internal March 1950 directive he argued that the CCP's assistance to Asian revolution was "one of the most important methods to consolidate the victory of the Chinese revolution in the international arena."[115]

Southeast Asia was indeed a potential front for invasion by foreign imperialists and a raiding base for remnant KMT groups near the still largely unpacified Southwest part of China (Yunnan and Guangxi Provinces, for example). The Soviets were simply not at risk from the same threats, and their relative attention to Korea and inattention to Vietnam demonstrated this. As Odd Arne Westad points out, Korea was not only more important to the Soviets than Vietnam, the Soviets all but managed North Korea's strategy and diplomacy as the nation that accepted Japanese surrender in Korea north of the 38th.[116] The North Korean leadership was therefore much closer to the Soviet regime than to the Chinese. The Vietnamese communists were a distant lot to Moscow, and, according to historian of Soviet foreign policy Ilya Gaiduk, Stalin was very skeptical about the internationalism in the Vietnamese communist movement, worrying that Ho's party had excessive contacts with the West. Stalin would eschew alliances with or aid packages for the Vietminh. Moreover, Stalin was not eager to run again the risks in another Korea-style conflict involving the United States and its allies.[117]

Westad reports that Stalin was more than relatively indifferent about Vietnam, he was nervous about Chinese support to the Vietnamese communists. In July 1949 Liu Shaoqi had promised CCP support for Ho Chi Minh's Vietminh and, as discussed earlier, Stalin had placed the CCP "in charge" of Asian revolution in a way that was fully consistent with a hierarchical alliance. But when Mao and his colleagues began reporting their plans for significant aid to the Vietminh, Stalin warned them against overconfidence, citing the danger that the United States could enter on the side of France. Such an escalation would pose real dangers for the communists inside and outside Vietnam. According to Westad's account, Mao took the lead on Vietnam pushing an aggressive plan of action there against a resistant, cautious Stalin. Although he backed the Vietnamese communists in principle, internally Stalin complained to his Soviet col-

---

[115] Liu quoted in Zhai, *China and the Vietnam Wars*, 21.
[116] Westad, *Decisive Encounters*, 317–320.
[117] Gaiduk, *Confronting Vietnam*, 5–11.

leagues about CCP overexuberance toward the Vietnamese and worried about the CCP's overconfidence that the United States would not support the French there.[118] In late January 1950, Beijing took the lead again in arranging for Ho's entourage to travel to Moscow, where they were greeted coldly by Stalin and warmly by Mao and Zhou Enlai, who were still in town for the negotiations regarding the Sino-Soviet treaty. Ilya Gaiduk states that the only reason that Stalin agreed to meet Ho at all was that he did not want to cause "bewilderment" in the visiting Chinese entourage in Moscow, who had arranged Ho's trip. According to Nikita Khrushchev's memoirs, Stalin was condescending and unpleasant to Ho after he arrived.[119]

The U.S. entrance into the Korean War and the blocking of the Taiwan Strait also energized Mao to increase the existing levels of support for Ho Chi Minh in Vietnam. Rather than turning his attention from Vietnam in his preparation for conflict in Southeast and Northeast China, Mao sharply increased his support of the Vietnamese communists. He did this for a range of motivations not fully shared by the Soviets: fear of future attacks from foreign and domestic foes in the Southwest as part of a three-pronged enemy offensive; his desire to prove his party's internationalism on the international stage; and his sincere belief in the viability and utility of Marxist-Leninist revolution in the developing world.[120]

Chinese secondary histories reveal that, despite the appearance of joint Sino-Soviet support for the anti-French struggle of the Vietnamese communists, in material terms only China provided support in the early 1950s.[121] China also sent military advisors to North Vietnam and Laos to assist the Vietnamese communists in their fight against the French military.[122] It is difficult to believe that the Soviets would have devoted such attention and expended so many resources on distant Vietnam even if they had not ceded "leadership" of Asian revolution to the Chinese.

For its part, Moscow initially lent more political support to the Vietnamese communist effort upon the outbreak of the Korean War, apparently sharing to some degree the Chinese calculations that Vietnam could become part of a three-pronged attack against the PRC. When the Korean War escalated, however, and the Truman administration successfully used the occasion to implement NSC-68, Stalin refused to take concrete actions to assist the Vietnamese communists.

---

[118] Westad, *Decisive Encounters*, 317.

[119] Gaiduk, *Confronting Vietnam*, 3–6.

[120] Chen, "China and the First Indo-China War," 90–92; and Chen, *China's Road to the Korean War*, 132.

[121] Guo, ed., *ZhongYue Guanxi Yanbian Sishi Nian*, 26–27, 55.

[122] Yang Kuisong, "Changes in Mao Zedong's Attitude toward the Indochina War, 1949–1973," Cold War International History Project (CWIHP) Working Paper No. 34 (February 2002), 5–7.

The bulk of the increased U.S. defense spending would be utilized in Europe and other areas far from East Asia, areas that the Soviets, like the Americans, deemed much more important than Southeast Asia. So, as Gaiduk argues, "The Korean War further discouraged Moscow from risking a further involvement in a war in the remote regions of Indochina or Southeast Asia."[123] While Vietnam was a vital interest to China, it was at best a secondary one for the Soviets.

Moreover, as Chinese historian Yang Kuisong argues regarding Beijing's aid for Ho Chi Minh, "[T]he decision to provide such aid reflected to some extent the CCP's geopolitical interests, but ideological considerations were predominant." He points out that Mao applied the same revolutionary advice he gave to Vietnam to other communist parties in Asia, not just those on his southern border. So, he concludes, "[I]t was in line with this tendency to support Asian revolution that the CCP offered assistance to the anti-French war led by the Vietnamese Communists."[124] As the uncontested leader of international communism and as someone still skeptical about rural revolution, Stalin also did not share Mao's need to demonstrate his revolutionary internationalism nor did he place great stake in the prospects for widespread Third World revolution. Again, for both realpolitik and ideological reasons, it is difficult to imagine that a more centrally controlled international communist movement under his control would have been as supportive of revolutionary forces in Southeast Asia as was Mao's independently motivated CCP in 1950–53.

---

[123] Gaiduk, *Confronting Vietnam*, 6–11.
[124] Yang, "Changes in Mao Zedong's Attitude toward the Indochina War," 5–7.

Chapter 4

THE BENEFITS OF COMMUNIST ALLIANCE COORDINATION AND THE CONTINUING COSTS OF U.S. ALLIANCE FORMATION, 1951–56

By the mid-1950s Mao had proved himself a full-fledged member of the international communist movement, and Sino-Soviet cooperation would be much closer than it was before the Korean War began. While hardly an unalloyed benefit for the United States and its allies, a relatively more coordinated and hierarchical international communist movement meant that peace deals were easier to negotiate in Korea and Indochina and that more aggressive local actors were less capable of dragging their more powerful allies into positions that would cause escalation of the existing conflicts.

With the important exception of the 1954–55 Taiwan Strait crisis, the period 1954–57 was one in which relations across the Cold War divide in East Asia were relatively stable. That crisis itself was precipitated in large part by the interaction of U.S. alliance politics in the region with the CCP's devotion to preserving the mission of one day "liberating" Taiwan and unifying the island politically with the mainland. The United States unwittingly sent threatening signals to China regarding future U.S.-Taiwan political and defense relations during the process of forming the Southeast Asia Treaty Organization. Mao responded to these signals with a preventive assault on the Taiwan-held offshore islands off of Fujian Province. These attacks failed and, if anything, served only to hasten the formation of a U.S.-ROC security alliance.

REACHING A KOREAN ARMISTICE: THE MIXED BLESSING OF GREATER COMMUNIST COORDINATION

By the first half of 1951, political and military conditions in the Korean War had become much clearer than they were in the summer and fall of 1950. The Chinese communists were in the war very deeply, as were U.S. and allied forces fighting under the UN flag. Although alliance coordination problems in the communist camp remained, particularly between Chinese military and political leaders and their North Korean counterparts, the level of coordination between the three communist governments had improved quickly and significantly after fall 1950. In particular the Sino-Soviet relationship was very tight and very cooperative once Beijing decided to enter the war in force. Differences between Kim Il-sung and Peng Dehuai on war strategy remained through at least the first half of 1951. As we would expect, Kim continued to be the most aggressive advocate of offensives and was often able to get at least limited backing in Moscow and Beijing, sometimes to Peng's great frustration.[1]

Coordination among the three communist leaderships in military and diplomatic affairs improved after China's massive entrance into the war. Unfortunately for the United States, however, the UN battlefield defeats and the general impression that the communists had the upper hand in December 1950 and early January 1951 meant that communist alliance coordination did not produce accommodation in negotiations, but rather continued belligerence. All things being equal, such coordination among one's enemies would have facilitated U.S. coercive diplomacy, but successful coercive diplomacy still required one's own demonstrated strength and resolve, and until the stabilization of the UN position in South Korea in early 1951, UN forces were unable to demonstrate sufficient military power to constrain their enemies. In other words, given the massive communist victories against UN forces in late November and early December 1950, the local interests of the North Korean communists in East Asia, the regional and ideological interests of the CCP, and the global interests of the Soviet Union as leader of the alliance all coincided. Moscow, Beijing, and Pyongyang all supported the idea of maximizing both North Korean gains and U.S. humiliation by crossing the 38th parallel with communist forces and thereby delegitimizing further the 38th parallel as a stable dividing line in Korea. In December 1950 and January 1951 the three allied communist capitals coordinated their positions and consistently opposed early armistice talks. They all agreed that any U.S. suggestion of a ceasefire was simply an attempt to buy time and regroup for future UN offensives against the North.

In this context, Moscow and Beijing decided that it would be best to ignore the concerns of the Chinese commander on the ground, Peng Dehuai, about his forces' need for rest. Instead Soviet and Chinese leaders chose to order attacks on UN and ROK forces south of the 38th parallel before those forces had a chance to regroup.[2] In an important meeting in Moscow in early December,

---

[1] For a discussion of the debates between Peng and Kim, see Shen Zhihua, "Sino-North Korean Conflict and Its Resolution during the Korean War," *Cold War International History Project Bulletin* (*CWIHPB*) nos. 14/15 (winter 2003–2004): 15.

[2] For Sino-Soviet coordination on the communist position at the UN regarding the conditions for an armistice, see "Roschin Conveying Message from Zhou Enlai to Soviet Government," December 7, 1950, in Weathersby, "New Russian Documents," document no. 47, p. 52. For the Chinese-language version, see "Weishen Guanyu Zhongguo Zhengfu Tingzhi Zai Chaoxian Junshi Xingdong de Tiaojian de Dianbao" [Roschin's Telegram Regarding the Chinese Government's Conditions for Halting Military Activities in Korea], December 7, 1950, in Shen, ed., *ChaoXian ZhanZheng*, 2:639.

Deputy Foreign Ministers Andrei Gromyko and Wang Jiaxiang discussed the desirability of attacking across the 38th parallel. They both decided that it was best to "strike while the iron is hot."[3] This policy line fit perfectly North Korea's strategic goals: to drive U.S. forces entirely off the peninsula as soon as possible.

Communist forces suffered terrible losses in the early months of 1951 after they briefly retook Seoul in January. Overextended and vulnerable in South Korea, the communists faced fierce counteroffensives from UN forces. Through the first half of 1951, however, all three communist capitals still ultimately wanted to drive the UN forces completely off the peninsula. Where they differed was how aggressively to pursue this goal. The North Koreans preferred the most consistently aggressive approach. For their part, the CCP leaders, particularly those in the field, became tactically much more cautious than the DPRK leaders, who seemed almost desperate to deal a death blow to UN troops before foreign anticommunist forces could consolidate a position in South Korea. After Chinese troops suffered heavy losses in January and February, Mao initially shifted to a strategy of rotating troops at the front, while still planning for a large-scale spring offensive after his forces had had a chance to rest and recuperate. By early March, Mao decided to adjust to a less aggressive, attritional strategy, at least for the foreseeable future. He would continue to prepare for a spring offensive, but seemed increasingly aware of the limits of Chinese power. In April, Mao would launch one last spring offensive, but it was limited in scope and designed largely to preempt an expected UN offensive. This campaign took heavy tolls on the communist side and forced them to give up some territory north of the 38th parallel.[4]

Following horrendous losses in the ill-advised offensives of January, February, and April 1951, in May 1951 the communist leaders in North Korea, China, and the Soviet Union recognized that the war was, at best, in deadlock. In fact, the communists believed themselves vulnerable to UN offensives should the anticommunist allies decide to recover more areas controlled by the communists in the near term. As for a major communist offensive, this would be out of the

---

The Soviet Politburo believed that a UN ceasefire would only allow the United States to "win time" to regroup; see "Lian Gong (bu) Zhongyang Zhengzhiju Guanyu Tingzhan Wenti Zhishi de Jueyi" [Politburo Resolution on the Directives Relating to the Question of a Ceasefire], December 7, 1950, in Shen, ed., *Chaoxian Zhanzheng,* 2:643.

[3] "Geweimike yu Wang Jiaxiang Huitan de Baiwanglu" [Memorandum of Meeting of Gromyko and Wang Jiaxiang), December 5, 1950, in Shen, ed., *Chaoxian Zhanzheng,* 2:635. The Chinese phrase for strike when the iron is hot is *chen shi da tie.* Chinese historian Niu Jun's coverage of these events suggests that Gromyko was reflecting a Politburo decision of the same day. See Niu Jun, "Zhanzheng Jubuhua yu Tingzhan Tanpan Juece: KangMei YuanChao Zhanzheng Juece Yanjiu Zhi San" [The Localization of War and the Decision for Ceasefire Talks: Three Studies of Decision-Making in the Korean War], paper presented at the Scholarly Conference on China in the 1950s, Fudan University, Shanghai, August 2004.

[4] Stueck, *The Korean War,* ch. 5; Zhang Shu Guang, *Mao's Military Romanticism: China and the Korean War, 1950–1953* (Lawrence: University Press of Kansas, 1995), chs. 6–7.

question anytime over the next year unless the sixteen Chinese divisions in Korea could be given time to rest and could be reequipped quickly through Soviet support.[5] Given these constraints, Mao opted for a piecemeal strategy designed to attack smaller enemy units and wear down enemy resolve.[6]

After they perceived that the war was deadlocked, Mao and Stalin would decide that the communist coalition should enter armistice talks if the United States were willing to pursue them. On June 1, 1951, George Kennan, a top advisor and former official of the Truman administration, told the Soviet ambassador to the UN, Jacob Malik, that the United States was interested in a ceasefire agreement. This move followed early feelers sent out to lower-level U.S. officials by Malik.[7]

Following consultations between Moscow and Beijing, Stalin and Mao decided in principle to pursue talks; Mao then called the most aggressive communist leader, Kim Il-sung, to the Chinese capital in early June to convince him of the wisdom of the plan, which included proposals for the 38th parallel as a ceasefire line. Mao made compromises of his own in this process, agreeing not to allow issues related to Taiwan or PRC representation at the UN to get in the way of talks. Stalin supported this approach, stating that "now, a ceasefire is a good thing to pursue."[8] The communist camp initiated its strategy later that month through a prepared statement by Ambassador Malik on June 23, 1951, suggesting that the communists might be willing to entertain a U.S. proposal for armistice talks. General Matthew B. Ridgway then proffered such a proposal for the UN in early July. The talks would begin that month in Kaesong, would break down from August to October, and would then restart at Panmunjom, where they would eventually produce an armistice agreement in July 1953.

Several factors had fostered the communist allies' decision to pursue peace talks. The Chinese and their North Korean allies suffered very heavy losses in

[5] See "Mao Zedong Guanyu Xiayi bu Zuozhan Jihua Zhi Shi Dalin Dian" [Mao Zedong's Telegram to Stalin Concerning the Next Step in the War Plan], January 15, 1950, in Shen, ed., *Chaoxian Zhanzheng*, 2:663. See also the February 7, 1951, letter from Mao to Zhou announcing the troop rotation system and the somber March 1 telegram to Stalin regarding the exhaustion of PLA forces in *Jianguoyilai Mao Zedong Wengao*, 2:104–5, 151–53. On June 5 Stalin approved of a protracted war strategy (*chijiu zhan* in the Chinese translation), because it would prove a training ground for Chinese troops and would lead to wavering in the Truman government and reduce American and British military prestige. For the English translation, see "Stalin to Mao Zedong," June 5, 1950, in Weathersby, "New Russian Documents," Document 65, pp. 59–60. For the Chinese translation, see "Shi Dalin Guanyu Fangyu Zuozhan Deng Wenti Zhi Mao Zedong Dian" [Stalin's Telegram to Mao Zedong Concerning the Question of a Defensive War, etc], June 5, 1951, in Shen, ed., *Chaoxian Zhanzheng* 2:784.

[6] Zhang, *Mao's Military Romanticism*, ch. 7; also see Chen Jian, "China's Changing Aims during the Korean War, 1950–1951," *Journal of American-East Asian Relations* 1, no. 1 (spring 1992).

[7] Stueck, *The Korean War*, 204–8.

[8] For an overview of this process, see Niu, "'Zhanzheng Jubuhua yu Tingzhan Tanpan Juece."

the UN counteroffensive with clearly negative implications for the broad and decisive offensive that Mao always seemed to be planning to launch some time just over the horizon. Another problem was that Soviet military assistance had proven more limited than Mao had hoped. Mao had always planned to depend on massive Soviet aid to equip his resting and rotating troops as they prepared for future offensives. In spring Mao had not been satisfied with the pace of Soviet weapons transfers, particularly given the quickly growing needs of his forces in the field.[9] On June 24, 1951, just before the opening of the armistice talks, Stalin rejected Mao's request for the Soviets to arm and outfit sixty Chinese divisions over time. Mao apparently wanted sixteen divisions (the number in Korea) outfitted within one year so that his forces could return to the offensive.[10] Stalin said that it was "impossible" and "inconceivable" that the Soviets could equip more than ten divisions within one year and said that the full Chinese request could not be met even within three years.[11] The quite-blunt late June telegram from Stalin to Mao was sent after the three communist capitals had already agreed to pursue armistice talks, but the limits on Soviet arms transfers had become clear before the decisions for an armistice in early June. The Soviet policy must have helped lock the Chinese communists into a position of sincerely pursuing a sustained ceasefire and accepting an outcome that fell far short of military "liberation" for all of Korea. Moreover, the limited nature of Soviet assistance must have helped determine the outcome of future debates on strategy later that year. Throughout the summer Chinese military elites debated the advisability of a sixth major offensive in August or September designed to drive UN forces back into South Korea and restore the 38th parallel as the military dividing line between the two sides. This aggressive strategy, unsurprisingly, was supported by Kim Il-sung, but was ultimately rejected by Mao as too risky.[12]

Apparently Soviet caution was a major constraining factor on Moscow's Asian allies. Soviet global concerns and desire to prevent escalation of the Korean War contributed to restraint on the Chinese and North Koreans in the war. This is what we would expect from a global leader of a revisionist alliance. General Xu Xiangqian, who was representing the PRC at discussions in Moscow in June 1951, recalled, "The Soviet leaders respected our victory in the Chinese revolution and supported the Chinese people's engagement in the Korean War.

---

[9] Stueck, *The Korean War*, 218–19.

[10] "Mao Zedong to Stalin," June 21, 1950, in Weathersby, "New Russian Documents," document no. 72, p. 62.

[11] "Stalin to Mao Zedong," June 24, 1951, in Weathersby, "New Russian Documents," document no. 73, p. 62. For the Chinese-language version, see "Shi Dalin Guanyu Xiang Zhongguo Tigong Jundui Zhuangbei Wenti Zhi Mao Zedong Dian" [Stalin's Telegram to Mao Concerning the Issue of Providing Equipment for the Chinese Military], June 24, 1951, in Shen, ed., *Chaoxian Zhanzheng*, 2:825.

[12] Zhang, *Mao's Military Romanticism*, ch. 7.

They were willing to provide certain assistance and help us speed up our army's modernization and standardization. The Soviet leaders, however, had their own worries and anxieties. I think they were afraid of provoking a war with the United States."[13]

By July 1951, then, the stage was set for armistice talks. The war had fallen into a deadlock—albeit an often quite violent one—that would be difficult for either side to break without major new sacrifices in the Korean theater. Not much significant territory would be gained or lost on either side of the 38th parallel between July 1951 and July 1953, when the armistice was signed (though the territory that was gained and lost was hotly contested at great costs to both sides).

The most important sticking point in reaching an armistice agreement was the dispensation of the POWs held by both sides. Many of the Chinese POWs in UN custody expressed a desire to remain outside the PRC, and the Truman administration refused to return them to China under those circumstances. For its part, the Truman administration demanded the return of all UN POWs.

PRC officials rejected the U.S. proposal as unacceptable in principle. The Chinese made it clear from the beginning that the POW repatriation issue was very important to Beijing's war strategy (but Mao did not initially seem to anticipate that it would become such a major sticking point in the negotiations).[14] Such an unequal exchange was viewed by the communists as the equivalent of a political defeat in the war. Communist elites worried about the future implications of such a major compromise for future coercive diplomacy in the standoff in Korea and over issues beyond the peninsula itself, including Taiwan. Moreover, many of the Chinese POWs who did not want to return to the mainland understandably wanted to travel to the "other" China—Chiang Kai-shek's Republic of China on Taiwan. This only made Beijing's acquiescence on this issue harder to obtain.

From the start, Moscow and Beijing consulted with each other in great detail and fully coordinated their position on armistice talks. The initial communist positions at the talks were actually relatively accommodating. Mao and Stalin agreed that the PRC would not link Taiwan and PRC UN representation with

[13] Marshall Xu Xiangqian, "The Purchase of Arms from Moscow," in Xiaobing Li, Allan R. Millett, and Bin Yu, *Mao's Generals Remember Korea* (Lawrence: University of Kansas Press, 2001), 143.

[14] Such coordination continued throughout the talks, and Stalin seemed very much in a veto position on any proposal, although Mao also took the initiative on many occasions. Shen, "Sino-North Korean Conflict," 19. For documentary evidence of such coordination, see "Mao Zedong to Stalin," June 30, 1951, in Weathersby, "New Russian Documents," document no. 78, pp. 63–64. For the Chinese version, see "Mao Zedong Guanyu Tingzhan Tanpan Deng Wenti Zhi Shi Dalin Dian" [Mao Zedong's Telegram to Stalin Regarding Armistice Talks, etc.) June 30, 1951, in Shen, ed., *Chaoxian Zhanzheng*, 2:835–36. For early references to the POW issues, see "Mao Zedong to Stalin," July 3, 1951 in Weathersby, "New Russian Documents," Document 84, p.66. For the Chinese version, see "Mao Zedong Guanyu Tingzhan Tanpan Zhong de Wo Fang Jianyi Wenti Zhi Shi Dalin Dian," [Mao Zedong's Telegram to Stalin Regarding the Issue of Our Side's Proposals during Ceasefire Talks], July 3, 1951 in Shen, ed., *Chaoxian Zhanzheng*, 2:853–54.

the ceasefire talks. In July 1951, the Chinese communists also decided, in consultation with Stalin, to leave off the agenda their most controversial, coveted, and elusive Korean War goal: the permanent removal of all foreign forces from Korea. The removal of foreign forces was considered not only an issue of principle but also a prerequisite for a lasting peace, particularly from the point of view of the North Koreans. Neither Mao nor Stalin, however, wanted this issue to get in the way of reaching a preliminary ceasefire. They proposed instead to discuss this issue after a ceasefire was reached.[15]

It is unclear just how much Kim wanted to resist these key decisions by Mao and Stalin, but he had little choice but to accede to such a strategy if his two protectors agreed that this was the way to proceed. It appears from available evidence in China and the Soviet Union that Kim Il-sung accepted quietly the advice of his two stronger allies about the desirability of negotiating a ceasefire. Kim participated in meetings in Moscow in June that helped bring that position about. There is some suggestion in the archival evidence that he did so with great dissatisfaction, however. According to an analysis of the Soviet representative to North Korea, Ambassador V.N. Razuvaev, Kim initially showed great frustration with the need to reach an armistice. Though Kim recognized the objective constraints on the communist forces, he remained angry at the Chinese, whom he believed to be leaning too hard toward the United States in seeking a ceasefire. The Soviet ambassador reports that the North Koreans saw the Malik statement at the UN as the "clearest manifestation of China's efforts to achieve a ceasefire and to throw off this burden of assisting Korea." Razuvaev also notes that the North Koreans were relatively slow to discuss Malik's statement at the UN in published reports and propaganda, a fact noted by at least one leading historian of the period as well.[16] Razuvaev also reported that Kim came home dejected after an early July visit to Beijing and thereafter begrudg-

[15] "Shi Dalin Guanyu Tingzhan Tanpan Celue Zhi Mao Zedong Dian," [Stalin's Telegram to Mao Zedong Regarding the Tactics in Ceasefire Negotiations], July 21, 1951, in Shen, ed., *Chaoxian Zhanzheng*, 2:907. In a footnote, Shen states that Stalin's telegram was a reply to Mao's telegram expressing tactics along these lines. After receiving Stalin's telegram, Mao gained Kim's agreement the following day. This was a reversal of Stalin's earlier position on this issue, in which he stated that the communists should resolutely demand both the 38th parallel as a ceasefire line and the full withdrawal of foreign forces from Korea in ceasefire talks. See "Telegram, Stalin to Mao Zedong," July 3, 1951, in Weathersby, "New Russian Documents," document no. 85, pp. 66–67, in which Stalin advises Mao and Kim to insist upon the total withdrawal of troops and to advance another proposal regarding the repatriation of refugees. This is in response to Mao's request for Stalin's opinion on the same day. See "Mao Zedong to Stalin," July 3, 1951, in Weathersby, "New Russian Documents," document no. 84, p.66.

[16] "Lazuwaerfu Guanyu Tingzhan Tanpan Zhong Chuxian de Zhengzhi Qingxu Ji ZhongChao Guanxi de Dianbao" [Razuvaev's Telegram Regarding the Political Emotions Arising from the Ceasefire Talks and Sino-Korean Relations], September 10, 1951, in Shen, ed., *Chaoxian Zhanzheng*, 3:1022–26; Stueck, *The Korean War*, 216.

ingly resigned himself to the ceasefire. Finally, the Soviet envoy reported that in a July 27 exchange with Mao regarding terms for a ceasefire, Kim found Mao's proposed position on a dividing line between the two enemy militaries to be too soft (ceding some territory previously controlled by North Korea to UN forces). Kim apparently complained in anger to his own colleagues that perhaps it "would be better to continue executing the war under conditions in which we do not receive help from China." According to Razuvaev, Mao apparently toughened his negotiating position on the issue of dividing lines the following day, to the great relief of Kim.[17]

It is not clear from available evidence what role intra-alliance differences played, if any, in the breakdown of armistice talks from August through October. From the standpoint of alliance management, the September 10 Razuvaev telegram, which was distributed to several top Soviet leaders in addition to Stalin, could not have been entirely unwelcome news in Moscow. The main point of the telegram was that residual Korean nationalist frustration regarding the need to negotiate a ceasefire and to divide Korea with a demilitarized zone (DMZ) was aimed at China, not the Soviet Union. Moreover, that frustration had been kept sufficiently in check to keep the North Koreans committed to the negotiation process. So, Stalin got to have his cake and eat it too. North Korean comrades would continue to pursue a path consistent with Moscow's wishes and thereby minimize the chance of escalation of war with the United States, but the North Koreans would steer their frustration about the need to compromise toward the Chinese, rather than the Soviets.

Razuvaev's story about Kim's frustration seems logical, especially when one compares Kim's alleged vexation with the similar anguish of South Korea's elites regarding a ceasefire pushed on them by the United States and others at the UN. Syngman Rhee and his colleagues, too, knew that an armistice agreement would deprive them of their own dream of unifying the peninsula under their control.[18] Kim did participate in the initial strategy to create the conditions for armistice talks, and we have no specific documentary evidence of his actively resisting or opposing the proposals of Mao and Stalin regarding these talks.[19] Razuvaev's

[17] "Lazuwaerfu Guanyu Tingzhan Tanpan Zhong Chuxian de Zhengzhi Qingxu Ji ZhongChao Guanxi de Dianbao," in Shen, ed., *Chaoxian Zhanzheng*, 3:1022–26. It is not clear if the reference to an early July trip to Beijing was an inaccurate reference to the early June trip discussed in Niu, "Zhanzheng Jubuhua," or a follow-on trip. The quotation is from p. 1024.

[18] See, for example, Stueck, *The Korean War*, ch. 6, esp. 211–15.

[19] The North Koreans likely had mixed feelings about any talks that might end a very painful war for the North but might also permanently deprive them of unification under their leadership. In his June 29, 1951, telegram on the issue, Kim expresses no opinion on whether or nor to start negotiations, merely asking Mao for guidance in how to respond to any overtures from UN commander Ridgway on that score. For his part, Mao believed that representatives should be sent for talks if they were offered, and proceded along these lines after Stalin agreed with his idea. For Kim's

account of Kim's frustration with a ceasefire agreement, however, is consistent with secondary histories discussing communist alliance politics in mid-1951. Citing the memoirs of Korean War generals, Shu Guang Zhang reports that in Beijing on June 3 and again later in the month Kim "vehemently demanded an immediate offensive to restore the 38th parallel," before ceasefire talks could be entertained.[20] Kim apparently remained the most aggressive member of the co-alition and was restrained only by his high degree of dependence on a relatively well coordinated and more cautious pair of allies in Beijing and Moscow. If these accounts are accurate, as seems likely, it is notable that when there was coordi-nation and cooperation between the two most powerful communist allies—the Soviet Union and China—there was little that the more aggressive Kim could do to change the outcome at the most fundamental level. Kim could only influ-ence events on the margins. For example, he did manage to affect how rigid the communists would be on issues like the location of the dividing line between the two enemy militaries. But his leverage at this point paled in comparison to the previous year, when he successfully manipulated a less cohesive coalition into a major war. In 1951 Kim was dragged into a more accommodating position by his more powerful allies and forced to accept in principle the negotiated division of his country by a demilitarized zone. As we will see, there are, therefore, direct parallels between summer 1951 in Korea and the negotiations leading up to the Geneva Accords on Indochina in 1954, discussed later in this book. In both cases, coordination in Moscow and Beijing on the alliance's grand strategy left the more radical local power little choice but to agree to a negotiated settlement that fell short of achieving its political and military ambitions in the near term.

There are some limits to this analogy between communist alliance dynam-ics in Korea from 1951 through 1953 and in Vietnam in spring 1954. Unlike the North Vietnamese communists following their victory against the French at Dien Bien Phu in spring 1954, North Korean forces in the second half of 1951 had not only just suffered a massive defeat, but they and large sections of the DPRK were also being continually punished by bombing from UN forces.

---

request for Mao's opinion (if not simply his instructions on how to deal with the matter), see "Mao Zedong to Stalin Transmitting 29 June 1951 Telegram from Kim Il-sung to Mao," June 30, 1951, in Weathersby, "New Russian Documents," document no. 77, p. 63. In another telegram Mao reports to Stalin that he advised Peng and Kim to consider the issue and to arrange to send appropriate representatives to the talks. See "Mao Zedong to Stalin," June 30 1951, in Weathersby, "New Russian Documents," document no. 78, pp.63–64. For the Chinese versions of these documents, see "Mao Zedong Zhuanfa Jin Richeng Guanyu Meiguo Dui Tingzhan de Fanying Zhi Shi Dalin Dian" [Mao Zedong's Transferring to Stalin of Kim Il-sung's Telegram Regarding U.S. Reaction to the [Idea of] a Ceasefire], June 30, 1951, in Shen, ed., *Chaoxian Zhanzheng*, 2:837; see also "Mao Zedong Guanyu Tingzhan Tanpan Deng Wenti Zhi Shi Dalin Dian" [Mao Zedong's Telegram to Stalin Regarding the Issue of Ceasefire Talks, etc.], June 30, 1950, in Shen, ed., *Chaoxian Zhanzheng*, 2:835.

   [20] Zhang, *Mao's Military Romanticism*, 157.

Under these constraints, from summer 1951 to early 1952, North Korea would reverse roles and become the main advocate of vigorously pursuing a peace settlement. By early 1952, as North Korean hopes of unifying the peninsula under Pyongyang's control dwindled and the economic and human costs of the war grew, the North Korean communists became interested in an armistice, if not a peace treaty. In February 1952 Kim told Mao that he had "no desire to continue the war."[21] As armistice talks remained deadlocked over the terms for the return of UN and communist POWs, North Korea morphed from the most aggressive member of the communist alliance in 1950–51 to its most conciliatory. The North Koreans' political goals were still expansive, but if they could not achieve them, they did not want to pay the heavy price of being the site for a bloody and economically costly sparring match between the two Cold War camps. For their part, both China and the Soviet Union had reasons to want to continue the fighting unless the United States offered sufficient concessions in negotiations, particularly on the POW issue. Mao's and Stalin's goals were to make the United States appear stymied in the war and to prevent the communists themselves from appearing to have lost prestige in the negotiation process. In other words, only an outcome that could be portrayed as a political defeat for the United States rather than for the communists was acceptable in Beijing and Moscow. Anything less, they believed, would compromise future coercive diplomacy in Korea, Taiwan, and elsewhere.[22] The United States was paying a direct, high price for stalemate in Korea, whereas the costs to its leading counterpart in the enemy camp, the Soviet Union, had become quite manageable.

In 1952 the Chinese and the Soviets were less eager to cut a deal with Washington than were the North Koreans and were therefore less willing to brook perceived injustices on the POW issue. But it is important to note that neither Beijing nor Moscow wanted escalation. At key junctures Moscow intervened to insure that neither China nor North Korea would take military initiatives that might cause an expansion of the war. As discussed earlier, in June 1951 Stalin did this consciously or unconsciously by limiting the amount of military aid he offered to the Chinese, thus limiting the option of future Chinese offensives. Similarly, in August 1952 Stalin would reject North Korean proposals for Chinese air offensives against South Korea. The North Koreans, under heavy bombardment themselves and seeing little progress in negotiations, had raised this proposal as

---

[21] Shen, "Sino-North Korean Conflict," 19.

[22] Mao emphasized the danger of harming communist countries' prestige while increasing that of the Americans in a telegram on July 15, 1952. Shen, "Sino-North Korean Conflict," 19. See also "Stalin to Mao Zedong," June 5, 1950, in Weathersby, "New Russian Documents," document no. 65, p. 59. For the Chinese translation, see "Shi Dalin Guanyu Fangyu Zuo Zhan Deng Wenti Zhai Mao Zedong" [Stalin's Telegram to Mao Zedong Concerning the Question of a Defensive War, etc.], June 5, 1951, Shen, ed., *Chaoxian Zhanzheng*, 2:784.

a way to increase communist negotiating leverage and to speed an agreement. Stalin not only rejected air strikes on the South, but he also went further, rejecting more broadly any new offensives, small or large, by the Chinese or the North Koreans while ceasefire negotiations continued.[23] Although the picture was mixed from the perspective of U.S. coercive diplomacy in 1951–53, the increase in Sino-Soviet-North Korean coordination after China entered the war made escalation easier to control and an armistice agreement more likely, even though it took many months of bloody warfare and ultimately the death of Stalin to bring about an acceptable compromise on the POW issue.

In our study, the 1952–53 period is a rare case in which the interests of the local power—in this case North Korea—were generally more moderate and less revolutionary than those representing the interests of the international movement as a whole. Still, this was only true because North Korea had been unable to convince a relatively unified pair of stronger allies to continue to pursue its more radical revolutionary agenda. Although they were more cautious than their North Korean comrades to avoid appearing overly conciliatory in talks, in general Beijing's and Moscow's position themselves were still comparatively moderate, ensuring that an armistice could be reached with the United States and its UN allies under the right political circumstances.

The August 1952 meeting between Zhou Enlai and Stalin in Moscow is instructive as to the alliance dynamics involved in the Korean armistice talks. The Chinese and the Soviets both opposed the North Koreans' push for a near-term armistice. They both believed that continued fighting within a controlled scope was worth the costs because it would serve the purpose of tying down U.S. energy and preventing the United States from asserting itself in other parts of the world.[24] To some degree, Beijing was still trying to impress the Soviets with Chinese internationalism almost two years after Chinese entrance into the war. Zhou said to Stalin that China's role in extending the conflict in Korea placed Beijing in the "vanguard," with Chinese troops at the front helping to stave off World War "for 10-15 years, assuming that they will succeed in containing the American offensive in Korea. The USA will not be able to unleash a third world war at all."[25] As one might expect of someone concerned about his leadership in the movement, Stalin granted China only partial credit for such an outcome. Stalin pointed out how weak the United States is in military affairs and questioned whether it would be able to launch another war even without the Korea problem. Stalin ridiculed the United States for being a bunch of "merchants" and "speculators" who did not know how to fight wars the way that the Soviets and

---

[23] "Stalin's Conversations," *CWIHPB* nos. 6/7, p. 13.
[24] Shen, "Sino-North Korean Conflict," 19.
[25] "Stalin's Conversations," *CWIHPB* nos. 6/7, p. 13.

their former enemies had. He mocked the U.S. inability to conquer a country like "little Korea" when it took the Nazis only weeks to subdue France. But at the same time Stalin did not want the Chinese to abandon the fight and capitulate in the negotiations. For this purpose, as in 1950, Stalin used a different tactic, appealing to Chinese parochial national interests to point out why a U.S. political victory in Korean War negotiations would be so costly. He said, "The Chinese comrades must know that if America does not lose this war, then China will never recapture Taiwan."[26]

For their part, the Chinese themselves wanted to make sure that they did not lose out politically in negotiations with the United States over prisoner swaps. By late 1952 Beijing did, however, appear willing to accept any face-saving way out of the problem offered by the UN. The problem was that in the talks to date, the United States had offered to return a much smaller percentage of Chinese POWs than North Korean POWs and demanded in exchange the full release of UN POWs (the United States proposed repatriating 80 percent of the North Koreans and only 32 percent of the Chinese). Beijing cared more than Pyongyang about these details for both straightforward and abstract reasons, including the belief that saving face and reputation for resolve were on the line. Moreover, despite the obvious costs of continuing the war, Beijing also was receiving some benefits. As North Korea's frontline ally, Beijing was receiving significant military assistance from the Soviet Union. This flow of weaponry and expertise that strengthened China's military overall reduced further Beijing's incentives for adopting a more accommodating negotiating position on the POW issue. CCP elites, of course, would not come out and say anything so cynical and self-serving to the Soviets regarding the links between Soviet aid and the CCP negotiating position, but this was precisely the Soviets' impression of the interests that drove Beijing's negotiating strategy at Panmunjom.[27] The logic was not foreign to Stalin. In June 1951 Stalin argued to Mao in typically cold-hearted fashion that an advantage of a protracted war in Korea for the communist camp was that the Chinese military could gain valuable experience against the U.S. and allied forces.[28]

As the leader of a more cooperative and integrated movement in East Asia than we saw in early 1950, Stalin played a mixed role in the armistice process. Through his ambassador, Jacob Malik, he initiated the process of armistice talks. But after talks began, Stalin encouraged continued belligerence, dismissing what he viewed as the excessive compromise toward the United States preferred by Pyongyang. By siding with the Chinese position over that of Pyongyang in 1952,

[26] Ibid.

[27] Shen, "Sino-North Korean Conflict," 19.

[28] See "Stalin to Mao Zedong," June 5, 1950, in Weathersby, "New Russian Documents," document no. 65, p.59. For the Chinese translation, see "Shi Dalin Guanyu Fangyu Zuo Zhan Deng Wenti Zhai Mao Zedong," in Shen, ed., *Chaoxian Zhanzheng*, 2:784.

Stalin fostered the continuation of conflict against the wishes of the increas-ingly dispirited North Koreans. Stalin concurred entirely with Mao's strategy of a protracted war for the purpose of securing a more advantageous long-term peace for the communist camp as a whole.[29] On the other hand, Stalin played a restraining role on the Chinese and the North Koreans, limiting their ability to launch large-scale offensives, and proscribing even smaller-scale offensives that might scuttle peace talks.[30] Stalin apparently liked extended conflict in East Asia as long as it was tightly controlled. More important still, Stalin was not at all averse to an armistice if it came at a sufficient price to the United States, and he clearly wanted to avoid escalation that would preclude such an armistice and perhaps spark a wider war.

Despite his stated desire for a tough position regarding the United States, in their August 1952 meeting Stalin agreed with Zhou that if the United States made any visible concessions on the POW issue, the communists should accept the concessions and move toward an armistice.[31] In a September 1952 meeting he even entertained Zhou's concept of a ceasefire prior to an agreement on the POW issue. Stalin, however, expected the United States to reject any such proposal.[32]

Stalin would not live to see an armistice. His death in March 1953 was ar-guably a major catalyst in pushing the communists toward a settlement. Still, the July 1953 Panmunjom armistice was probably compatible with Stalin's core goals during the Korean War.[33] Stalin basically preferred an armistice to a for-mal peace treaty, and such a treaty is lacking to this day. Stalin, like Mao, also wanted to find face-saving ways for the communists to cope with U.S. refusals to simply turn over all Chinese POWs to the PRC. The combatants would eventu-ally find one by employing the good services of a third party, India, to vet the POWs' wishes as to repatriation. This solution was fully consistent with what Zhou and Stalin had agreed upon in summer 1952. The Panmunjom Armistice appeared to accord with Stalin's wishes because it kept U.S. forces on station and on alert in the region but greatly reduced the downside dangers of escalation to a broader regional or global war, something the Soviets, with bigger fish than Korea to fry, could not allow.

From the point of view of the United States, Sino-Soviet coordination was not an entirely positive phenomenon. But by increasing greatly the transpar-

---

[29] Historian William Stueck suggests that one reason why Stalin wanted a lengthy negotiation process is that continued war would tie up Chinese resources and prevent China from increasing its prestige relative to that of the Soviets by more actively pursuing revolution in Southeast Asia. Stueck, *The Korean War*, 220–21.

[30] "Stalin's Conversations," *CWIHPB* nos. 6/7, p. 13.

[31] Ibid., 10–13.

[32] Ibid., 17–19.

[33] For an in-depth analysis of Soviet views and debates on armistice negotiations, see Stanley, *Paths to Peace*, ch. 4.

ency of the situation, it made U.S. coercive diplomacy more effective and thereby reduced the likelihood of catastrophic surprises such as Peng Dehuai's brilliant counteroffensive in northernmost Korea in November 1950. Stalin sided with the Chinese in disagreements with the North Koreans regarding whether or not to vigorously pursue a ceasefire in 1952. Stalin also very much appreciated Mao's logic about the utility of an extended Korean conflict for the world communist movement, an argument probably created by Mao for precisely this effect, given Mao's apparent desire to use Korea to erase any doubts in Moscow about the CCP's internationalism. Still, both Mao and Stalin insisted that an armistice was a desirable outcome and it was only in this international context that Pyongyang desired an armistice even more urgently than its allies. In other words, if either Stalin and Mao had held out the possibility of major offensives to unify the peninsula in the foreseeable future and/or if Kim thought he could manipulate one or both of his allies into supporting such an offensive, Kim likely would have been the most, rather than the least, aggressive member of the alliance, as he had been from 1949 through most of 1951. In this sense, Stalin's caution as the leader of the most powerful member of the communist alliance, and as the leader of a state with global, not just regional, interests, was a sobering influence on the strategy and tactics of the communist alliance for both the Chinese and North Koreans. That influence was very important for the prospects of the two enemy camps reaching a settlement and avoiding further escalation.

THE GENEVA CONFERENCE: THE BENEFITS OF SINO-SOVIET UNITY FOR THE ANTICOMMUNIST CAMP

By the time the Korean War ended in July 1953, Mao had certainly proven himself a true member of the movement to Stalin and his successors. Although Soviet suspicions about Mao's becoming a rival in Asia may have continued, nobody seemed particularly concerned about his Titoism and lack of devotion to internationalism.[34] During the Korean War and in its immediate aftermath, Mao continued to earn his red stars by training the Vietnamese communist forces engaged in the anticolonial struggle with the French and by supplying those Vietnamese forces with an impressive quantity of guns and ammunition. At a time when the Soviets paid little attention to Indochina, Chinese technical and logistical support was critical to the Vietnamese communists' offensives in northernmost Vietnam during 1953–54 and the impressive victory at Dien Bien Phu in spring 1954. As discussed in the previous chapter, the Soviets under Stalin

---

[34] Foreign Minister Chen Yi recalled that China's Korean War entry erased the fears of Stalin and his successors about Mao's Titoist tendencies. Zhang, *Mao's Military Romanticism*, 249.

and his immediate successors had been relatively indifferent to Vietnam, worried about the implications for Soviet policy toward France—particularly about the political consequences of collaborating with Paris's colonial enemies—and generally unhelpful to the Vietnamese communists' armed struggle in Southeast Asia. Moscow was also nervous about China's activism in Indochina. China, on the other hand, offered almost unlimited amounts of ammunition to Vietnamese artillery gunners in 1953–54, thus providing the Vietnamese with the capability to coerce the French garrison at Dien Bien Phu into submission.[35] According to one estimate, during the campaign at Dien Bien Phu, China would transfer to Vietnamese troops "200 trucks, 10,000 barrels of oil, over 100 cannons, 3,000 pieces of various types of guns, 2,400,000 bullets, over 60,000 artillery shells, and about 1,700 tons of grain."[36]

The number of Chinese military advisors and the amount of material aid flowing to Vietnam increased sharply after the Panmunjom armistice.[37] Through the military advisors he dispatched to Vietnam and Laos, Mao urged an aggressive anti-French stance by the Vietnamese forces to secure the northern part of Vietnam for the communists. The creation of a stable, friendly regime on his border to replace a colonial European power allied with the United States would provide a strategic buffer for Mao's new nation. Beijing had an exaggerated view of the coordination between the United States and France in Southeast Asia.[38] But what was just as important, assisting the successful Vietnamese communist campaign against the French established and solidified his revolutionary internationalist credentials in the immediate region and beyond.[39] Mao's CCP tried to build on its new self-image as leader of revolution in the developing world against the United States and its allies.

The Chinese assistance that helped the Vietnamese communists secure victory at Dien Bien Phu against the French also helped create the strategic backdrop for the peace settlement at the Geneva Conference in 1954. In mid-June of that year a more accommodating new government came to power in France under Pierre Mendes-France and seemed eager to reach an agreement. With the signing of the Geneva Accords on July 21, Vietnam would be divided at the 17th parallel with an agreement, ultimately unfulfilled, to carry out nationwide elections in North and South Vietnam in 1956. The French and the Vietminh would also agree (in principle) to withdraw forces from Laos and Cambodia. The various

---

[35] Guo, *ZhongYue Guanxi Yanbian Sishi Nian*, 38.

[36] Chen, "China and the First Indo-China War," 103–4.

[37] Chen, "China and the First Indo-China War," 98–104.

[38] For example, see "Zhou Enlai to Mao Zedong and Others, Regarding the Situation of the First Plenary Session, May 9, 1954," *Cold War International History Project Bulletin* (*CWIHPB*) no. 16 (fall 2007/winter 2008): 17.

[39] Yang, "Changes in Mao Zedong's Attitude toward the Indochina War."

signatories agreed that these would become neutral countries, militarily allied neither with the communist nations nor the Western powers and containing no foreign military basing.[40]

The agreements that were reached in July seemed a compromise much more consistent with U.S. interests than would be continued warfare or escalation in Vietnam.[41] Secretary of State Dulles had been very worried about the type of agreement that would come out of the Geneva Conference. He refused on principle to sign the accords, since they granted sovereignty to the communists in northern Vietnam. Fearing that any accords would be a near total capitulation to the communists, Dulles had hoped that the negotiations would break down. He personally preferred a policy of opposing all compromises by the British and the French and limiting U.S. activity at the conference. Eisenhower intervened and reversed Dulles's position, fearing that discord between the Western allies would only serve the interests of the communists. The president insisted that Dulles back the negotiating positions of the new French leader, Mendes-France, albeit indirectly, and encourage active diplomacy in Geneva by the U.S. representatives to the talks, Walter Bedell Smith. In fact, Dulles traveled to Paris in early July just before the accords were reached. His goal was to coordinate U.S. and French positions toward the talks. Eisenhower's intervention made sense. When the Geneva Accords were finally reached, even Dulles was pleasantly surprised. The Geneva Accords seemed to secure a territorially contiguous, noncommunist southern Vietnam and, thereby, provide the basis for containing communism throughout the rest of Indochina and Southeast Asia more broadly.[42] As historian George Herring sums it up, "Although notably hostile toward the Geneva agreements in public, he and Eisenhower were not entirely displeased, seeing in them the means to accomplish their larger goals."[43] Those larger goals included creating, for the purposes of containment, an alliance of regional actors that was "free of the taint of French colonialism."[44]

U.S. interests and concerns were well represented in the negotiation process. We know from declassified Chinese documents that Anthony Eden and

[40] Zhai, *China and the Vietnam Wars*, ch. 2; Zhang, *Mao's Military Romanticism*, 249; Gaiduk, *Confronting Vietnam*, ch. 3.

[41] For the evolution of the Eisenhower administration's attitudes toward the conference from May to July, see The Pentagon Papers: The Gravel Edition, 1:108–46, available at http://25thaviation.org/id966.htm.

[42] Richard H. Immerman, *John Foster Dulles: Piety, Pragmatism, and Power in U.S. Foreign Policy* (Wilmington, Del.: Scholarly Resources, 1999), 94–95; also see George C. Herring, "The Indochina Crisis: 1954–55," in Richard H. Immerman, ed., *John Foster Dulles and the Diplomacy of the Cold War* (Princeton: Princeton University Press, 1990), 213–33.

[43] Herring, "The Indochina Crisis," 219.

[44] Ibid., 223. The quotation is from the diary of James Hagerty, Eisenhower's press secretary.

other British officials would share U.S. views with Zhou Enlai during the nego-tiations.[45] Once an agreement was reached, even in public comments President Eisenhower had positive things to say about the accord, stating on July 21 that he was "glad that an agreement has been reached at Geneva to stop the bloodshed in Indochina." While adding that "the agreement contains features we do not like," he also pledged that "the U.S. will not use force to disturb the settlement."[46]

Given the common interests of France, Great Britain, and the United States in a negotiated settlement, the most interesting question is not why they accept-ed it, but why the communists did (particularly the Vietnamese communists). In understanding why, we see how a relatively unified and stable movement was easier for anticommunist actors to contain through coercive diplomacy and ne-gotiations than was the internally contentious but externally more aggressive communist alliance of the 1960s. The only actors whose motivations remained the same over the two decades were the Vietnamese communists, who wanted to use armed struggle to unify the nation under their leadership as soon as possible. The Vietnamese communists were dependent, however, on the material and po-litical support of Beijing and Moscow for their revolutionary activities. For their part, in the mid-1950s the Chinese and the Soviets were preparing for a "peace offensive" toward the Western powers and noncommunist regimes, particularly in the Third World. Neither nation wanted war in the short-term, particularly if it involved the United States. Beijing was still recovering from the anti-Japanese War, the Civil War, and the Korean War. It also did not want to encourage a tighter U.S.-led encircling alliance in the region.[47] In the 1950s Soviet leaders, like U.S. leaders, were focused more on global "strongpoints" like Europe and the Middle East than on revolution in Southeast Asia. In 1953–54 Moscow still hoped to turn French opinion against the European Defense Community and German remilitarization by appearing moderate on Indochina.[48] Moreover, the Soviets were in an economic rebuilding period, and a political transition period following the March 1953 death of Stalin. The Soviets therefore saw anything but peace in Vietnam as draining of critical resources better used elsewhere. Escalation in Vietnam could lead to increased U.S. military activity locally and

---

[45] See, for example, "Minutes of Conversation between Zhou Enlai and Anthony Eden, July 13, 1954," in *Cold War International History Project Bulletin* (*CWIHPB*) no. 16 (fall 2007/winter 2008): 63.

[46] White House Statement by the President, July 21, 1954, Papers of John Foster Dulles, Firestone Library, Princeton University, Princeton, N.J., box 80, reel 31, "Correspondences with Dwight D, Eisenhower."

[47] Qu Xing, "ZhongYue Zai Yinzhi Zhan Wenti Shang de Zhanlüe Yi Zhi yu Celüe Chayi," [Sino-Vietnamese Strategic Coordination and Tactical Differences on the Question of the Indochina War], in Zhang Baijia and Niu Jun, eds., *Lengzhan yu Zhongguo de Zhoubian Guanxi* (Beijing: World Knowledge Publishers, 2002), 303–6.

[48] Gaiduk, *Confronting Vietnam*, ch. 3.

globally, as it had in Korea. It is fairly clear that the Soviets in the mid-50s saw Vietnam as a relatively unimportant issue and were relatively willing to allow China to play a leading role there, as long as the conflict would not escalate. After all, in a period in which Soviet relations with the Chinese communists were still good, there was little reason for Moscow to compete too jealously with the CCP for the loyalties of Ho Chi Minh and his movement.[49]

As leading historian Chen Jian writes, "The 1954–55 period shined as a golden age of the Sino-Soviet alliance."[50] There were many good reasons for Beijing to coordinate its foreign policy with Moscow. Beijing still relied heavily on Moscow to break out of its diplomatic isolation. Moscow helped arrange the PRC's entrance onto the great power stage by guaranteeing PRC participation in the 1954 Geneva Conference. At Beijing's request, the Soviets also gave their novice allies training in how to tackle challenges at international conferences.[51] Moreover, China was implementing its first Five Year Plan (1953-57) with massive assistance and guidance from the Soviets. Accordingly, in the mid-1950s China was clearly looking for a breathing spell in its efforts to spread revolution and challenge the Western powers in East Asia. Zhou Enlai apparently told the Soviets that he needed to disabuse the Vietnamese comrades of the notion that one could simply assume that China would enter into a war in Southeast Asia as it had in Korea to counter any future U.S. escalation. In addition to the direct costs of such a repeat Sino-American military conflict, Zhou appeared concerned that it would assist the United States in building an encircling alliance around China from Japan to Indonesia to India.[52] Neither the Soviets nor the Chinese were looking to spread revolution in a risky fashion in East Asia. So, Beijing and Moscow both had strong incentives to cooperate in establishing a peaceful international environment.

The PRC had its first opportunity to play a major role at an international conference at Geneva. At the conference, the Soviets and Chinese were working largely from the same playbook, and coordination between Beijing and Moscow would never be tighter. In fact, before the conference started Zhou Enlai made

---

[49] See, for example, Ang Cheng Guan, *Vietnamese Communists' Relations with China* (Jefferson, N.C.: McFarland, 1997), ch. 1.

[50] Chen Jian, *Mao's China and the Cold War* (Chapel Hill: University of North Carolina Press, 2000), 62. Communications between Moscow and Beijing in March and April suggest a strong consensus between the two capitals about how to proceed at Geneva. See, for example, "Telegram, Zhou Enlai to CCP CC Chairman Mao Zedong. CCP CC Vice Chairman Liu Shaoqi, and the Central Committee of the CCP, Concerning Soviet Premier Georgy Malenkov's Conversation with Zhou Enlai About the Vietnam Issue," April 23, 1954, *Cold War International History Project Bulletin* (*CWIHPB*) no. 16 (fall 2007/winter 2008): 15.

[51] See "From the Journal of [Soviet Foreign Minister Vyacheslav M.] Molotov and PRC Ambassador [to the Soviet Union] Zhang Wentian, March 6, 1954," in *Cold War International History Project Bulletin* (*CWIHPB*) no. 16 (fall 2007/winter 2008): 86–87.

[52] Gaiduk, *Confronting Vietnam*, ch. 1; and Zhai, *China and the Vietnam Wars*, 53–54.

two separate trips to Moscow for consultations. Beijing knew that the Soviets had been responsible for PRC inclusion in the conference, overcoming the resistance of the United States. Moreover, the relatively inexperienced CCP leadership wanted practical advice from the Soviets. After his second trip to the Soviet capital Zhou flew directly to the Geneva Conference. While in Moscow, the Vietnamese and Chinese comrades met in plenary session with Soviet Foreign Minister Vyacheslev Molotov in order to coordinate policies. Beijing would join the Soviets in encouraging the Vietnamese communists to consolidate victories in the North and postpone aggressive actions in the South.[53]

In Moscow, the Soviets seemed to throw wet blankets on their Chinese and Vietnamese colleagues in a preemptive fashion. They told them not to expect to gain too much from the conference and to avoid providing the United States pretexts to escalate the war. The Soviets clearly wanted an armistice agreement and did not want the aspirations of the Vietnamese or the Chinese to get in the way.[54] As Ilya Gaiduk points out, in managing the alliance, the Soviets were more concerned about the views of the Chinese communists than the Vietnamese communists, and were heartened that the former seemed much more willing than the latter to entertain a peace settlement based on the territorial division of Vietnam.[55]

Although there was a large degree of commonality in the Chinese communist and Soviet positions, the Chinese had somewhat greater aspirations for the conference than the Soviets. As major players in the Indochina War, they knew that they would be instrumental in pounding out an armistice agreement. As rookies in international conferences, they were determined to prove themselves effective both as international actors and as allies to the Vietnamese communists. The Chinese, while cautious and basically in agreement with the Soviet desire for an armistice, also wanted to assure some gains for the Vietnamese communists.

As we would expect from our study of Kim Il-sung and the Korean War, the Vietnamese communists were the most aggressive in pursuing gains for themselves at the expense of the French. They wanted to avoid any agreement that

[53] Robert K. Brigham, *Guerrilla Diplomacy: The NLF's Foreign Relations and the Viet Nam War*, (Ithaca: Cornell University Press, 1998), 4. Also see Chen, "China and the First Indo-China War," 106–11; and Chen Jian, "China's Involvement in the Vietnam War, 1964–69," *China Quarterly* 142 (June 1995): 357. Sino-Soviet coordination began as early as early March 1954. See "Telegram, PRC Ambassador to the Soviet Union and Vice Foreign Minister Zhang Wentian to the PRC Foreign Ministry, Zhou Enlai and the CCP Central Committee, Reporting the Preliminary Opinions of Our Side on the Geneva Conference to the Soviet Side," March 6, 1954, in *Cold War International History Project Bulletin* (*CWIHPB*) no. 16 (fall 2007/winter 2008): 13–14; for Soviet documentary evidence of the same coordination, see "From the Journal of [Soviet Foreign Minister Vyacheslav M.] Molotov and PRC Ambassador [to the Soviet Union] Zhang Wentian," March 6, 1954, in *CWIHPB* no. 16 (fall 2007/winter 2008): 86–87.

[54] Zhai, *China and the Vietnam Wars*, 52–53.

[55] Gaiduk, *Confronting Vietnam*, 18–19.

reduced their ability to unify Vietnam under their control in the near future. For related reasons, they also wanted to maintain and increase their influence in the ongoing domestic political struggles in Laos and Cambodia.

In principle, both major allies of Ho Chi Minh's movement supported a peace deal that would include the partition of Vietnam into clear and distinct communist and noncommunist sections and the removal from Laos and Cambodia of foreign forces (including Vietnamese communist forces). In discussions with their Vietnamese comrades, they were quite insistent on a settlement that emphasized a stable armistice in Vietnam and a geographic consolidation of Vietnamese communist victories to date rather than a crazy-quilt map of communist strongholds surrounded by territory held by anticommunist forces. The latter concept was initially pushed by the more aggressive Vietnamese communists as part of their plan to continue armed struggle aimed at unification in the nearer term.[56]

The Vietnamese communists would also be asked to compromise on policies toward Laos and Cambodia, where Ho Chi Minh's party had exerted military and political influence. The Vietnamese communists were attempting to control strategically important territory to be used for logistical and other support for revolution in Southern Vietnam and for the spread of communist revolution into the rest of Indochina more generally. The Chinese entourage at Geneva, led by Zhou Enlai, would broker a deal between Britain and France on one side and the Vietnamese communists on the other for the removal of all foreign forces from Laos and Cambodia. This was important because the Vietnamese communists had played a strong hand in spreading revolution to Laos in particular. As Zhou Enlai would learn from discussions with noncommunist representatives to the conference from Laos and Cambodia, this revolutionary activity by Vietnamese communist forces and their local communist allies went far beyond opposing French occupation and included destabilizing any local anticommunist political forces in the kingdom.[57] In this period of consolidation, Beijing was much more interested in preventing escalation of the conflict or providing any pretext for a long-term U.S. military presence on China's border in Laos or northern Vietnam than it was in spreading revolution beyond North Vietnam.[58]

[56] Ibid., 18–19. For the earliest evidence that a clear dividing line was the goal of the Chinese entourage, see "Preliminary Opinions on the Assessment of and Preparation for the Geneva Conference," (drafted by Zhou Enlai), March 1954, in *Cold War International History Project Bulletin* (*CWIHPB*) no. 16 (fall 2007/winter 2008): 12–13.

[57] Zhai, *China and the Vietnam Wars*, 56–57.

[58] In the same spirit of bridge-building with nationalist, nonrevolutionary regimes in the region, in 1955 Beijing would try to foster better relations between communist and noncommunist anticolonial regimes at the Bandung Conference. Beijing would also open ambassadorial-level talks with the United States in Warsaw. The only exception to this trend was the 1954–55 Taiwan Strait

Despite the more expansive goals and aggressive strategies of the Vietnamese communists, Ho Chi Minh and his advisors eventually acceded to the Chinese and Soviet insistence on accepting a clear geographical boundary between communist and noncommunist Vietnam.[59] Even before the dramatic final victory at Dien Bien Phu on May 8, 1954, the Vietnamese communists felt that they were ascendant and that, with the further backing of the Chinese and Soviets, they could continue armed struggle toward near-term unification.[60] The Vietminh leadership, therefore, was extremely reluctant to accept division of Vietnam and came around to this position in June only under persistent Soviet and, especially, Chinese pressure.[61] Moreover, they resisted both the details of the dividing line in the final accord (which they believed to be drawn too far north) and the two-year delay in elections designed to abolish the dividing line and unify the nation.[62] Finally, the Vietnamese communists, originally the "Indochina" com-

---

crisis, discussed later, which had more to do with Beijing's peculiar nationalist fetish with Taiwan and with changes in the U.S. alliance structure in the region than it did with intra-alliance politics in the communist alliance.

[59] See "Preliminary Opinions on the Assessment of and Preparation for the Geneva Conference," *CWIHPB* no. 16, pp. 12–13.

[60] For a discussion of the early meetings in March 1954 between the Soviets and the Vietnemese communists on the issue of a negotiated partition of Vietnam, see Gaiduk, *Confronting Vietnam*, 18–19.

[61] Ibid., 37–38; also see Yang, "Changes in Mao Zedong's Attitude toward the Indochina War," 9–10.

[62] Chen, *Mao's China and the Cold War*, 62. Gaiduk, *Confronting Vietnam*, chs. 2–3. For recent documentary evidence of the importance of the demarcation line and the timing of elections to the Vietnamese, see "Minutes of Zhou Enlai's Meeting with [Pierre] Mendes-France," July 17, 1954, in *Cold War International History Project Bulletin* (*CWIHPB*) no. 16, (fall 2007/winter 2008): 68–69. Perhaps anticipating problems with the division of Vietnam and the holding of future elections, in a later meeting Zhou volunteered to Mendes-France that "in Vietnam, there are two regrouping areas with two governments. Within a specific period they control their respective areas. But the regrouping areas in Vietnam are only a provisional solution, and this does not harm reunification." "Minutes of Conversation between Zhou En-lai, Pierre Mendes-France, and Eden," July 19, 1954 in *CWIHPB*, no. 16 (fall 2007/winter 2008): 77. It is clear that the setting of dates for elections for any time after 1955 was the "last resort" position of the Chinese and Vietnamese at Geneva, and Ho Chi Minh's accession to that compromise position was won by the Chinese in early July. See "From the Journal of Molotov: Secret Memorandum of Conversation with Zhou Enlai and Pham Van Dong," July 16, 1954, *CWIHPB*, no. 16 (fall 2007/winter 2008): 95. After accepting the idea of a dividing line, the Vietnamese communists apparently initially suggested a line between the 13th and 14th parallels. See "From the Journal of Molotov," 89. Also see "Rineiwa Huiyi Qijian Zhang Wentian Dashi yu Yingguo Daibiao Kaqiya de Tanhua Jiyao" [Notes from the Meeting between Amb. Zhang Wentian and British Representative (Harold) Caccia], July 19, 1954, 5:45–6:30 p.m., document 206-00005-10(1) Foreign Ministry Archives, PRC, Beijing. The author found this document during research in summer 2004. This document was apparently subsequently translated into English in *CWIHPB* no. 16 (fall 2007/winter 2008): 77–78 but is given a different document number and has the time of 5:45–6:00 p.m., instead of 5:45–6:30 p.m., which seems a bit short for such a meeting. For evidence that China and Vietnam had not yet reached a consensus on the details of the division of zones at the Geneva Conference as late as June 1954, see "Telegram CCP Central Committee to Wei Guoqing, Qian Xiaoguang and Convey to the Vietnamese Workers Party Central Committee,

munists, also had long planned to extend their influence in Laos and Cambodia, areas of strategic concern to both the United States and China as well.

The Vietnamese communists would be compelled by their allies to accept division of the Vietnamese nation at the 17th parallel and the delay of planned elections, intended to end this state of affairs, until 1956. Ho Chi Minh's movement would also need to accept, in principle, if not in practice, the notion that all foreign troops, including Vietnamese communist troops, should be withdrawn from Laos and Cambodia. A final compromise was international oversight of the peace agreement, including monitors in Vietnam, Laos, and Cambodia to police against aggression, revolutionary activities, or irredentism from either side. The solidarity and moderation exhibited by their Chinese and Soviet allies left a materially poor Vietminh no chance to play one capital off the other or to gain the level of material and political support necessary to continue the fight against the southern regime and to spread communist revolution to nations bordering Vietnam.[63]

For their part, the French settled for an agreement only reluctantly. Only the dramatic defeat at Dien Bien Phu, followed by the rise of a new more accommodating French government under Mendes-France on June 16, would allow France to accept many of the peace terms, and the details of a peace settlement still mattered in Paris. In order to facilitate an agreement, in negotiations with the British and French representatives, Zhou Enlai and his Chinese colleagues would all but admit that Vietnamese forces were in Laos, despite the fact that the Vietminh denied their existence. After learning much more about the political realities on the ground in Southeast Asia and recognizing the independence and nationalism of states in Laos and Cambodia, China then played a major role in pressuring the Vietnamese to agree to withdraw forces from those countries as payment in return for pledges of neutrality vis à vis Laos and Cambodia from local governments, from European governments, and, indirectly through the British representatives in Geneva, from the United States.[64]

Regarding the Meeting Between the Premier and Comrade Ding," June 20, 1954, *CWIHPB* no. 16 (fall 2007/winter 2008): 48–49.

[63] Chen, "China and the First Indo-China War"; Qiang Zhai, "China and the Geneva Conference of 1954," *China Quarterly* no. 129 (March 1992): 103–22; Gaiduk, *Confronting Vietnam*, ch. 3; the Chinese diplomatic historian Niu Jun puts it this way: "Moscow concluded that the CCP had lost its appetite for engaging the imperialists at any given chance and thought it a better ally for it." See his chapter, "The Sino-Soviet Alliance and the United States," in Odd Arne Westad, ed., *Brothers in Arms: The Rise and Fall of the Sino-Soviet Alliance* (Washington, D.C.: Woodrow Wilson Center, 1998), 171.

[64] Zhai, *China and the Vietnam Wars*, 56–57. After initially arguing to the incredulous British in May that the leftist resistance in Laos was purely a local movement that predated even the CCP, the Chinese representatives later all but recognized explicitly the large Vietminh presence outside of Vietnam. See "Rineiwa Huiyi Qijian Zho Enlai Zongli yu Faguo Zongli Jian Waizhang Mengdia-Foulangsi de Huitan Jiyao" [Notes from the Meeting Between Premier Zhou Enlai and French

As the most active diplomatic actor in Geneva on the communist side, Zhou Enlai pursued the following goals for the PRC: a consolidated territory for the Vietnamese communists in the northern part of Vietnam, free of French military presence; prevention of escalation of the war and direct involvement of the United States; and the removal of French forces from Cambodia and Laos without having them replaced by U.S. military forces or alliances.[65] Zhou knew that achievement of these goals required persuading the Vietnamese communists to make the necessary compromises on their near-term goals in southern Vietnam and in Laos and Cambodia.[66] The communists and anticommunists involved in the negotiations needed to settle on an international advisory body that was acceptable to all and that could verify compliance with whatever agreement was made.

Judging from the secondary accounts and from the recently opened PRC Foreign Ministry Archives, Zhou worked tirelessly during the talks and during the recess. He shuttled between the headquarters of the various entourages in Geneva, brokering a deal between the French and Vietnamese communists. To

---

Prime Minister and Concurrent Foreign Minister, Mendes-France during the Geneva Conference], June 23, 1954, in document no. 206-00006-06(1); "Rineiwa Huiyi Qijian Zhang Wentian Dashi yu Yingguo Daibiao Kaqiya de Tanhua Jiyao" [Notes from the Meeting Between Ambassador Zhang Wentian and British Representative (Harold) Caccia during the Geneva conference], July 18, 1954, Document 206-00005-11(1); and "Rineiwa Huiyi Qijian Zhang Wentian Dashi yu Yingguo Daibiao Kaqiya de Tanhua Jiyao" [Notes from the Meeting Between Ambassador Zhang Wentian and British Representative (Harold) Caccia during the Geneva conference], July 19, 1954, 5:45–6:30 p.m., document no. 206-00005-10(1)—all from the Foreign Ministry Archives, PRC, Beijing. For the original statements about the purely local nature of the Laotian resistance, see, for example, "Rineiwa Huiyi Qijian Zhou Enlai Huifang Yingguo Waijiao Dachen Ai Deng de Tanhua Jilu" [Records of the Discussions during the Visit of Zhou Enlai to British Foreign Secretary Eden during the Geneva Conference], May 20, 1954, doc. 206-00005-02(1) in Foreign Ministry Archives, PRC, Beijing. By June 16, Zhou seemed to recognize that Vietnamese "volunteers" might still be in Laos. See "Rineiwa Qijian Zhou Enlai Waizhang Liu Yue Shiliu Ri Fang Yingguo Waijiao Dachen Ai Deng de Tanhua Jilu" [Record of the Talks During Foreign Minister Zhou Enlai's Visit to British Foreign Secretary Eden on June 16, 1954, during the Geneva Conference], June 16, 1954, document no. 206-00005-05(1), Foreign Ministry Archives, PRC, Beijing.

[65] For evidence of this concern, see "From the Journal of Molotov: Secret Memorandum of Conversation with Zhou Enlai and Pham Van Dong," July 16, 1954, in *CWIHPB* no. 16, p. 96.

[66] For a good rundown of Chinese concerns late in the negotiating process, see "Rineiwa Huiyi Qijian Zhang Wentian Dashi yu Yingguo Daibiao Kaqiya de Tanhua Jiyao" [Notes from the Meeting between Ambassador Zhang Wentian and British Representative (Harold) Caccia], July 19, 1954, 5:45–6:30 p.m., document no. 206-00005-10(1) Foreign Ministry Archives, PRC, Beijing. For evidence of Beijing's oft-stated concern that Laos and Cambodia would not remain neutral after French and Vietnamese withdrawal but might become U.S. allies, perhaps hosting U.S. military bases, see "Rineiwa Qijian Zhou Enlai Waizhang Liu Yue Shiliu Ri Fang Yingguo Waijiao Dachen Ai Deng de Tanhua Jilu" [Record of the Talks during Foreign Minister Zhou Enlai's Visit to British Foreign Secretary Eden on June 16, 1954, during the Geneva Conference], June 16, 1954, document no. 206-00005-05(1), pp. 4–5, Foreign Ministry Archives, PRC, Beijing. In that meeting Zhou stated that China could live with several outcomes for Laos and Cambodia as long as the two nations did not become U.S. military bases as had Thailand.

do so he apparently put a good deal of pressure on the latter to accept peace and a dividing line between north and south even as he represented the Vietnamese communist positions in discussions with envoys from London and Paris. As the communist side's "honest broker," Zhou met frequently with British Foreign Secretary Anthony Eden, who played a similar broker's role for both France and the United States, especially before the rise of the new Mendes-France government in mid-June (Mendes-France was eager for an agreement and would decide to negotiate directly with Zhou Enlai at Geneva).[67] Zhou gathered information from the noncommunist representatives of Laos and Cambodia about the popular legitimacy of their regimes and the destabilizing influence of Vietnamese communist infiltration into Laos in particular. He then apparently pressured the Vietnamese communists, who publicly denied the presence of their troops in other countries, to agree in principle to withdraw any "remaining" troops from those countries.[68]

[67] For documentary evidence in China of the British and the Chinese playing the honest broker roles for both, see, for example, "Rineiwa Huiyi Qijian Zhou Enlai Huifang Yingguo Waijiao Dachen Ai Deng de Tanhua Jilu" [Records of the Discussions during the Visit of Zhou Enlai to British Foreign Secretary Eden during the Geneva Conference], May 20, 1954, document no. 206-00005-02(1); and "Rineiwa Huiyi Qijian Zho Enlai Zongli yu Yingguo Waijiao Dachen Ai Deng 7 Yue 13 Ri de Tanhua Jilu" [Record of the July 13 Discussion Between Premier Zhou Enlai and British Foreign Secretary Eden during the Geneva Conference], July 13, 1954, document no. 200006-00005-07; and "Rineiwa Huiyi Qijian Zhang Wentian Dashi yu Yingguo Daibiao Kaqiya de Tanhua Jiyao" [Notes from the Meeting between Amb. Zhang Wentian and British Representative (Harold) Caccia during the Geneva Conference], July 19, 1954, 5:45–6:30 p.m., Document 206-00005-10(1)—all in the Foreign Ministry Archives, PRC, Beijing. On several occasions Eden would reassure Zhou that the United Kingdom had no interest in seeing Laos or Cambodia become military bases for any third country after a peace settlement in Indochina. In such instances, Eden seemed to be speaking not only for himself, but also, indirectly, for the United States. See, for example, "Rineiwa Qijian Zhou Enlai Waizhang Liu Yue Shiliu Ri Fang Yingguo Waijiao Dachen Ai Deng de Tanhua Jilu" [Record of the Talks during Foreign Minister Zhou Enlai's Visit to British Foreign Secretary Eden on June 16, 1954, during the Geneva Conference], June 16, 1954, document no. 206-00005-05(1), p. 4–5, Foreign Ministry Archives, PRC, Beijing.

[68] See "Rineiwa Huiyi Qijian Zhou Enlai Guanyu yu Lao Waizhang Shaneinekun de Tanhua Neirong Zhi Zhonggong Zhongyang de Baogao" [Zhou Enlai's Report to the CCP Central Committee Regarding Meetings During the Geneva Conference with Laotian Foreign Minister Sananikone], June 23, 1954, Document No. 206-00046-28, PRC Foreign Ministry Archives, Beijing. On the same day as this report, Pham Van Dng met with the Laotian Foreign Minister. At those talks, the Laotian Royal government and the DRV discussed Laotian government talks with the opposition, recognition of the Pathet Lao movement, and removal of French and Vietnamese forces from Laos. No consensus was reached at the meeting. But the frank discussion of Vietnamese bases outside of Vietnam should be considered a major breakthrough in the talks. See "Rineiwa Qijian Li Kenong Guanyu Bubao Yuenan Fu Zongli Fan Wentong yi Laowo Waizhang Tanhua Yaodian Zhi Zhong Gong Zhongyang de Baogao" [Li Kenong's Report to the CCP Central Committee on the Supplemental Report of the Important Points Made at the Meeting of Vice Premier Phan Van Dong and the Laotian Foreign Minister], June 29, 1954, document no. 206-00046-38 (1) in Foreign Ministry Archives, PRC, Beijing. Note: Li's Report to Beijing was dated June 29, 1954, but it was regarding a Vietnamese-Laotian meeting on June 23, 1954.

During the intersession in the talks in late June through early July, Zhou went on a whirlwind tour of South Asia and Southeast Asia. In July he also met with China's Vietnamese communist allies in the southern Chinese city of Liuzhou to prepare the final negotiating positions of the communist camp for the July Geneva Accords signed later that month. Zhou showed a great deal of initiative, and he apparently achieved more for the Vietnamese communists in negotiations than the Soviets had expected. But he kept the Soviets fully informed of his strategy in the negotiations, and one does not get a sense from the archives that Zhou was acting in ways that ran against Soviet interests or intentions.

Only selected documents regarding Chinese-Vietnamese discussions at the Geneva Conference have been released in China to date, but there is some evidence in the recently opened PRC Foreign Ministry archives that the Vietnamese communists initially resisted the compromises they were being asked to make by their foreign comrades and by the British and French. The tenor of the meetings between Zhou and British leaders, in particular, suggests that Beijing had been wrestling with a difficult partner in Hanoi throughout the negotiations.[69] The best evidence that the Vietnamese communists' arms were twisted by China and the Soviet Union at Geneva comes from recently available Chinese documents from the late 1960s. These documents demonstrate that the Geneva Accords were a bitter pill for the Vietnamese communists. Saigon's decision to refuse to honor the commitment to national elections in 1956 (a decision Washington supported) was infuriating but expected in Hanoi. The legacy of Geneva therefore, would become a bit of an embarrassment for the Chinese communists in the 1960s. At that time, Beijing elites were accusing Soviet leaders publicly and privately of antirevolutionary revisionism and selfish nationalist chauvinism in international politics. It appeared that the PRC in 1954 had been guilty of the same sins of moderation of which Beijing was accusing Moscow in the 1960s. As will be discussed in chapters 5 and 6, Vietnam was one of the countries

---

[69] On two occasions in his July 13 meeting Zhou suggested that the Vietnamese were demanding more than they were being offered on issues such as a dividing line in Vietnam. At one point Zhou told Eden that he met with Ho Chi Minh in China in July during the recess in the Geneva Conference and discussed many issues relating to Southeast Asian peace, including Laos and Cambodia. Implying quite a bit of initial resistance from the Vietnamese leadership, he said that "eventually" (*zuihou*) the two were able to reach a consensus view on how to proceed. In the same meetings, Eden tried to reassure Zhou that, while the United States was not happy with the settlement in the making, its concerns were driven more by mistrust that the PRC would colonize Southeast Asia than by a desire to gain strategic advantage against China. The meeting concluded with Eden stating that just as he was leaving Washington from his trip there during the recess, President Eisenhower used the term "peaceful coexistence" in a public statement. See "Rineiwa Huiyi Qijian Zhou Enlai Zongli yu Yingguo Waijiao Dachen Ai Deng 7 Yue 13 Ri de Tanhua Jilu" [Record of the July 13 Discussion Between Premier Zhou Enlai and British Foreign Secretary Eden during the Geneva Conference], July 13, 1954, document no. 200006-00005-07, in Foreign Ministry Archives, PRC, Beijing.

that China was trying most to impress during the Sino-Soviet rift in the 1960s, and Hanoi's leaders were fully aware of Beijing's history on this score. In October 1968 Chinese Foreign Minister Foreign Chen Yi berated Le Duc Tho, the head of Hanoi's committee to oversee the southern campaign, for agreeing to peace talks with Washington earlier that year. He did so by referring to the mistakes at Geneva. Le Duc Tho acidly reminded Chen that Vietnam only made that mistake in 1954 because "we listened to your advice." Chen Yi fired back, "You just mentioned that in the Geneva Conference, you made a mistake because you followed our advice. But this time, you will make another mistake if you do not take our words into account."[70] In November 1968 when Mao was criticizing the Vietnamese communists for being too soft by negotiating with the Americans, he was quite frank about Beijing's differences with the Vietnamese communists in 1954. Mao stated that at Geneva, "it was difficult for President Ho to give up the South, and now, when I think twice, I see that he was right." Zhou Enlai quickly chimed in, blaming the compromises at the Geneva Conference on the Soviets. Mao, however, insisted, "I think we lost an opportunity."[71] Whether the Soviets or the Chinese had exerted the most restraint on the Vietnamese communists in 1954, it is clear that, long after 1956, the Vietnamese communists continued to blame both the Soviets and the Chinese for the alliance's moderation at Geneva.[72]

While differences in the alliance remained, the communist alliance in East Asia was never more unified and well coordinated than it was at Geneva.[73] The most important factor was that China and the Soviet Union were in basic agreement that an armistice needed to be pursued vigorously and that their dependent

---

[70]   Chen Yi and Le Duc Tho, October 17, 1968 in Odd Arne Westad, Chen Jian, Stein Tonneson; Nguyen Vu Tung, and James Hershberg, eds., "77 Conversations Between Chinese Foreign Leaders of the Wars in Indochina," Cold War International History Project (CWIHP), Working Paper No. 22, p.136 (hereafter Westad et al., eds., "77 Conversations"). One internal Party history reports that there was a subtle difference in attitudes about Geneva in Moscow and Beijing. It argues that Moscow had real faith in the accords and the possibility of peaceful coexistence followed by peaceful transformation to a unified, communist Vietnam. Beijing, on the other hand, saw the peace settlement as a breathing spell and a tactic in a long-term armed struggle against the "imperialists." Guo, *ZhongYue Guanxi Yanbian Sishi Nian*, 65. For a rather different portrayal of events, stating that Vietnamese communists agreed with Beijing and Moscow all along at Geneva, see Yang Gongsu, *Zhonghua Renmin Gongheguo Waijiao Lilun yu Shixian* [The Theory and Practice of PRC Diplomacy], a limited edition Beijing University textbook, 1996 version, p. 336; internally circulated. Yang argues that China put no pressure on Ho Chi Minh to accept the Geneva Accords and that Zhou and Ho agreed all along that accepting temporary division of Vietnam was the best strategy.

[71]   Mao Zedong and Pham Van Dong, Nov. 17, 1968, in Westad et al., eds., "77 Conversations," no. 39, pp.137–53.

[72]   See the statements by Luu Van Loi to Chester Cooper in James G. Blight, Robert S. McNamara, and Robert K. Brigham, *Argument Without End: In Search of Answers to the Vietnam Tragedy* (New York: Public Affairs, 1999), 79–80.

[73]   This basic point is outlined well in Gaiduk, *Confronting Vietnam*, chs. 2–3.

ally, the Vietnamese communists, must be convinced of this basic fact. The lack of rivalry between Mao and his Soviet counterparts in this period meant that the Vietnamese communists could not play one communist ally off the other as they would in the 1960s. Neither major ally shared the Vietnamese communists' intense interest in taking the South nor its desire to destabilize and spread its influence in Laos and Cambodia. The United States and its anticommunist allies, who wanted a negotiated settlement of the conflict in Indochina, were the clear beneficiaries of this enemy unity and the restraint that it fostered.

But it is, of course, possible that alliance cohesion was not the key variable here: the national interests of China and, to a lesser degree, of the Soviet Union in securing a friendly buffer state for China and preventing escalation of conflict in Indochina may provide a complete explanation. That realpolitic explanation would be fine if the same nations had maintained that position into the 1960s. The clear national interest is hard to deduce purely from geography and the distribution of power in any case, but neither changed between the mid-1950s and the 1960s, and yet the communist camp's behavior in the two periods was markedly different. For reasons I outline in the next chapter, shifts in U.S. policy in the interim do not provide sufficient explanation for this change. It is clear that the key change from the mid-1950s to the 1960s was one from alliance unity and consensus, at least between China and the Soviet Union, to alliance rivalry and internal discord, which not only prevented effective restraint on the Vietnamese communists by their stronger allies, but also served to catalyze overall foreign assistance to the Vietnamese communists thus fostering their belligerence against the South.

U.S. REGIONAL ALLIANCE FORMATION AND THE MAKING OF THE 1954 TAIWAN STRAIT CRISIS

In the early 1950s, the United States had set up a series of bilateral treaties in East Asia and one trilateral treaty in the case of ANZUS (the defense treaty of Australia, New Zealand, and the United States). In August 1951, the United States signed a bilateral defense treaty with the Philippines. In September 1951 the U.S.-Japan Security Treaty was signed. In the same month, ANZUS was formed. After an armistice was reached in Korea, Washington entered into another bilateral defense treaty with Seoul. During the Geneva Conference, Dulles made it clear that he wanted to set up a broader, multilateral anticommunist defense pact in Southeast Asia to stem the spread of communism there. Although he privately wanted the Geneva talks to collapse when he first introduced the idea of such an alliance, he claimed publicly that the creation of a pact like this should assist in the process of a peace settlement in Indochina. In an April press conference in London, he answered a query about an East Asia defense pact and its relationship to the Geneva Conference as follows: "Well, I would think that the discussions about the collective defense should proceed more or less independently of the Geneva talks. There is an interconnection? Of course; but I have felt that the Geneva talks were more apt to succeed if there was some strong alternative to a failure to accomplish the result we wanted by negotiation. The very existence of that alternative, will, I think, make it more likely that there will be successes in Geneva."[74]

Dulles's concept would come to fruition with the signing of the Manila Pact in September 1954, creating the Southeast Asia Treaty Organization (SEATO). Signatories included the United States, Great Britain, Australia, New Zealand, France, the Philippines, and Pakistan. None of the three noncommunist states of Indochina—Laos, Cambodia, and South Vietnam—could join because of restrictions in the Geneva Accords, but the SEATO treaty carried a "special protocol" that covered those areas and thereby called for their protection against aggression.[75] This protocol was actually the very essence of the treaty, which had as a central purpose to enforce the geographic delineations of the Geneva Accords and prevent the rest of Indochina and Southeast Asia from falling to the communists.[76]

As was the case with the development of the U.S.-Japan security partnership in the period 1948–51, the process of creating SEATO produced certain destabilizing effects in relations across the Cold War divide. The Eisenhower administration was quite public about its desire to form such a pact in the spring and summer of 1954. The notion of an encircling alliance of anticommunist states would, in and of itself, be of great concern to the Chinese communists, as was expressed consistently by Zhou Enlai and his colleagues during the Geneva talks.[77] What made the prospect of much more concern to Mao and other CCP

[74] Dulles Papers, box 79, "Reel: RE CHINA, PEOPLE'S REPUBLIC OF," Secretary Dulles Press Conference, London, April 13, 1954, National Archives.

[75] For the Chinese insistence at Geneva that no countries in Indochina could be included in a future U.S.-led treaty, see "Minutes of Conversation between Zhou Enlai and Anthony Eden," July 17, 1954, in Cold War International History Project Bulletin (CWIHPB) no. 16 (fall 2007/winter 2008): 65–68.

[76] For a brief description of the treaty, see Paul H. Clyde and Burton F. Beers, The Far East: A History of Western Impacts and Eastern Responses, 1830–1975 (Englewood Cliffs, N.J.: Prentice-Hall, 1975), 500–501. For analysis of Dulles's strategy in creating the alliance, see Herring, "The Indochina Crisis: 1954–55," 219–33.

[77] It is interesting that Zhou calls ANZUS an understandable reaction to the threat of future Japanese aggression but sees no similar justification for a Southeast Asian Pact to include Laos, Cambodia, and Vietnam. See "Minutes of Zhou En-lai's Meeting with [Pierre] Mendes-France," July 17, 1954, CWIHPB no. 16, pp. 68–69. In another conversation Anthony Eden at Geneva, Zhou would say that "The ANZUS Pact is directed against the possible resurgence of Japanese militarism, just as the Sino-Soviet alliance, and therefore is somewhat justified. This is because these countries face the menace of Japanese militaries. But the problem in Southeast Asia is of a different nature." See "Minutes of Conversation Between Zhou Enlai and Anthony Eden," July 17, 1954, CWIHPB no. 16, p. 67.

elites was the possibility that the United States might conclude a defense treaty with Chiang Kai-shek's Taiwan and, perhaps, link Taiwan into the SEATO structure. In an important July 27, 1954, document on strategy toward the United States, SEATO, and Taiwan, the Chinese Communist Party Central Committee noted recent discussions between Washington and Taipei about a defense treaty. The Chinese top leadership did not believe a final decision had yet been made but stated that "if the United States and Jiang [Chiang] sign such a treaty, the relationship between us and the United States will be tense for long period, and it will become more difficult [for the relationship] to turn around. Therefore, the central task of our struggle against the United States at present is to break up the U.S.-Jiang treaty of defense and the Southeast Asian treaty of defense."[78] The worst outcome for China would have been the linking of the two treaties because this would have potentially involved more countries than just the United States in relations across the Taiwan Strait and solidified the defense relationship between Chiang's Taiwan and the United States. Such an outcome could have prolonged the division of the mainland and Taiwan and, under certain circumstances, allowed Chiang Kai-shek to drag the United States into a war to recover the mainland. None of these prospects were particularly bright ones for Mao. Chiang Kai-shek had been lobbying vigorously in the United States in 1954 for a U.S.-ROC defense pact and had been pressuring a reluctant Eisenhower administration to agree to one. Chiang's government employed not only confidential government-to-government channels in this campaign, but also congressional lobbying and media connections, so Beijing did not need sophisticated intelligence assets to be fully aware of the problem. According to well-connected PRC scholars, Mao was concerned that the development of SEATO and a U.S.-ROC alliance would augur a more stubborn "Taiwan problem" if the PRC did not take preventive action in the near term.[79]

---

[78] "Telegram, CCP Central Committee to Zhou Enlai, Concerning Policies and Measures in the Struggle Against the United States and Jiang Jieshi after the Geneva Conference," July 27, 1954, *CWIHPB* no. 16, p. 83. Zhou Enlai would present these ideas to the Soviet allies on July 29, but he would gain their agreement in large part by painting the Chinese response to this growing threat as limited to public statements about tensions in the Strait and the buildup of Chinese defenses along the coastline to ward off "violations of the maritime or air boundaries of China." See Memorandum of Conversation between Soviet Premier Georgy M. Malenkov and Zhou Enlai," July 29, 1954, *CWIHPB* no. 16, pp. 102–3.

[79] Nancy Bernkopf Tucker, *Taiwan, Hong Kong, and the United States, 1945–1992*, (New York: Twayne Publishers, 1994), 38–40. Also see Nancy Bernkopf Tucker, "John Foster Dulles and the Taiwan Roots of the 'Two Chinas' Policy," in Immerman, ed., *John Foster Dulles and the Diplomacy of the Cold War* (Princeton: Princeton University Press, 1990), 240–42. See Zhang Baijia, "The Changing International Scene and Chinese Policy Toward the United States," in Robert S. Ross and Jiang Changbin, eds., *Re-examining the Cold War: U.S.-China Diplomacy, 1954–73* (Cambridge: Harvard University Asia Center, 2001), 49; and Gong Li, "Tensions in the Taiwan Straits: Chinese Strategies and Tactics," in Ross and Jiang, eds., *Re-examining the Cold War*, 145–47.

Beijing was gravely concerned that inclusion of the Republic of China on Taiwan in the alliance system would strengthen and encourage Chiang militarily and politically. Moreover, this linkage between the United States and Taiwan would occur at a time when China was recovering from the Korean War and Chiang still had strong irredentist claims on the mainland that could be fulfilled only with American support. In the past when the United States had upgraded its military relations with Chiang's Republic of China on Taiwan, Chiang quickly became more belligerent, accelerating harassment of the mainland and the seizing of tiny offshore islands just off the mainland coast.[80] More important still, inclusion of Taiwan in an alliance would seem to lock the United States permanently in the Chinese Civil War with long-term implications for cross-Strait unification and the mainland regime's security, as the United States would be providing assistance to anticommunist forces on Taiwan who wanted to use military means and subversion to overthrow Mao's regime.[81]

In Mao's eyes, a related dangerous pattern was a diplomatic trend in the region, partially supported by Beijing's own diplomacy in Korea and Vietnam: the settlement of Asian civil wars by internationally recognized DMZs and political separation lines that created the appearance of two, geographically distinct legitimate governments in Korea and Vietnam. Such an outcome, if it were going to become a precedent, clearly ran counter to Mao's desire to unify both sides of the Taiwan Strait under CCP leadership and to gain universal diplomatic recognition of Beijing as the sole legitimate government of all of China.[82] He did not want a permanent or semi-permanent division of his own nation and, therefore, wanted to take actions to prevent such an outcome. Ironically, then, Beijing's moderate stance in Korea in mid-1953 and Vietnam in mid-1954 created precedents that would lead to Chinese national security concerns and would, eventually, encourage CCP belligerence in the Taiwan Strait in 1954. In order to stem these developing trends in the region, Mao decided to launch concerted artillery attacks on the KMT-controlled offshore islands of Quemoy and Matsu (Jinmen

---

[80] For example, after the Americans committed to defending Taiwan in late June 1950, Chiang took the opportunity to increase attacks on Shanghai and to seize islands off of the Zheijiang and Fujian coast. Xu Yan, *Jinmen Zhizhan* [The Battle over Quemoy] (Beijing: Chinese Broadcast Television Press, 1992), 159.

[81] He Di, "The Evolution of the People's Republic of China's Policy toward the Offshore Islands," in Warren Cohen and Akira Iriye, eds., *The Great Powers in East Asia, 1953–1960* (New York: Columbia University Press, 1990), 222–45; and Thomas Stolper, *China, Taiwan, and the Offshore Islands* (Armonk, N.Y.: ME Sharpe, 1985), 19–27; Gerald Segal, *Defending China* (Oxford: Oxford University Press, 1984), ch. 7; Gong Li, *Tensions in the Taiwan Straits: Chinese Strategies and Tactics*; in Ross and Jiang , eds. *Re-examining the Cold War*, 145–47.

[82] See Zhang Baijia, "The Changing International Scene and Chinese Policy toward the United States," 50.

and Mazu) in September 1954, thereby sparking a major crisis with both Taiwan and the United States over the next several months.

In his sometimes surprisingly frank 1992 open-source book, *Jinmen Zhi Zhan* (The Battle for Quemoy), Colonel Xu Yan of the National Defense University sums up the foregoing factors behind Mao's decision when he writes,

> Beginning in July 1954, the Central Committee, [and] Mao Zedong at the same time placed "liberation of Taiwan" and the coastal islands question in prominent positions, demanding that the PLA increase its struggle in the coastal regions. This decision made at this time was the result of comprehensive consideration of multiple factors [such as] the international situation, the struggle across the Taiwan Strait, and domestic political mobilization and economic construction, etc. The main reason for the prominence of the Taiwan issue was the result of the 1954 Geneva Conference... [Mao's] first consideration was the international strategic situation's influence on the question of reunification of the motherland. At that time, the separation of Korea into North and South had been fixed and completed, the result of the ceasefire in Indochina was also a dividing line [separating] north and south Vietnam. The United States was also mustering together Britain, France, Australia, the Philippines, New Zealand, Thailand, and Pakistan in preparation for signing the Southeast Asia Treaty Organization whose purpose was to contain China. At the same time it [the U.S.] was plotting the solidification (*gudinghua*) of the separation of the two sides of the Taiwan Strait. The need to stress the "liberation of Taiwan" was an expression of the CCP's resolute position on unification of the motherland, and of the smashing of plots to divide China. In addition to this ... under conditions of armistice in Korea and the Geneva conference's decision for a cease-fire, [Mao] was unwilling to allow the peaceful influences to slacken the national people's fighting spirit (*quanguo renmin douzhi*). And that it is not even to mention that at that time the United States military was still illegally occupying the Chinese territory of Taiwan, constituting a major threat to the new China; and the Taiwan KMT also had continuously harassed the mainland."[83]

Xu goes on to state that Mao's decision in summer 1954 did not represent a decision for large-scale military attack across the Strait (*da guimo de duhai zuozhan*) but rather Mao's efforts to affect real or perceived negative trends in international politics and domestic politics by the coercive use of force.

According to Xu, in addition to these international political objectives, there was also a domestic dimension to the operations in 1954–55. As a revolutionary, Mao emphasized the psychology of struggle as necessary for his populace to meet the domestic goals set by the Communist Party. The land reform cam-

---

[83] Xu, *Jinmen Zhi Zhan*, 173–74.

paign and the pacification of the West was carried out during the Korean War and, according to Xu, Mao hoped that the successful implementation of the First Five Year Plan (1953–57) would be assisted by reminding the population that the international environment was hostile and that struggle was needed for the PRC to meet its goals. Although I see less consistent hard evidence of this in the 1954–55 case than in the later case of 1958, the goal of domestic mobilization would clearly become a theme in the 1958 Taiwan Strait crisis, itself a result of long-term strategic thinking in Beijing, albeit of a radically different sort.[84] It is noteworthy that one important Central Committee document from July 1954 supports all of Xu Yan's points, with some emphasis on the domestic component. In a telegram to Zhou Enlai in Geneva, the Central Committee called for preparation for struggle against the prospect of a U.S. defense treaty with Chiang, perhaps linked to a future SEATO, by stating, "The introduction of the task is not just for the purpose of undermining the American-Jiang plot to sign a military treaty; rather, and more importantly, by highlighting the task we mean to raise the political consciousness and political alertness of the people of the whole country; we mean to set up our people's revolutionary enthusiasm, thus promoting our nation's socialist reconstruction."[85]

As in Korea in late 1950, in late summer 1954 Mao believed international forces were shifting against the PRC and that brash action needed to be taken to alter the trend lines. In this case, Mao used limited force to send a message to the United States and to Chiang about the costs of forming a formal alliance. Mao's policy backfired and, if anything, only hastened the formation of a formal U.S.-ROC alliance. Discussing this case, He Di, formerly a leading scholar at the Chinese Academy of Social Sciences, argues that "Mao Zedong had long believed in using warfare for political objectives."[86]

---

[84] Ibid., 159. Xu's argument contradicts directly a leading work on the offshore islands problem in the West. Thomas Stolper writes that, given Mao's domestic agenda, "external crisis then would only be a distraction." Stolper, *China, Taiwan, and the Offshore Islands*, 18. The Chinese historian Niu Jun argues that strategic and military considerations dominated Mao's thinking in the 1954–55 crisis as it did in 1958. See Niu, "San Ci Taiwan Haixia Junshi Douzheng Juece Yanjiu" [A Study of Decision-making during the Three Military Conflicts across the Taiwan Strait], unpublished manuscript presented at the conference on "China in the 1950s," Fudan University, Shanghai August, 2004.

[85] "Telegram, CCP Central Committee to Zhou Enlai, Concerning Policies and Measures in the Struggle Against the United States and Jiang Jieshi after the Geneva Conference," July 27, 1954, *Cold War International History Project Bulletin (CWIHPB)* no. 16 (fall 2007/winter 2008): 84. I am grateful to Alastair Iain Johnston for calling this document to my attention.

[86] He, "The Evolution of the People's Republic of China's Policy toward the Offshore Islands," 226. Unfortunately, one rarely if ever finds a single critical word about Mao's military strategy, even in internally circulated readings, and so the 1954–55 crisis is not portrayed as a failure but as a reasonable and carefully crafted use of coercive diplomacy on Mao's part.

In January the U.S.-ROC Mutual Defense Treaty would be ratified. The PLA upped the ante and attacked and seized the offshore island of Yijiangshan off of Zhejiang, and in February the PLA took the Tachens (Dazhens), also off of Zhejiang, after they were evacuated by ROC troops. According to He Di, the assaults on the Yijiangshan and the seizing of the abandoned Dazhens, were of more military significance than political significance as they provided good practice for a weak PLA amphibious capability. The islands were very far from Taiwan, weakly defended, but had provided a base for engaging in irritating harassment and blockade activities launched by the ROC. But Mao had apparently recognized that full-scale invasion of the most important ROC offshore garrisons on Quemoy and Matsu was impractical given the strong fortifications on the island and the weak state of PLA amphibious forces. The motivation for the artillery attacks on those islands, which sparked the crisis with the United States, was apparently largely political.[87] It is hardly clear that Mao would have wanted to seize all of the offshore islands even if he believed the PLA to be more capable. As part of Fujian province, the offshore islands of Quemoy and Matsu provided a notional political bridge between Taiwan and the mainland. Cutting that bridge, especially with no hope of taking Taiwan itself soon thereafter, would serve the perceived American goal of permanently wresting Taiwan away from the mainland. Although He Di traces the long history of PLA preparation for invading the small islands first before the larger islands, at least the timing of the January 1955 attacks still seems politically motivated.

Despite Mao's ire at the United States, he clearly wanted to avoid provoking a war with the superpower. Apparently one of the great constraints on the scope of Chinese actions—the attacks on Quemoy and Matsu in September 1954, the shelling of Yijiangshan in November 1954, the taking of Yijiangshan in January 1955, and the recovery of the abandoned Dachens in February 1955—was Mao's consistent desire to avoid escalation by engaging American forces directly.[88]

Beijing was not driven in this period by its desire to impress upon the Soviets Mao's revolutionary credentials, as it was in 1949–51. In fact, the crisis made the Soviets quite nervous by all accounts, and there was little indication in Washington or Beijing that the Soviets wanted to risk World War III over Taiwan.[89] Even though Sino-Soviet relations seemed to be at their apex during the Geneva Conference and the 1954–55 Taiwan Strait crisis, there was a clear difference

[87] See ibid., 226.

[88] Ibid., 222–31.

[89] Zhou Enlai would present Chinese concerns about Taiwan and SEATO to the Soviet allies on July 29, but suggested only rhetorical and defensive countermeasures by the PRC. See "Memorandum of Conversation between Soviet Premier Georgy M. Malenkov and Zhou Enlai," July 29, 1954, *CWIHPB*, no. 16, pp. 102–3.

of opinion in Moscow and Beijing over the importance of solving the Taiwan issue on Beijing's terms. This is why in the middle of the crisis, January 1955, Mao decided that China needed its own nuclear weapons and developed an indigenous program, albeit with large amounts of Soviet support in the program's early phases.[90] Nor was Mao driven by any sense of competition with the Soviet Union for ideological leadership of the international movement, a pretense that Mao would only begin to develop after two or three years of Khrushchev's rule.

Mao seemingly miscalculated by failing to appreciate Washington's reluctance to meet Chiang Kai-shek's requests for a bilateral alliance in 1953–54. In so doing, he underestimated how PRC belligerence might encourage U.S. leaders to accede to Chiang's entreaties, rather than dissuade them from doing so. To return to the terms discussed in chapter 1, Dulles and Eisenhower feared entrapment by a smaller ally, Chiang's KMT, just as the Truman administration had feared entrapment by Syngman Rhee's ROK in 1945–53. Washington had no desire to be dragged into the Chinese Civil War and was, therefore, rather nervous about signing a defense pact with Taipei, particularly anything that would offer an unconditional defense commitment, including under circumstances in which Chiang had provoked the mainland by launching attacks of which Washington did not approve in advance. So, for most of 1954 the Eisenhower administration resisted signing a defense pact. Although it is difficult to say if and when a U.S.-ROC defense pact would have been signed in any case, Mao's military actions in the Taiwan Strait, beginning in late summer and early fall, only enabled and accelerated the process of U.S.-ROC alliance formation.[91]

[90] John Wilson Lewis and Xue Litai, *China Builds the Bomb* (Stanford: Stanford University Press, 1988), ch. 3. See Shen Zhihua, "Yuanzhu yu Xianzhi: Sulian yu Zhongguo de Hewuqi de Yanzhi" [Assistance and Limitations: Sino-Soviet Nuclear Weapons Research and Development], *Lishi Yanjiu* [Historical Studies] (March 2004): 114 for Mao's decision to build nukes. This article details the impressive amount of assistance that the Soviets provided the PRC in the nuclear field from 1954 to 1958. During the crisis, Ambassador Chip Bohlen wired back to the State Department, "I find it difficult to believe that the Soviet Government would be prepared to run serious risk of involvement in major war over Chinese claims to Formosa." See "U.S. Amb (Bohlen) to State, October 2, 1954," in *Foreign Relations of the United States, Vol. 14: China and Japan*, 1:674.

[91] Tucker, "The 'Two Chinas' Policy,'" 238–42; and Tucker, *Taiwan, Hong Kong, and the United States*, 37–40. Robert Accinelli, "'A Thorn in the Side of Peace': The Eisenhower Administration and the 1958 Offshore Island Crisis," in Ross and Jiang, eds. *Re-examining the Cold War*, 107–12, 118–19. The reasons for U.S. concern about entrapment were clear. In September Chiang continued to insist that the only way for the United States to deal with communism in East Asia over the long run was to assist him in retaking the mainland. See Rankin to Department of State, September 9, 1954, in *Foreign Relations of the United States, Vol. 14: China and Japan*, 1:581–82. Concerns about entrapment and the need to maintain the ability to veto ROC offensives were expressed repeatedly in State Department deliberations about a mutual defense pact with Taiwan in fall 1954. See, for example, Memorandum of Conversation by Secretary of State, October 10, 1954 and October 15,

Even when the U.S.-ROC Mutual Defense Treaty was created in December 1954 and ratified in January 1955, U.S. concerns about entrapment were still very real, as demonstrated by a secret addendum to the treaty stating that the United States need not honor its commitment to Taiwan's defense if the latter were to launch an attack on the mainland without prior consultation with and approval from Washington. As Dulles explained to the NSC in discussing the proposed treaty in October, "The treaty should be defensive in nature and this aspect should be accepted by the ChiNats. It would not be consistent with our basic policy of non-provocation of war were the United States to commit itself to the defense of Formosa, thus making it a 'privileged sanctuary,' while it was used, directly or indirectly, for offensive operations against the ChiComs."[92] A second aspect of the treaty designed to avoid entrapment was the vague wording regarding the U.S. commitment to defend offshore islands held by the KMT other than the main islands of Taiwan and Penghu. By not mentioning the offshore islands nearest the mainland by name, instead referring to "such other territories as may be determined by mutual agreement," the United States suggested it would help defend the smaller offshore islands closer to the mainland, such as the Tachens, Quemoy, and Matsu, only if the threat to those islands also posed a threat to the main islands of Taiwan and Penghu.[93] This fear of entrapment would create conditions that helped undercut deterrence in the second Taiwan Strait crisis of 1958, which is discussed in chapter 5.

One reason why Mao likely overestimated Washington's desire to ally with Taiwan formally and underestimated the tensions in U.S.-ROC relations was the

---

1954; and Memorandum of Conference with the President (John Foster Dulles) White House, October 18, 1954, *Foreign Relations of the United States, Vol. 14: China and Japan*, 1:724, 734, 770. For negotiations of the secret defensive protocol and why, for domestic political reasons, the KMT demanded that it be kept secret, see Memorandum of Conversation between Chinese Foreign Minister, Chinese Ambassador, Chinese Emb Min, Robertson, and McConaughy, *Foreign Relations of the United States, Vol. 14: China and Japan*, 1:870–71. For detailed research on the negotiations between Washington and Wellington Koo, the ROC ambassador to the United States, on how to handle the status of the offshore islands, see Jin Guangyao, "Gu Weijun yu Mei Tai Guanyu Yanhai Daoyu Jiaoshe (1954.12-1955.2)," [Wellington Koo and the U.S.-Taiwan Negotiations Regarding the Offshore Islands], a paper presented to the Scholarly Conference on China in the 1950s, Fudan University, Shanghai, August 2004.

[92] Report of Secretary of State to NSC, October 28, 1954, *Foreign Relations of the United States, Vol. 14: China and Japan*, 1:811.

[93] See Tucker, "The 'Two Chinas Policy,'" 242–243. The restrictions on ROC offensives without U.S. preapproval were actually in effect even before the treaty was signed. As early as April 1954, the Eisenhower administration had conditioned the transfer of F-84 jet fighter-bombers on such a pledge from Taipei. See Immerman, *John Foster Dulles*, 121. Also see, "Memorandum of Conversation by the Secretary of State," October 10, 1954 *Foreign Relations of the United States, Vol. 14: China and Japan*, 1:724. For Dulles's discussion of how the restraint clauses in the treaty were negotiated in mid-October, see "Memorandum of Conversation, by Key," October 18, 1954, in *Foreign Relations of the United States, Vol. 14: China and Japan*, 1:774.

general trend in the Eisenhower administration's East Asia policy in 1954 during and just after the Geneva talks. Although the creation of SEATO was designed to create stability and deter communist aggression in the region, the process of creating SEATO helped undercut U.S. coercive diplomacy by undermining assurances that Chinese passivity would protect PRC core interests. Washington, Taipei, and other regional actors seemed to respond to the PRC peace offensive with policies that appeared to the Chinese communists to threaten Beijing's national security interests. In 1954, because of real and perceived U.S. alliance policies in the region, Beijing saw a closing window of opportunity or an opening window of vulnerability, depending on one's perspective, in which it could either prevent such an alliance from forming or, at a minimum, make sure that, if one was formed, everyone involved would understand that the project would not be cost free.

Aside from Beijing's nationalistic reaction in the Taiwan Strait, Chinese foreign policy would be relatively moderate in the middle 1950s and fully in tune with Soviet designs for a breathing spell in the Cold War. This was demonstrably true before and during the Geneva Conference in 1954. Espousing the "Five Principles of Peaceful Coexistence," in April 1955 Zhou Enlai would play a leading role at the Bandung Conference of African and Asian states in Indonesia, a precursor of the Non-Aligned movement. Although such a conference was designed to challenge the world leadership of the European powers and the United States, what was notable for our purposes was Beijing's distinct lack of commitment to Third World communist revolution. A major theme of the conference was noninterference in the internal affairs of other states, a theme that does not accord well with leadership of world revolution. After the 1954–55 Taiwan Strait crisis wound down in spring, the PRC also opened up ambassadorial talks with the United States in Geneva.[94] Hardly anything concrete was accomplished at these meetings, which would last through 1957, but they were, by their very existence, a sign of reduced U.S.-PRC tensions. Finally, under Zhou Enlai's leadership in the 1955–57 period, China would adopt a posture of seeking peaceful liberation of Taiwan, if such a process were possible. This was as close as Mao's government would come to Khrushchev's own policy line of peaceful transformation to socialism and peaceful coexistence with the capitalist world.[95] In a nutshell, in the mid-1950s Mao's PRC and Khrushchev's Soviet Union seemed mostly to be on the same page in international affairs, and aside from the Taiwan

---

[94] Chen Jian, *Mao's China and the Cold War*, 191–92; Xia, *Negotiating with the Enemy*, ch. 4; Kenneth T. Young, *Negotiating with the Chinese Communists: The United States Experience, 1953–1967* (New York: McGraw Hill, 1968).

[95] Chen Jian, *Mao's China and the Cold War*, 170–171; also see Yang, "Changes in Mao Zedong's Attitude toward the Indochina War, 1949–73," 13–15.

Strait crisis, a big beneficiary of the cohesion in the communist alliance was the United States and its regional security partners. In comparison to the earlier period of communist alliance formation (1950–51) and the later period of communist intramural tensions and rivalry (1958–69) the communist movement was less aggressive when Mao's China was a secure but subordinate member of a Soviet-led alliance. Where conflicts existed, as in Indochina in 1954, peace was also easier to negotiate than it would be in the 1960s.

Chapter 5

THE SINO-SOVIET SPLIT AND PROBLEMS
FOR THE UNITED STATES IN ASIA, EUROPE,
AND THE AMERICAS, 1956–64

Coordination and comity in the communist camp was unparalleled during 1953–57. In 1956, the Soviets and Chinese would demonstrate their high degree of cohesion by cooperating fairly actively in addressing problems within the communist camp in Eastern Europe and in North Korea. This period of alliance unity was, however, relatively short-lived as ideological differences, distrust, and jealous rivalries for international leadership between Stalin's successor, Nikita Khrushchev, and Mao Zedong divided the communist alliance in the years leading up to escalation of the Vietnam War. One can date the true split in the Sino-Soviet alliance in various ways, but in the years 1958–59 relations between the two communist capitals were already quite poor. By 1960 the Soviets would suddenly withdraw all economic advisors from China. In the following years, Mao would launch an intense ideological campaign against the Soviets within the international communist movement.

The Sino-Soviet rivalry would carry real costs not only for Beijing and Moscow but also for the United States and its friends and allies in the region, who were attempting to contain the spread of communism there. Before the United States would escalate its involvement in Vietnam, Beijing was supporting military revisionism by its communist allies in Southeast Asia. Mao's radicalism ran against the Soviet foreign policy line of peaceful coexistence and peaceful transformation at the time. Eventually, Beijing would catalyze the Soviets into a harder stance designed to compete for prestige within the communist camp with Mao's party. The biggest beneficiaries of the dispute were third-party communists in Vietnam, Cuba, and Germany, who were able to manipulate the rivalry to gain increased Soviet and Chinese support for their revisionist causes and to reject advice from Moscow to adopt more accommodating policies toward the enemy bloc.

THE GATHERING IDEOLOGICAL STORM IN SINO-SOVIET RELATIONS

Peaceful Coexistence, Peaceful Transformation, and Eastern Europe

Mao had become increasingly concerned about the Sino-Soviet relationship since Khrushchev's February 1956 de-Stalinization speech. In both form and content, the speech was unwelcome to Mao. The speech was made without prior consultation with Beijing, something Mao believed to be entitled to as the longest ruling Communist Party chief in the international communist movement. Mao had often praised Stalin in China publicly. Khrushchev's criticism of Stalin's "cult of personality" could easily be turned on Mao himself in the CCP context. Khrushchev's push for peaceful transformation to socialism and peaceful coexistence with the West and his mismanagement (according to Mao) of his alliances in Eastern Europe, leading to crises and interventions in Hungary and Poland in 1956, all made Mao suspicious of Stalin's successor.[1]

Mao would publicly back Khrushchev on the international stage in 1956–57, thus somewhat extending what scholars have referred to as the "honeymoon period" in the alliance.[2] In 1956 CCP elites would also coordinate closely with Soviet officials to counsel Kim Il-sung on how to manage the prospect of severe factionalism within Kim's party.[3] Despite this outward coordination between the two parties there was trouble brewing as Mao was already privately quite frustrated with the Soviet leadership.[4]

These early misgivings between Mao and Khrushchev were reciprocal. According to one knowledgeable Chinese specialist on Russian affairs, when Mao offered to help the Soviet leader solve his Eastern European problem through Chinese mediation, Khrushchev was grateful but felt real jealousy (*jidu*) toward Mao. This jealousy would grow in November 1957 when Mao gained quite a bit of attention from third-country communists in Moscow during the celebrations of the fortieth anniversary of the Bolshevik Revolution in Moscow. According to this scholar, before the summer and fall of 1958, Mao did not challenge Khrushchev openly, but to the contrary supported him often against international

---

[1] For the Chinese view on peaceful transformation and the deStalinization speech more generally see, Han Nianlong and Xue Mouhong, chief eds., *Dangdai Zhongguo Waijiao* [Contemporary Chinese Diplomatic Relations] (Beijing: Social Science Press, 1987), ch. 10; internally circulated. For an impressive historical review of the history of Sino-Soviet tensions, Lorenz M. Luthi, *The Sino-Soviet Split: Cold War in the Communist World* (Princeton: Princeton University Press, 2008).

[2] For such a portrayal, see, for example, Shen Zhihua, "Khrushchev, Mao, and the Unrealized Sino-Soviet Military Cooperation," unpublished manuscript written for the Parallel History Project on China and the Warsaw Pact, October 2002, p. 2. The manuscript can be found on line at www.isn.ethz.ch/php/documents/collection_11/texts/Zhiua_engl.pdf.

[3] See James F. Person, "We Need Help From the Outside: The North Korean Opposition Movement of 1956," Cold War International History Project (CWIHP) Working Paper No. 52 (August 2006).

[4] According to Wu Lengxi, head of the New China News Agency, Mao would, on occasion, express that frustration to Soviet ambassadors and envoys in Beijing. The most dramatic reporting from Wu regards the Polish crisis of fall 1956. Wu reports that Mao threatened Moscow via Ambassador Yudin that China would back the Poles and publicly condemn the Soviets if they invaded Poland with military force. See Wu Lengxi, *Shinian Lunzhan: 1956–66, ZhongSu Guanxi Huiyilu* [Ten-Year Theory Battle: 1956–66, A Memoir of Sino-Soviet Relations] (Beijing: Central Documents Publishing House, May 1999), 1:34–39.

challengers like the Hungarians and the Yugoslavians and domestic foes like the so-called anti-Party Group. In this account, rather than chastising Khrushchev, Mao saw him as an inexperienced "little brother" in the international communist movement who needed tutelage and assistance. Predictably, this condescending attitude did not win Khrushchev's affections for the Chinese leader.[5]

Khrushchev also apparently viewed the Chinese leadership with a similar mix of condescension, jealousy, and derision. The new authority of the Chinese Communist Party was demonstrated when, in early 1957, Zhou Enlai was sent as an emissary between Moscow and the East European capitals to smooth over remaining tensions.[6] When Zhou discussed his findings in Eastern Europe with his Soviet comrades in a way that was somewhat critical of the Soviet Union, Khrushchev apparently bristled at what appeared to him to be insubordinate and impolite preaching.[7]

The mutually disrespectful and distrustful attitudes of the two leaderships and the perceptions they created also help explain the row in the summer of 1958 over Soviet proposals for a joint naval fleet and for Soviet-owned radio stations based in China for communication with Soviet submarines. What appeared like a helpful gesture and normal allied cooperation to Khrushchev seemed a "chauvinist" insult and a threat to Chinese national security by Mao. Mao's bitter reaction was therefore surprising and worrisome to Khrushchev.[8]

A major sticking point in the relationship between Khrushchev and Mao was the former's apparent commitment to peaceful transformation of noncommunist states to communism through legal and parliamentary methods, as well as his related policy of peaceful coexistence and cooperative arms control agreements with Cold War adversaries in the United States and Europe. Beijing did not challenge Moscow openly in the period 1956–57, and its own policy line in the region, outlined at the end of chapter 4, was largely consistent with the Soviet line. But Mao clearly saw such peaceful gestures to the West as temporary and

[5] Discussion with two leading experts on Mao's relations with the communist world, Shanghai, August 2004.

[6] Shen Zhihua, "Bo Xiong Shijian yu Zhongguo: Zhongguo Zai 1956 Nian 10 Yue Weiji Chuli Zhong de Juese yi Yingxiang," [China and the Polish-Hungarian Incident: China's Role and Influence in Handling the October 1956 Crisis], a paper presented to the Scholarly Conference on China in the 1950s, Fudan University, Shanghai, August, 2004.

[7] Luthi, The Sino-Soviet Split, 67.

[8] Li Yueran, Waijiao Wutai Shang de Xin Zhongguo Lingxiu [New China's Leaders on the Diplomatic Stage] (Beijing: Liberation Army Press, 1989), 170; Shen, "Khrushchev, Mao, and the Unrealized Sino-Soviet Military Cooperation," 13–26. For documentation on Mao's suspicions and anger about the Soviet offers of a joint fleet and radio stations on Chinese soil, see "Discussions with Soviet Ambassador Yudin," July 22, 1958, in Mao Zedong Waijiao Wenxuan [Selected Diplomatic Documents of Mao Zedong] (Beijing: Central Literature Publishing House, 1994), 322–33; and "Comments on the Request by the Soviet Union to Build a Log-Wave Radio Transmitter in this Country," June 7, 1958 in Jianguo Yilai Mao Zedong Wengao, 7:265–66.

tactical measures, while at least in Mao's opinion, Khrushchev seemed wedded to these ideas in principle.[9] Despite Zhou's rhetoric about eventual peaceful unification with Taiwan in 1955–56, after the Korean War and Taiwan Strait crisis of 1954–55 Mao had trouble believing that the problems with Taiwan could really be resolved peacefully in the PRC's favor. By 1957, peaceful unification seemed less and less likely.[10]

Communist parties in China and North Vietnam were involved in national unification struggles, and so their leaders naturally objected to Khrushchev's notion of "peaceful transformation" to socialism. Saigon's 1956 refusal to abide by the Geneva Accords and allow nation-wide elections suggested the naïveté of Khrushchev's faith that parliamentarian socialist transformation was possible.[11] The enemies' reversal on the election pledge was a bitter pill, but the more globally oriented Soviets were tone deaf to Vietnamese concerns, and continued to view Southeast Asia as an opportunity to promote further détente with the West. In fact, the Soviet leadership was so inconsiderate of the Vietnamese communists' national unification concerns that in late 1956 and early 1957 Moscow briefly sponsored legislation at the UN that would allow both Vietnams to enter the organization. This move apparently was made without prior consultation with Hanoi and was one that Zhou Enlai would describe to his Vietnamese comrades as a "sell-out."[12] Although the Soviets would quickly withdraw the proposal under pressure from allies, much of the damage to its relations with the Vietnamese was already done.[13] Just before Mao decided to launch an artillery assault on Quemoy and Matsu in summer 1958, he expressed his frustration with Soviet moderation in international relations and the need to demonstrate to the Chinese people and the world that the PRC could stand up to the Americans. Vice Foreign Minister Zhang Wentian's comments on Mao's message to a small group of leading foreign policy elites on June 17 are revealing on this score. Zhang noted that Mao discussed the need for a policy of confrontation with the United States and the need to build up confidence at home in the long-term struggle against imperialism. Zhang's chronicle reads as follows:

[9] See Han and Xue *Dangdai Zhongguo Waijiao*, ch. 10. See also Wu, *Shinian Lunzhan*, vol.1, ch. 2; Luthi, *The Sino-Soviet Split*, 62–70.

[10] This was a position that Mao would adhere to until his death in 1976, despite occasional assurances to the Americans that he sought peaceful unification in the early 1970s. For Mao's discussions with the Nixon and Ford administrations on this score, see Romberg, *Rein in at the Brink of the Precipice*, chs. 3–4. Also see Xia, *Negotiating with the Enemy*, chs. 7–8.

[11] For the Vietnamese communists' view on "peaceful transformation," see Ang, *Vietnamese Communists' Relations with China*, 26; and Zhai, *China and the Vietnam Wars*, 79–80.

[12] Gaiduk, *Confronting Vietnam*, 85; Ang, *Vietnamese Communists' Relations with China*, 49–50.

[13] After a meeting with Ho, Khrushchev would withdraw support for the proposal. Ang, *Vietnamese Communists' Relations with China*, 58.

Our foreign policy and that of the Soviet Union and other socialist countries are basically consistent [*yi zhi de*, sometimes translated as "identical"], but there are also some differences. Generally speaking, they are somewhat afraid of the United States. Khrushchev says not to fear, but in [his] actions [he] expresses that he is somewhat afraid [*you xie pa*], afraid of the United States, afraid of West Germany. This psychology of dread [*kongju de jingshen zhuangtai*] is greater in other socialist countries. In the past, out of consideration of the Soviet Union, we did not discuss the Chairman's [Mao's] thinking very clearly in our propaganda. Now we need to give a greater role (*jiayi fahui*) to the Chairman's thinking. In international relations and foreign policy we need to openly set our direction as an example (*gongkai de shuli qi women de fangxiang*).[14]

Frustration with Khrushchev's moderation was not limited to Asian communists. The East Germans also disagreed with Khrushchev's moderate policies on Berlin, requesting permission to close the border to West Berlin throughout the 1950s.[15]

The Great Leap Forward and the 1958 Taiwan Strait Crisis

On the domestic front, problems like the drying up of Soviet capital aid and the declining utility of Soviet advice in solving various bottlenecks in the more backward Chinese economy would also encourage Mao to reduce his dependence on Soviet advice and strategies. Only someone of Mao's ideological ilk might have tried to solve his economic problems with a disastrous, radical, and largely utopian set of policies like the Great Leap Forward (1958–61). So we must recognize the ideological and psychological roots of the Great Leap Forward, which was hardly a rational response to objective factors. But one can, at least, fairly state that the reconsideration of economic policy during the First Five Year Plan that lay at the foundation of the Great Leap was itself rooted in some real problems.[16] As I have argued in much more detail elsewhere, fears and suspicions regarding his relations with Moscow only made Mao even more radical at home and somewhat more aggressive abroad. The Great Leap itself had international dimensions as a way for Mao's China to gain more independence from and strength in its relations with the Soviet Union and within the inter-

---

[14] *Zhang Wentian Nianpu* [Chronicle of Zhang Wentian] (Beijing: Chinese Communist Party History Publishers, 2000), 2:1098–99.

[15] Hope M. Harrison, *Driving the Soviets Up the Wall, Soviet-East German Relations, 1953–61* (Princeton: Princeton University Press, 2003).

[16] Christensen, *Useful Adversaries*, ch. 6; Roderick MacFarquhar, *The Origins of the Cultural Revolution. Vol. 2, The Great Leap Forward 1958–60* (New York: Columbia University Press, 1983); and David Bachman, *Bureaucracy, Economy, and Leadership in China: The Institutional Origins of the Great Leap Forward* (Cambridge: Cambridge University Press, 1991).

national communist movement. The Great Leap was also an early indication of Mao's desire to assert his own model of revolutionary and social change as an alternative to the post-Stalin Soviet model. This would become more obvious during the 1960s, particularly during the Cultural Revolution, but the initial impetus can be found in the late 1950s, as Mao became increasingly disillusioned with Stalin's successors and increasingly confident about his own rightful place as a co-equal or superior to his Soviet counterparts. In many senses, the Great Leap was Mao's first attempt to market himself as a natural leader of the international communist movement.[17]

Internationally, the domestic requisites of the Great Leap made Mao more aggressive, rather than less so. Mao used conflict in the Taiwan Strait to mobilize his public around nationalist themes during the more radical phases of the Great Leap and demonstrate his independence from the Soviets internationally.[18] In the late spring and early summer of 1958 in the earliest phases of the Great Leap policies and in the leadup to the Taiwan Strait crisis, Mao would attack revisionism in the socialist states' foreign policy and call for a more aggressive foreign policy posture.[19] So, even in its nascent stages the competition for leadership in a hierarchical revisionist alliance caused problems for the United States in its coercive diplomacy toward the communist camp.

No matter how quixotic it proved to be in the end, practical implementation of the Great Leap required great sacrifice from the vast majority of Chinese citizens. Mao's need to sell the program to his population also encouraged him to escalate tensions with the United States and to manipulate a crisis in the Taiwan Strait for purposes of domestic mobilization. This logic was an element of the first Taiwan Strait crisis but was a much more important element in the 1958 crisis. Unlike 1954–55, before the latter crisis there were few new or pressing security concerns related to Taiwan's behavior or U.S. alliance behavior in the region that might have motivated Mao to provoke a crisis with the KMT and its U.S. allies. But Mao believed that he needed tensions for domestic mobilization purposes, and he provided them in late August by ordering the shelling of the KMT-held offshore islands of Quemoy and Matsu. Mao did this without first consulting with the Soviets, who almost certainly would have opposed such an action.[20] In this

---

[17] For a recent discussion of the international roots of the Great Leap, see Shen Zhihua, "Sulian dui 'Da Yue Jin' He Renmin Gongshe de Taidu ji Jieguo" [The Soviet Attitudes about the "Great Leap Forward" and the People's Communes and Their Results], *Zhonggong Dangshi Ziliao* [Chinese Communist Party History Research Materials], 1:120; also see Christensen, *Useful Adversaries*, ch. 6.

[18] Christensen, *Useful Adversaries*, ch. 6.

[19] See Zhang, "The Changing International Scene and Chinese Policy Toward the United States," 55; and *Zhang Wentian Nianpu*, 2:1098–99.

[20] For a thorough study of Chinese and Soviet documents that concluded Mao never informed Khrushchev in advance of the attacks, see Shen Zhihua, "Yi Jiu Wu Ba Nian Paoji Jinmen Qian, Zhongguo Shi Fou Gaozhi Sulian" [Did China Notify the Soviet Union in Advance Before Shelling

instance, there is little doubt that lack of coordination, disagreements, and a budding rivalry in the communist alliance made China more aggressive, not less so. The shelling produced a major nuclear crisis with the United States that lasted several weeks. So, we see again how a period of little intra-alliance coordination and a high degree of discord in the communist camp critically complicated U.S. containment strategies in East Asia. What was bad for the Soviets was not necessarily good for the United States, at least not in the near term.

As discussed in chapter 4, entrapment fears in the Eisenhower administration were real, and an unconditional commitment to Taiwan combined with Chiang's own interests in provoking conflict with the mainland could have undercut the assurance side of the U.S. deterrence equation. The United States had no interest in provoking China into attacking the offshore islands. Many scholars assert that one contributing factor in Mao's decision to attack the ROC forces on Quemoy and Matsu was his desire to send a signal about what he saw as dangerous trends in the U.S.-ROC alliance, in particular the stationing of nuclear-capable Matador missiles on Taiwan in 1957.[21] Given that the attacks on Quemoy and Matsu occurred a full year after the Matador missiles were introduced and were timed perfectly for domestic mobilization purposes during the most radical phases of the Great Leap Forward and for demonstrating independence from Moscow at a time when relations with the Soviets were worsening, it is hard to sustain the argument that Mao's concerns about trends in the U.S.-ROC alliance were the most important reason for the 1958 artillery attacks. That having been said, Mao's more general concern about the U.S. and ROC getting too comfortable with one another militarily likely did factor into Mao's decision to stir the pot in cross-Strait relations in 1958. By attacking the offshore islands Mao could remind the United States of the costs and dangers of its alliance with Taipei and test U.S. resolve toward the protection of the offshore islands in particular.[22]

Quemoy in 1958?], *Zhonggong Dangshi Yanjiu* [Chinese Communist Party History Studies] no. 3 (2004): 35–40. For other authoritative accounts suggesting Mao never informed the Soviets, see Feng Xianzhi and Jin Chongji, eds., *Mao Zedong Zhuan* [Biography of Mao Zedong] (China Central Documents Publishing House, 2003), 1:855. Wu Lengxi argues that Mao and Khrushchev "did not speak a single sentence" about the shelling of Quemoy and Matsu in late July and early August. Wu, *Shinian Lunzhan*, 1:186.

[21] Deterrence and punishment of Chiang Kai-shek and the United States is routinely listed as one of the contributing factors in Mao's decision. See, for example Xu Yan, *Jinmen Zhi Zhan*; Zhang Shu Guang, *Deterrence and Strategic Culture: Chinese-American Confrontations 1949–58* (Ithaca: Cornell University Press, 1992), ch. 8; and Li Jie, *Zhongsu Lunzhan de Qiyin: Guocheng Ji Qi Yingxiang* [Origins of the Sino-Soviet Ideological Conflict: Process and Its Influence], paper presented at the October 1997 Conference on Sino-Soviet Relations and the Cold War, Chinese Central Committee Documents Research Office, Beijing, p. 6; Gong Li, "Tension Across the Taiwan Strait," 156–58. Also see, Niu, "San Ci Taiwan Haixia Junshi Douzheng Juece Yanjiu," 11. Niu quoted Mao as saying in late June 1958 that an assault on the offshore islands would "directly attack Chiang Kai-shek and indirectly attack the United States."

[22] On this last point, see Niu, "San Ci Taiwan Haixia Junshi Douzheng Juece Yanjiu," 17; and Gong Li, "Tension Across the Taiwan Strait," 157–58.

In 1958, Mao was careful to limit the nature and geographic location of his artillery attacks against Quemoy and Matsu to minimize the chance of escalation to war with the United States. Mao wanted a crisis, not a war. For example, he placed constraints on air operations in support of the artillery attacks and never attempted to take the islands by force.[23] But the artillery assaults on Quemoy and Matsu turned out to be riskier than Mao had originally expected, as the U.S. responded quite vigorously by deploying two aircraft carrier battle groups and running supplies to the islands by sea and air.[24] Mao had reason for his overconfident assessment that the U.S. response would likely be quite limited and that he could attack the islands in a limited fashion without much likelihood of escalation. After all, the U.S. commitment to the offshore islands' defense had intentionally been left ambiguous in the U.S.-ROC Mutual Defense Treaty. As was discussed in chapter 4, in order to avoid entrapment, the Eisenhower administration decided not to make firm commitments to ROC-held islands other than Taiwan and Penghu (the Pescadores).[25] This ambiguity gave the United States more freedom of maneuver and put some constraints on Chiang Kai-shek's ability to drag the United States into war, but it also helped undercut the credibility of U.S. deterrent threats against such limited coercive efforts aimed at the ROC by the PLA. So, in this sense as well, U.S. alliance politics and dilemmas contributed indirectly but significantly to the breakdown of coercive diplomacy in 1958.

This is not to say that the recipe for peace and stability from an American perspective was less ambiguity as well as clearer and less conditional alliance commitments to Chiang's regime. At the broadest level, there was no simple way out of the alliance security dilemma for the United States. Many policies that might have increased the credibility of a deterrent threat could have undercut assurances, and policies in place that were designed to prevent entrapment and bolster assurances undercut the credibility of threat.[26]

A second complicating factor related to lack of cohesion and coordination in the U.S.-ROC alliance was Washington's inability to settle the crisis by encouraging the KMT to withdraw its overextended and highly exposed troops from the offshore islands. The KMT opposed this because of its long-term goal of returning to the mainland. The offshore islands provided notional bridges to the mainland because of their geographic location and, more abstractly, because they were universally recognized as part of the mainland province of Fujian. A second related and more immediate problem for the KMT was its domestic legitimacy on Taiwan. In the 1950s the KMT was fully dominated by mainland

---

[23] Ye Fei, *Ye Fei Huiyilu* [Memoirs of Ye Fei] (Beijing: Liberation Army Press, 1988), 650–56.

[24] See, for example, Niu, "San Ci Taiwan Haixia Junshi Douzheng Juece Yanjiu," 8.

[25] Accinelli, "'A Thorn in the Side of Peace,'" 107–113, 118–119.

[26] This is one reason why one common translation of "dilemma" in Chinese is *liang nan*, or "double-sided trouble."

Chinese who had escaped to the island after the KMT's defeat in the Chinese Civil War on the mainland. Without the argument that Taiwan was part of a unified China and the KMT government was the sole legitimate government of that larger Chinese nation of which Taiwan is a part, Chiang's KMT would have had very little legitimacy on Taiwan itself. The United States was clearly worried about the damage to KMT morale and stability on Taiwan if the offshore islands and the more general offensive long-term mission that they represented were somehow lost under military pressure from the Chinese communists.[27]

For their part, the CCP elites recognized this KMT position and appreciated it, as it prevented permanent Taiwan independence from the mainland, something Beijing correctly believed that Washington would have preferred in the 1950s if it could have had its druthers. For this reason, when late in the crisis Washington was urging Chiang to remove his forces from Quemoy and Matsu as part of a ceasefire in the crisis, Mao ordered the otherwise confusing policy of offering to bring supplies to the islands to feed the besieged garrisons there so that they would not leave under such U.S. pressure.[28] So, the domestic politics of the Chinese Civil War exacerbated greatly the U.S.-ROC alliance security dilemma and complicated efforts at crisis management and de-escalation.

On the communist side of the equation, mutual tensions and misgivings in Sino-Soviet relations were both a cause and effect of the 1958 crisis. The Soviets' request earlier in the year for a joint naval fleet and a submarine radio station on Chinese soil triggered China's postcolonial nationalist sensitivities and raised tensions to a fever pitch, as Mao berated the Soviet ambassador to China about Soviet colonial designs on China.[29] After this bitter encounter Khrushchev hurriedly planned a secret trip to Beijing to attempt to smooth relations a bit. At those tense summit meetings in late July and early August, Mao apparently never disclosed his plans for shelling of Quemoy and Matsu, which were well under way. As Khrushchev was preparing to return home from his secret mission to China, the CCP also publicly disclosed the Soviet leader's visit. This fact made some Chinese scholars opine that Mao's berating of the Soviet ambassador might have been a ruse to draw Khrushchev to Beijing. By then leaking the news of the secret meeting Mao could signal to the world Soviet support for Mao's China. This disclosure then might have been seen as a useful deterrent against

---

[27] Accinelli, "'A Thorn in the Side of Peace,'" 109–12.

[28] See Gong Li, "Tension across the Taiwan Strait," 163. Gong cites Mao's statements at an October 3 Politburo Standing Committee Meeting that the KMT and CCP agree on opposing a "two China" solution to cross-Strait relations. Also see Niu, "San Ci Taiwan Haixia Junshi Douzheng Juece Yanjiu," 17. Niu claims that Mao initially wanted to drive KMT forces off the island and then settled for this strategy once the U.S. had clarified its commitment to Taiwan by sending so many forces to the area. Also see Christensen, *Useful Adversaries*, ch. 6.

[29] Shen, "Yi Jiu Wu Ba Nian Paoji Jinmen Qian, Zhongguo Shi Fou Gaozhi Sulian"; and Niu Jun, "San Ci Taiwan Haixia Junshi Douzheng Juece Yanjiu," 18–19.

robust U.S. intervention following the planned attacks on the offshore islands of Quemoy and Matsu. Mao launched those attacks just three weeks after Khrushchev's departure.[30]

The Chinese diplomatic historian Niu Jun argues that while it is not entirely clear if Mao manipulated Khrushchev to come to Beijing in the first place, "at a minimum after Khrushchev arrived he intentionally used [the situation] in order to facilitate the creation of an impression internationally that the Soviet Union supported the attack on Quemoy and Matsu."[31] If this is true, then Mao's approach backfired. The Eisenhower administration indeed believed that the Soviets had approved of the attacks on Quemoy and Matsu for precisely this reason. But this only strengthened U.S. resolve to prevent the offshore islands from falling under military pressure. As Robert Accinelli points out in his study of Eisenhower administration policy, Dulles's belief that the Soviets were behind the attacks meant that they were a broad test of U.S. resolve and could not have been based in particularistic Chinese concerns about U.S. behavior toward Taiwan or the offshore islands. This conclusion only stiffened Washington's backbone.[32] So, if one believes that Mao initially wanted to rid the island of KMT forces and only changed his strategy when the United States responded very vigorously to the Chinese challenge, as does Niu, then Mao's manipulative alliance tactics sent the wrong kind of coercive signal, provoking a tough U.S. reaction rather than deterring one.

Chinese scholars may be correct that Mao leaked the news of the meeting with Khrushchev as part of a deterrence strategy aimed at the United States. But Mao's concern about Soviet bullying and domination as expressed to the Soviet ambassador and later to Khrushchev himself was likely fully sincere. In any case, Sino-Soviet relations would get much worse within weeks of the summit. The large-scale artillery assault on the offshore islands of Quemoy and Matsu were launched without prior consultation with Moscow. Moreover, the attack sparked a major nuclear crisis with the United States at a time when Moscow was pushing a line of peaceful coexistence. The crisis would last several weeks. In the first week or two of the crisis Mao would launch the most radical phases of the Great Leap Forward, a revolutionary economic plan that was already in ill repute in Moscow and would only gain more derision from Soviet leaders as it intensified. For their part, Beijing elites, however fairly, would come to view Soviet support during the crisis as limited and late in the offing. Moreover, in areas where the Soviets offered concrete support, Mao still treated these offers with suspicion.

[30] Off-the-record discussion with Chinese scholars, August 2004. For written versions of this portrayal of a conspiring Mao, see Shen, "Yi Jiu Wu Ba Nian Paoji Jinmen Qian, Zhongguo Shi Fou Gaozhi Sulian"; and Niu, "San Ci Taiwan Haixia Junshi Douzheng Juece Yanjiu," 18–19.

[31] Niu, "San Ci Taiwan Haixia Junshi Douzheng Juece Yanjiu," 18–19.

[32] Accinelli, "A Thorn in the Side of Peace," 119–20.

For example, the Soviets offered to send fighters and bombers with missiles to Fujian province weeks after the Taiwan Strait crisis broke out, but Mao refused because the Soviets would not yield command of the units over to China.[33] Mao's ambassador in Moscow reported having warned Mao at the time that Khrushchev might try to use the occasion of the crisis to increase Soviet control over China and cross-Strait issues.[34]

According to the historian Shen Zhihua, during the crisis Mao expressed his independence from the USSR in a manner that truly upset Khrushchev. In late September a U.S.-built "Sidewinder" air-to-air missile fell on Chinese territory. This was the U.S. military's most advanced air-to-air missile and had been transferred to the ROC air force during the crisis. The Soviets naturally wanted to get their hands on the missile right away for research, but the Chinese demurred. After a delay of weeks they would eventually turn the system over to the Soviets but only after removing perhaps the most important component of interest to the Soviets, the infrared sensor.[35]

In Mao's eyes, what was perhaps more important still in 1958–59 than fears of Soviet chauvinism, was Soviet criticism in 1959 of his domestic "Great Leap" development plan. The Soviet leadership viewed the plan as unrealistic and as diverging from core Marxist principles. This criticism only reinforced Mao's view that the Soviet leadership was domineering, overly bureaucratic, insufficiently revolutionary, and rife with Russian nationalist chauvinism.[36] At the July 1959 Lushan Plenum when he lashed out at domestic critics of the Great Leap Forward such as Peng Dehuai, Mao also focused his ire on Khrushchev's criticism of the concept of people's communes, a core component of the Great Leap. The Soviet leader had criticized the communes in July 1959 during a visit to Poland and, to add insult to injury, his speech was subsequently published in *Pravda*.[37] When Mao purged Peng Dehuai, he accused him of being too closely linked to the USSR.[38]

---

[33] For details, see Wu, *Shinian Lunzhan*, 1:187. Wu reports that the offer for Soviet air units was made in mid-October or later. Other sources report that Khrushchev offered to send Soviet interceptor squadrons to Fujian in mid-September. See John Lewis and Xue Litai, *China's Strategic Seapower: The Politics of Force Modernization in the Nuclear Age* (Stanford: Stanford University Press, 1994), 17; and Liu Zhihui, *Chashang Chibang de Long* [Dragon with Wings], (Beijing: Liberation Army Arts Press, 1992), 150–52.

[34] The Chinese ambassador to the Soviet Union, Liu Xiao, warned Beijing that Khrushchev might try to use the occasion of the crisis to bring the Taiwan question under Soviet control and under "Soviet-Western and in particular Soviet-American world hegemony." See Liu Xiao, *Chushi Sulian Banian* [Eight-Year Mission to the Soviet Union] (Beijing: Central Archives and Historical Records Press, 1986), 70.

[35] Shen, "Yuanzhu yu Xianzhi," 127.

[36] Shen, "Sulian dui 'Da Yue Jin' He Renmin Gongshe de Taidu ji Jieguo," 118–39.

[37] Ibid., 134.

[38] Luthi, *The Sino-Soviet Split*, 130.

In a nutshell, by the late 1950s Mao deemed the current Soviet leaders un-
worthy of international leadership in the communist movement, thus setting
the stage for the Sino-Soviet competition for leadership that would catalyze the
revolutionary and belligerent tendencies of the movement overall for the next
ten years. Of course, he saw himself and his own party as quite worthy of that
leadership. As Shen argues, by the end of 1959 Mao clearly wanted "China to
become the symbol (*biaolü*) and model (*bangyang*) of the international communist movement."[39] He opines that Mao chose to launch his ideological conflict
(*lunzhan*) with the Soviets on the issues of revolution and war, but his real goal
was to have China take the lead in the international communist movement more
generally, including responsibility for communicating theories of economic
management and transitions from socialism to communism. Regarding the
budding Sino-Soviet dispute in 1958–59, Chinese scholar Yang Kuisong writes,

> [W]hat irritated Mao the most was Soviet unwillingness to carry on revolution. For
> Mao, revolution, whether it was the class struggle or the anti-imperialist variety, was
> not only the focal point of his life experience but also the key to the success of the
> Chinese revolution. In his mind the negation of revolution, particularly violent revo-
> lution [as in the case of Khrushchev's concept of peaceful transformation], meant
> the negation of the universal applicability of the Chinese revolutionary model and
> the rejection of the 'unique contribution' that he had made to Marxism-Leninism.
> That was why in 1958 Mao ordered the distribution among high-level party cadres of
> quotations on continuous revolution by Marx, Engels, Lenin, and Stalin. He wanted
> those cadres to understand the nature of the Sino-Soviet dispute.[40]

Mao's insistence on the superiority of his approach to class struggle and socialism obviously posed a major challenge to the Soviet Communist Party's self-
image as leaders of the international communist movement.

Toward the Break: The Collapse of Sino-Soviet Military
and Economic Cooperation

Soviet military assistance would continue to flow to China, even in the nuclear
field, after the 1958 Taiwan Strait crisis began. In fact, in response to the crisis,
the Soviet Union increased its assistance to the Chinese Air Force in particular. The Soviets, however, began backing out of their earlier commitments to
full nuclear cooperation, including the transfer of a test nuclear weapon. The
Soviets saw the Chinese as adventuristic and unreliable in part because they
launched the Taiwan Strait crisis without first informing the Soviet allies and in

---

[39] Shen, "Sulian dui 'Da Yue Jin' He Renmin Gongshe de Taidu ji Jieguo," 138.
[40] Yang, "Changes in Mao Zedong's Attitude toward the Indochina War," 18.

part because the Soviets were confused about the PRC's apparent lack of desire to seize the islands once they were subject to an artillery blockade.[41] In June 1959, the Soviets officially scrapped the nuclear agreement altogether, as relations between Mao and Khrushchev worsened further.[42] Wu Lengxi, the head of the influential Xinhua News agency at the time, recalls that the cancellation of the agreement was a major event in Sino-Soviet relations as Chinese elites began to wonder if the Soviets were aligning with the West and the United States against China.[43] It was in this context that Khrushchev's peace summitry with Eisenhower in September 1959 at Camp David and discussions of a nuclear-free Asia were viewed as a sell-out in China. The Soviet discussions about a nuclear-free Asia must have looked less worrisome to Mao in 1958 when Khrushchev was still actively and secretly supporting the Chinese nuclear weapons program. This might help explain why China publicly supported the concept in early 1958. But especially after the abrogation of the agreement to transfer a test weapon in mid-1959, Beijing had to worry more seriously about Soviet-American collusion to keep smaller powers under their control and about Soviet complacency and satisfaction with Moscow's place as an international superpower following Sputnik and other Soviet technological achievements. Of particular concern was the progress achieved in Geneva by Soviet, Eastern Bloc, U.S., and Western European scientific delegations who were attempting to create a verifiable nuclear test ban treaty. On August 21, 1958, two days before the PRC launched the artillery barrage on Quemoy and Matsu, the scientists announced that they could design a system of verification that could cover the entire globe to police violations of a future test ban. Khrushchev, who helped organize the conference, did not invite Chinese representatives but included China explicitly in the areas that any treaty would cover.[44]

[41] On Khrushchev's negative reaction to the shelling of Quemoy and Matsu, see Liu Xiao, *Chushi Sulian Ba Nian* [Eight-Year Mission to the Soviet Union] (Beijing: Chinese Communist Party History Research Materials Publishing House, 1986), 72. Liu was Ambassador to the Soviet Union in 1958–59. Also see Shen, "Yuanzhu yu Xianzhi," 126–28.

[42] For continued Soviet support after the 1958 summit and the Taiwan Strait crisis, see Shen, "Khrushchev, Mao, and the Unrealized Sino-Soviet Military Cooperation," 23–25. For the temporarily positive effect of the Taiwan Strait Crisis on Sino-Soviet military cooperation, particularly Soviet aid to the Chinese air force, see Shen, "Yi Jiu Wu Ba Nian Paoji Jinmen," 40. For the temporary and increasingly wary Soviet cooperation on nuclear matters with China after the crisis, leading to the abrogation of the agreement to transfer a test weapon, see Shen, "Yuanzhu yu Xianzhi," 110–31. Also see, Viktor Gobarev, "Soviet policy Toward China: Developing Nuclear Weapons, 1949–69," *Journal of Slavic Military Studies* 12, no. 4 (December 1999): 19–27.

[43] Wu, *Shinian Lunzhan*, 1:207–8.

[44] On the Geneva meetings and the negative impact on Sino-Soviet relations, particularly when Mao shelled Quemoy and Matsu, thus running counter to the spirit of détente that Khrushchev had worked to create with the West, see Shen, "Yuanzhu yu Xianzhi," 126–27. On the general issue of arms control and the effects on Sino-Soviet relations, see Shen, "Khrushchev, Mao, and the Unrealized Sino-Soviet Military Cooperation," 25.

Mao and Khrushchev then had a terribly tense summit in October 1959, setting the stage for a full-scale rift in the alliance and the withdrawal of Soviet advisors with their blueprints in July 1960.[45] At the October summit, one of the issues that aggravated bilateral relations was Khrushchev's suggestion that China adjust its policy toward Taiwan so as to improve relations with the United States. Khrushchev's apparent siding with the United States following the Camp David summit, particularly on an issue that the CCP considered one of domestic, not international, politics, led to a strong rebuke from both Zhou Enlai and Mao.[46]

THE EFFECTS OF THE SINO-SOVIET DISPUTES ON THIRD-PARTY COMMUNISTS IN ASIA

In July 1964 relations between the Chinese and Soviets would become so poor that Mao began considering the possibility of armed conflict between the two communist states in discussions with the CCP Politburo.[47] It would take nearly five more years for Sino-Soviet relations to deteriorate to that degree and, when they did, the United States was a major beneficiary. But while the Sino-Soviet rivalry was escalating (1960–69) and the two communist nations' competition for the hearts and minds of the Vietnamese was heating up (1964–69), the biggest beneficiaries would clearly be the Vietnamese communists. Until the early 1970s the big losers of this Sino-Soviet competition were Saigon and Washington, who were facing a determined group of Vietnamese enemies who were receiving much more material assistance from abroad and much more international political backing for the continued spread of revolution to the South than they received in the mid-1950s, when the Sino-Soviet house was in much better order and the movement as a whole was, therefore, more transparent, more willing to negotiate, and more moderate overall.

Vietnam 1959–61

Sino-Soviet tensions and the rebellion against Khrushchev's policy line of peaceful coexistence and peaceful transformation perhaps affected Southeast Asia most dramatically. After months of consulting with Beijing and keeping Moscow in the dark, in May 1959 Hanoi would agree to assist the South Vietnamese

---

[45] For detailed coverage of this summit, see the memoirs of Li Yueran, one of Mao's Russian interpreters, see Li, *Waijiao Wutai Shang*.

[46] Wu, *Shinian Lunzhan*, 1: 222–23.

[47] Wang Zhongchun, "The Soviet Factor in Sino-American Normalization," in William C. Kirby, Robert S. Ross, and Gong Li, eds., *Normalization of U.S.-China Relations: An International History* (Cambridge: Harvard University Asia Center, 2005), 149.

communist cadres in national unification by all means necessary. The National Liberation Front (NLF) was born. North Vietnam's secrecy and prevaricating toward Moscow and its relative openness with the CCP demonstrated how much more conservative Moscow was than Beijing about revolution in the South.[48] When the Soviets discovered the North Vietnamese designs on the South, they were surprised and upset at the provocative behavior of their Vietnamese comrades. As Gaiduk writes, "[T]he Soviet Union, as a great power, had to take into account geopolitical considerations, not only ideological preferences."[49]

The Chinese Communists, on the other hand, promised political, economic, and military support for the NLF effort in 1959. It is notable that this backing occurred well before the United States began increasing the number of military advisors in South Vietnam in the early 1960s and almost six years before the United States would send in a large number of ground troops. In the late 1960s Beijing's relatively active support for belligerence in Vietnam in comparison with the Soviets' desire for restraint would be affected at least in part by national security concerns in Beijing about U.S. activity in Southeast Asia that were not fully shared in more distant Moscow. But it is very difficult to sustain an argument that Chinese support for the NLF in the period 1959–64 was driven by standard realpolitik concerns about friendly borders. After all, North Vietnam, the bordering state, was solidly under Ho's control. U.S. military activity in South Vietnam was still very limited (with hundreds, not even thousands, of advisors on the ground). Moreover, the 17th parallel separating North and South Vietnam was distant from the Chinese border. All of these factors work against an argument suggesting that Chinese support for revolution in this period was driven primarily by defensive national security concerns.

In August 1959 the Vietnamese communists would announce to the Chinese communists that they were concerned about potential collusion at the upcoming Eisenhower-Khrushchev summit at Camp David and lack of international support for their revolutionary efforts. In one of the darkest moments in its economic history—the Great Leap disaster—China responded with a promise of $500 million dollars in military and other assistance.[50] In 1959 Beijing also supplied Vietnam with 57 million British pounds in preferential loan packages. This outstripped anything coming from a relatively indifferent though much richer Moscow.[51] Ilya Gaiduk reports that, from 1956 to 1960, overall Chinese economic

---

[48] Gaiduk, *Confronting Vietnam*, 111–14.

[49] Ibid., 114.

[50] Ang, *The Vietnamese Communists' Relations with China*, 122–30. Ang offers reports from British observers in Hanoi about the conspicuously boisterous celebration of China's National Day in Hanoi in 1959.

[51] Ibid., 102.

assistance to Vietnam was double that of the much wealthier Soviet Union.[52] Meanwhile, Sino-Soviet relations were deteriorating further.

This picture of a cozy Hanoi-Beijing condominium and a distant USSR would apparently change briefly in early 1960 either because the Chinese would not or could not promptly carry through on their promises of aid. By early 1960 China was getting little press in Hanoi and the Soviets were clearly being wooed.[53] From available evidence, it is unclear if this brief, public pro-Soviet tilt occurred because Hanoi was upset at China for failing to keep its word or whether Hanoi was trying to seduce the Soviets into adding Soviet aid to the pot of Chinese aid that it had already received.

The Chinese communists did not stand idly by in the face of Hanoi's tempo-rary tilt toward Moscow. Instead, they attempted to outbid their richer and more powerful Soviet rivals. In January 1960 the Chinese foreign ministry agreed to give Hanoi extensive loan packages. Moreover, Zhou Enlai wanted to insure that military aid to Vietnam was given unconditionally and *gratis*. He asked that Chi-nese diplomats never casually mention military aid when discussing economic loans. Beijing was clearly trying to avoid treating Vietnam the way that the Sovi-ets had treated China after the Korean War. The Soviets had created great bitter-ness in Beijing by demanding repayment from China for military transfers dur-ing wartime.[54] By spring and summer 1960 the Chinese were very active again in the Indochina struggle. Zhou Enlai traveled to Phnom Penh in May. In July Sihanouk would reveal an informal offer of Chinese defense assistance in case of foreign violations of Cambodian territory.[55] Also, in May, Ho Chi Minh traveled to China, and Chinese and Vietnamese leadership entourages shuttled between Beijing and Hanoi. Beijing fully backed Hanoi's plan for a protracted military and political struggle to "liberate" the South. Although the Soviets, like the Chi-nese, would offer economic aid packages at this time, the Soviets, unlike the Chi-nese, continued to push for a peaceful settlement of the Vietnamese Civil War.[56]

In June 1960 Zhou Enlai briefed an expanded session of the CCP Politburo about China's key foreign policy differences with the Soviets, including attitudes about: 1) the nature of imperialism and the causes of war; 2) arms reduction; 3) peaceful coexistence; 4) support for the revolutionary struggle of oppressed peo-ples, and so on.[57] In the same month CCP representatives criticized the Soviet

---

[52] Gaiduk, *Confronting Vietnam*, 90.

[53] Ang, *The Vietnamese Communists' Relations with China*, 139.

[54] January 21, 1960, entry in *Zhou Enlai Nianpu*, 2:281.

[55] Ang, *The Vietnamese Communists' Relations with China*, 142–43. Ang says that according to U.S. intelligence reports in July, Zhou had even promised fighter planes to Sihanouk.

[56] Guo, *ZhongYue Guanxi Yanbian Sishinian*, 67. Ang, *The Vietnamese Communists' Relations with China*, 148–149; Zhai, *China and the Vietnam Wars*, 83.

[57] June 8, 1960, entry in *Zhou Enlai Nianpu*, 2:325.

Union strongly, albeit not directly by name, for the first time at an international communist conference (in Bucharest).[58] As discussed further in the pages that follow, initially indirect public attacks on proxy targets would escalate into direct attacks on the Soviet leadership in Chinese Communist Party publications, speeches, and in talks at international conferences. This would become a pattern over the next several years.[59] Clearly upping the ante in the global competition with the Soviets, in July 1960 Zhou Enlai agreed to increase grain transfers and related loans to distant Albania.[60] (Within weeks of this decision, all Soviet advisors would be withdrawn from China). In an internal speech in August 1960, Zhou Enlai argued that the best way for China to deal with the growing split within the movement was for it to try fostering international mobilization and solidarity of the socialist camp and guide it away from "revisionism" and "semi-revisionism."[61] In December 1960, Beijing was the first foreign government to recognize officially the Vietnamese National Liberation Front (NLF), designed to foment revolution in the South. Despite the terrible famine caused by the Great Leap Forward, in January 1961 China agreed to preferential loans to Hanoi worth 142 million RMB.[62] In that same month, Zhou Enlai committed to long-term assistance to Albania on the basis of their common fight against "imperialism and revisionism."[63] Playing to Chinese strategy, Albanian leaders would take explicitly negative positions toward Khrushchev in meetings with Chinese elites in 1961–62 and declare that Mao and his CCP should be the leader of the international communist movement. Vice-Premier Li Xiannian would not reject this goal for the CCP, stating only that Mao believed the process of securing CCP leadership of the international movement would take ten years.[64]

[58] On Bucharest, see Luthi, *The Sino-Soviet Split*, 167–74; Shen, "Khrushchev, Mao, and the Unrealized Sino-Soviet Military Cooperation," 25.

[59] In a most impressive piece of multinational research in several languages, Luthi, *The Sino-Soviet Split*, chs. 5–6, offers a complete review of the beginnings of the polemical battles between the Chinese and Soviets within the international communist movement.

[60] July 26, 1960, entry in *Zhou Enlai Nianpu*, 2:335. The entry mentions both financing for grain shipments and discussion of the Sino-Soviet dispute with the Albanians. China would prove unable to meet Albania's full requests later in the year because of "the calamity," better known in the West as the Great Leap Forward. See December 25, 1961 entry in ibid., 2:447–48. Despite its geographic distance from Eastern Europe, China supplied impressive levels of military aid to Albania from 1961 to 1978, including 752,000 guns, 11,000 pieces of artillery, 890 tanks and armored vehicles, and 180 airplanes. See "Jianguo Hou Wo Dui Wai Junshi Yuanzhu Qingkuang," 34.

[61] August 14, 1960, entry in *Zhou Enlai Nianpu*, 2:340.

[62] Ang, *Vietnamese Communists' Relations with China*, 179–83, 196–98.

[63] "Memorandum of Conversation with Comrade Zhou Enlai," *Cold War International History Project Bulletin* (*CWIHPB*) no. 16 (fall 2007/winter 2008): 189.

[64] See the minutes of Enver Hoxha's meeting with Li Xiannian in "Report on the Second Meeting with the CCP Delegation to the Fourth Congress of the Albanian Party," February 25, 1961 in *Cold War International History Project Bulletin* (*CWIHPB*) no. 16 (fall 2007/winter 2008): 200–216. In a June 1962 meeting with Zhou Enlai, an Albanian Communist Party leader, Hsyni Kapo, would

For their part, Soviet officials seemed to understand the implications of the Chinese initiative to actively support revolution in the developing world. In a June 27, 1960, conversation with his Soviet counterpart Stepan Chervonenko, Albanian Ambassador to the PRC Mihal Prifti pointed out that "more and more Maoism is being touted as the Marxism of the 20th Century." Chervonenko replied,

> I think the Chinese comrades accept that the October Revolution was truly an event of historical proportions on a worldwide scale. But they think that its influence has been larger over the European countries, while the Chinese Revolution, according to their opinion, also of worldwide importance, is more important for the peoples of Asia, Africa, and Latin America.... As a result, the peoples of these countries and the entire world's workers' and communist movement should take lessons from the Chinese Revolution. They should walk in its path and not in that of the October Revolution. Furthermore, after Stalin's death, someone should be at the helm of the entire workers' and communist movement. This person is Mao Zedong and the CCP.[65]

Prifti also reported to Tirana that "when I asked why China, facing such difficulties [at home], was [still] helping Mongolia and Vietnam with such large sums, he [Chervonenko] answered that this was due to the fact that China wanted to control these countries.[66]

The Chinese strategy of supporting armed revolution in noncommunist former colonies began winning internationalist support as early as November 1960, especially in Asia.[67] This caused great consternation in Moscow and debate ensued in the Kremlin about how to answer the China challenge. Still, as James Richter argues, the split within the Communist Party of the Soviet Union (CPSU) on international affairs in late 1960 led more to paralysis in Moscow than it did to a concerted effort to reverse the peaceful coexistence line and increase support for revolution.[68] So in a high-level meeting with the Vietnamese in June 1961 the Chinese were enthusiastic about the Vietnamese desire to lib-

---

call Mao the "eminent leader of the international communist movement." See "Memorandum of Conversation, Hsinyi Kapo and Ramiz Alia with Zhou Enlai," June 27, 1962, in *CWIHPB*, no. 16 (fall 2007/winter 2008): 248.

[65] Memorandum of Conversation between Albanian Ambassador to the PRC Mihal Prifti and Soviet Ambassador to the PRC Stepan V. Chervonenko," June 27, 1960, *CWIHPB*, no. 16 (fall 2007/winter 2008): 187.

[66] Ibid., 189.

[67] For an Albanian and Chinese recap of the growing support for China's position among some Asian delegates at the November 1960 Moscow Conference, see "Memorandum of Conversation, Comrade Abdyl Kellezi with Comrade Zhou Enlai," April 20, 1961, in *Cold War International History Project Bulletin* (*CWIHPB*) no. 16 (fall 2007/winter 2008): 219–26.

[68] James G. Richter, *Khrushchev's Double Bind: International Pressures and Domestic Coalitions* (Baltimore: Johns Hopkins University Press, 1994), 137–39; Ang, *The Vietnamese Communists' Relations with China*, 1–2.

erate the South by force, while two weeks later the same Vietnamese entourage would get only a lukewarm response from the Soviets.[69]

The Soviets were not yet major players in Vietnam in the early 1960s despite Vietnamese efforts to change that. As Russian historian Ilya Gaiduk points out, the Chinese had supplied much more assistance than the Soviets to the Vietnamese communists in the 1950s and had maintained much closer relations with Hanoi than had the Soviets, who had weak connections in Hanoi and had provided only "negligible" military assistance. With some exceptions noted later, this trend continued into the early 1960s, even after the dispatch by the Kennedy administration of large number of U.S. military advisors to the South. Even as late as January 1964, when Vietnamese communist leaders, including General Secretary Le Duan, visited Moscow, the Soviets limited their support largely to rhetoric. By one estimate, from 1962 to the beginning of 1965 the Soviets provided only $100 million dollars worth of military aid to the Vietnamese communists. Instead of providing additional assistance up front, the Soviets complained about Chinese influence in Hanoi and insisted that further Soviet assistance would require a change of attitudes in Hanoi toward the Sino-Soviet split. Moreover, they consistently seemed much more eager than their Chinese counterparts to seek a negotiated settlement rather than escalation in Vietnam.[70]

Laos 1959–64

The degree to which the Soviets could compete with the Chinese in supporting revolution was limited by geography and by Khrushchev's efforts to improve relations with the new Kennedy administration and avoid escalation of problems in Laos and Vietnam. In late 1961 while the Soviets pushed for compromise in Laos to avoid escalation of that war to one that might involve the Americans, the Chinese pushed for a harder line by the leftist Pathet Lao against the Western-leaning Laotian conservatives. As late as 1957, in the period of relative Sino-Soviet harmony, Beijing had lined up with the Soviets to dissuade Vietnam from scuttling previous peace deals by infiltrating North Vietnamese forces into Laos. Beijing cited the danger of U.S. escalation.[71] But beginning in 1959 Beijing would start to support armed resistance by Laotian leftists and China would turn a blind eye to North Vietnamese infiltration of areas just west of the 17th parallel dividing line between the North and South, asking only that

---

[69] Ang, *The Vietnamese Communists' Relations with China*, 196–98.

[70] Ilya V. Gaiduk, *The Soviet Union and the Vietnam War* (Chicago: Ivan R. Dee, 1996).

[71] Gaiduk, *Confronting Vietnam*, 127.

the Laotians avoid providing evidence of such infiltration to international in-spectors.[72] In August and November 1960 in meetings with Ho Chi Minh, Mao would encourage armed struggle by the North Vietnamese in both Laos and South Vietnam. He said there was nothing wrong with killing people and urged his Vietnamese comrade not to fear reprisals from "the reactionaries" because their killing of people in response just mobilizes the people against the reaction-aries. In these meetings Ho Chi Minh had expressed some reservations about the threat of escalating violence, and Mao and his colleagues were trying to en-courage Ho's belligerence. Mao's approach therefore was in direct contrast with the Soviet strategy toward the region at the time.[73] In 1961–62, China would support a much more belligerent stand by the Pathet Lao than would Moscow, who, according to Gaiduk, "more than once assured the Americans of their de-sire to eliminate the Laotian problem as an obstacle on the way to agreement on more important international matters."[74] Put succinctly, "the Chinese and the North Vietnamese were not as concerned as the Soviet Union about the danger of broadening the conflict."[75]

Unlike South Vietnam or Cambodia, Laos borders China, and since the late 1950s the United States had been actively supporting anticommunist forces in the civil strife there. For that reason, it is more difficult to parse out the factors of national security as opposed to revolutionary fervor in this case. In some senses this is a chicken and egg problem. U.S. support for anticommunists in Laos followed U.S. detection of direct and indirect North Vietnamese interven-tion there, an intervention that was fully backed by Beijing, but not by the Sovi-ets. In any case, the Chinese feared that too much compromise by the Pathet Lao and sympathetic "neutralists" might allow for continued U.S. influence in Laos and thereby threaten not just the Chinese border but the plan for expanding revolutionary activities in South Vietnam, which required secure logistics lines through Laos to facilitate Hanoi's support for communists in the South. In the early 1960s the Chinese would persuade both the Vietnamese communists and sympathetic Laotian forces to take a tougher stand against conservative forces in Laos, thereby complicating peace negotiations.[76] As one recent Chinese history of the period states, the Soviets in 1962 stated that they supported the moderate Laotian neutralists, and in fact they did. The Chinese on the other hand said in public that they supported the moderate neutralists but in reality they sup-

[72] Yang, "Changes in Mao Zedong's Attiude toward the Indochina," 17; Gaiduk, *Confronting Vietnam*, ch. 7.

[73] Yang, "Changes in Mao Zedong's Attitude toward the Indochina War," 20.

[74] Gaiduk, *Confronting Vietnam*, 151.

[75] Ibid., 153.

[76] Ang, *The Vietnamese Communists' Relations with China*, 208–25.

ported the revolutionary parties and urged a tough stand.[77] In discussions with the Albanians in 1962, Deng Xiaoping criticized the Soviets for considering "the agreement for the creation of a coalition government in Laos as an example of how to achieve agreements with Western powers through talks." As an alternative, Deng clearly presented the Chinese strategy of supporting a coalition government as a temporary measure to keep the United States out of the country while the Pathet Lao strengthened its position there.[78]

To assure peace and stability, the relatively conservative Soviets were willing to settle for an internationally enforced Austrian-style neutrality in Laos or even for a geographic division of the nation along the lines of Korea or Vietnam. In fact, one of the reasons why the Soviets saw a divided Laos as acceptable was that such a solution might shore up the unstable situation in the rest of Indochina, including in a divided Vietnam. But preserving stability in Vietnam was hardly a concern shared by Khrushchev's increasingly radical Asian allies. The Chinese and the Vietnamese rejected these solutions and acted to undermine them at every step because they would harm revolutionary forces already in place in Laos, including those sympathetic to North Vietnamese efforts against the South and willing to assist and protect Hanoi's logistic operations.[79] Gaiduk writes,

> Whereas Moscow regarded the settlement in Laos as a necessary precondition for the stabilization of the situation in Indochina, Hanoi wanted to use the Laotian crisis to instigate precisely the processes the Soviets wanted to prevent. The North Vietnamese were not completely opposed to solution of the problem of Laos, but they supported only a solution that allowed them to continue their activities in South Vietnam. ... The situation was complicated because the Chinese shared most of Hanoi's views."[80]

By January 1962, fighting had broken out again in Laos between the Pathet Lao–Neutralist forces and the royalist forces loyal to Vientiane. Moreover, North Vietnamese forces would join the fight in significant numbers for the first time since 1954.[81] China supported the increased militancy of the Pathet Lao because it might result in increases in the relative strength of anti-American, pro-Hanoi

---

[77] Quan Yanchi, *Gongheguo Mi Shi* [Secret Diplomacy of the Republic] (Hohhot, PRC: Inner Mongolia Publishers, April 1998), 29–30.

[78] "Memorandum of Conversation Deng Xiaoping, [CCP Liaison Department Director] Wang Jiaxiang, Hsyni Kap, and Ramiz Alia," June 19, 1962, in *Cold War International History Project Bulletin* (*CWIHPB*) no. 16 (fall 2007/winter 2008): 241.

[79] Gaiduk, *Confronting Vietnam*, ch. 8.

[80] Ibid., 167.

[81] Ang, *The Vietnamese Communists' Relations with China*, 208–225. Robert McNamara would recall the Vietnamese military actions in Laos in 1962, labeling them "difficult to understand." In the process McNamara belies a false belief in the notion that these acts were a result of misperceptions and poor governmental coordination rather than a systematic attempt by Hanoi with Beijing's support to ensure that Laos could be used to support revolutionary activity in South Vietnam. See Blight et al., *Argument Without End*, 124–25, 130–36.

forces in any future coalition government. Toward this end, China sent weapons to the Pathet Lao. At the same time, Beijing elites were privately critical of the Soviet position for being too accommodating to pro-American forces and insufficiently protective of the interests of revolutionary parties. Beijing also supported the dragging out of negotiations so as to allow the Pathet Lao to consolidate its hold over strategically important areas in the country.[82]

Eventually a National Union would be formed in Laos according to U.S. and Soviet wishes. The Chinese would publicly support the July 1962 Geneva Accords on neutralization of Laos, but the hands of the Pathet Lao and the North Vietnamese had already been sufficiently strengthened to guarantee that Laos would be a conduit for support of revolution in South Vietnam. The Chinese communists knew that the North Vietnamese had no intention of honoring their end of the agreement by removing all of their forces from Laos. In fact, in November 1961 Mao encouraged Ho Chi Minh to continue fighting in Laos with North Vietnamese forces, but suggested that Hanoi should keep up the pretense that it had no forces in the country.[83] By January 1964 Mao was encouraging Hanoi to send several thousand more soldiers into Laos.[84] North Vietnamese violations of the 1962 Geneva Accords were quite consequential in the Indochina War. Robert McNamara reports that the Kennedy administration became so untrusting of Hanoi after those violations that hope for future peace agreements regarding South Vietnam were severely harmed.[85]

Vietnam 1962–64

China's foreign aid skyrocketed in the early 1960s, particularly aid to Vietnam, and this occurred at a time when the PRC was suffering from severe economic disasters at home. In 1961 Mao made further commitments to transfer enough weapons to equip 230 battalions of Vietnamese forces.[86] After backing the NLF in August 1959 with the promise of aid, including military aid, Beijing apparently had some difficulty fulfilling that promise with actual deliveries. But despite China's domestic hardships, the aid began flowing in during the early 1960s. In 1962 China supplied 90,000 guns of various types to the "Southern Campaign."[87] In October 1962, Mao and Zhou met with a military entourage from North Vietnam led by General Giap. Zhou would promise "comprehensive" assistance

---

[82] Zhai, *China and the Vietnam Wars*, 104–10; Gaiduk, *Confronting Vietnam*, 168.

[83] Yang, "Changes in Mao Zedong's Attitude toward the Indochina War," 21. Yang cites the minutes of meetings between Mao and Ho Chi Minh on November 14, 1961.

[84] Yang, "Changes in Mao Zedong's Attitude toward the Indochina War," 28.

[85] Blight et al., *Argument Without End*, 110; and Brigham, *Guerrilla Diplomacy*, 26.

[86] Yang, "Changes in Mao Zedong's Attitude toward the Indochina War," 21–22.

[87] Guo, *ZhongYue Guanxi Yanbian Sishinian*, 69; and Ang, *The Vietnamese Communists' Relations with China*, 229.

to North Vietnam, including military, political, and economic aspects.[88] In late 1962 and early 1963, the Vietnamese communists argued in Beijing that the United States might respond to communist military action in the South with escalation and attacks on the North.[89] For a combination of ideological and national security concerns apparently not shared in Moscow, China responded impressively. According to an internal Party history, from 1962 to 1966 China supplied 270,000 guns, 5,400 cannon and artillery pieces, 200 million rounds of ammunition, 900,000 artillery shells, over 700 tons of TNT, 200,000 military uniforms, and 4 million meters of cloth.[90]

The picture of China's Vietnam strategy in 1962 appears to be a mix of ideological and national security concerns, but the former factor seems more important than the latter. In 1962 the United States increased the number of military personnel in Vietnam from 3,000 to 11,000. But China's general support for the NLF and promises of military aid long preceded this escalation. Moreover, however large they had become, U.S. forces in Vietnam still had a training and advisory role and were not a combat force capable of invading North Vietnam, let alone China. Even if they had been so capable, it does not naturally follow that a logical Chinese defensive countermeasure would have been support for revolution in the South, rather than simply the bolstering of North Vietnamese defense. It is important to note that from a realpolitik point of view, Beijing already had a buffer in North Vietnam at the time. The spread of revolution to the South in the 1960s was not the same type of goal for China as was the eviction of the French from the North in the early 1950s. It can hardly be seen as a defensive measure or as one purely related to China's parochial national interest.

It is true that Mao had already launched a study of the vulnerability of China's coastal industries to enemy attack in spring 1964, a study that would eventually lead to the costly and radical "Third Front" program, in which large amounts of China's defense industrial complex would be moved deep into the interior. In line with speeches he made in 1964 about the threat of U.S. aerial attack on China, Mao launched the study with the argument that one never knew when imperialists might attack. But Mao did not implement the study's findings until August 12, 1964, after the alleged Gulf of Tonkin Incident and the launching of punitive strikes against the North by U.S. naval air power, which was then followed by the truly large-scale introduction of hundreds of thousands of ground

[88] October 5, 1962, entry in *Zhou Enlai Nianpu*, 2:500.

[89] Qu Aiguo, Bao Mingrong, and Xiao Zuyue, "*Zhongguo Zhiyuan Budui YuanYue KangMei Junshi Xingdong Gaishu*" [A Narrative of the Military Activities of the Chinese Volunteer Units in the Assist Vietnam Oppose America War] (Beijing: Junshi Kexue Chubanshe, 1995), 40. Chen, "China's Involvement in the Vietnam War," 359.

[90] Guo, *ZhongYue Guanxi Yanbian Sishinian*, 69.

forces beginning in 1965. [91] Most important, Mao was clearly surprised by the U.S. attacks and the responses that followed, demonstrating his relatively low threat assessment prior to U.S. escalation. According to Chinese Communist Party historian Li Danhui, one reason that Mao was surprised was that he had come to believe in the early 1960s that the United States was too reliant on its allies to directly intervene in regional conflicts in the near term. He thought that the most likely direct combatants in wars against enemies like China would be U.S. proxies such as Japan, not the United States itself.[92] If this is correct, then the U.S. tightening of its relations with its allies in the 1950s unintentionally undercut deterrence by falsely signaling U.S. over-reliance on those allies.

The history of 1962 provides us with a good test of the thesis that Mao's ideological leanings and his desire to compete for ideological leadership of the international communist movement, not China's national interests in accord with more straightforward realpolitik calculations, was driving his radical policies toward the Soviets, relations with other communist parties, and Third World revolution. In the first half of 1962 Vice Foreign Minister Wang Jiaxiang put forward a proposal for reorientation of PRC foreign policy that would be in accord with almost any objective analysis of China's needs. His program, often referred to as the "three reconciliations and one reduction" (*sanhe yishao*), called for a reduction of Chinese foreign aid for revolution abroad during China's recovery period from the Great Leap disaster and reconciliation with at least some foreign governments. Arguably with good reason, Wang considered it very dangerous for the PRC to maintain poor relations with both the Soviet Union and the United States at the same time as it was trying to pull itself out of domestic disaster. He therefore wanted Beijing to improve relations with one superpower or the other.[93] For the first half of 1962, Wang's moderate stance enjoyed tacit or explicit support from several top leaders, including Zhou Enlai and Peng Zhen. Sino-Soviet relations even appeared to be thawing a bit in those months after a bit of a truce in the nastiest interparty attacks during 1961.[94]

---

[91] For coverage of Mao's speeches on the threat from the United States in 1964, see Zhai, *China and the Vietnam Wars*, 140–41. On the development of the Third Front policy and its relation to threat assessments, see Zhai Qiang, "China and Johnson's Escalation of the Vietnam War, 1964–65," a paper presented at the conference entitled "New Evidence on the Cold War in Asia," Hong Kong, January 1996, pp. 22–25. For an excellent overview of the system, see Barry Naughton, "The Third Front: Defense Industrialization in China's Interior," *China Quarterly* no. 115 (September 1988): 351–86. Roderick MacFarquhar, *The Origins of the Cultural Revolution, Volume 3: The Coming of the Cataclysm, 1961–66* (Oxford and New York: Oxford University Press and Columbia University Press, 1997), 369. I am grateful to Dali Yang and David Bachman for useful comments on the Third Front.

[92] Li Danhui, "ZhongSu Guanxi yu Zhonggguo de YuanYue KangMei" [Sino-Soviet Relations and the Chinese Aid Vietnam, Resist America [Effort]] in *Dangshi Yanjiu Ziliao* no. 251 (June 1998): 3–4.

[93] MacFarquhar, *The Origins of the Cultural Revolution*, 3:270–72.

[94] Ibid., 3:270–272. On the superficial and short-lived reduction in tensions in 1961–62, see Luthi, *The Sino-Soviet Split*, 197–98. Luthi points out that Moscow even transferred designs for a Mig-21 fighter plane in 1961–62.

In August 1962 Mao would harshly criticize Wang and demand that China stay on a revolutionary path that encouraged revolution abroad and criticized Soviet revisionism in the international communist movement. By early 1963 the CCP was fully devoted to a long-term ideological struggle with the Soviets, support for concepts of armed revolution, and opposition to the concepts of peaceful coexistence and peaceful transformation to socialism. At that time Beijing elites would berate their Vietnamese comrades for not siding more clearly with the Chinese in the growing Sino-Soviet dispute.[95]

Given Mao's relatively limited threat assessment regarding U.S. escalation in Southeast Asia, from 1959 through the first several months of 1964 the spread of revolution and the improvement of his own party's position within the international communist movement was a much more important goal in Mao's policy toward Indochina than bolstering the security of his own nation against attack by support of a local ally against a common enemy. In fact, Beijing had more than just revolution in Southeast Asia in mind at the time. As historian Chen Jian puts it, in Beijing's broader ideological struggle with the Soviets, Vietnam had "become a litmus test for true communism."[96]

Obviously, Chinese assistance to the Vietnamese communists raised the costs of containing them for the United States and its local partners. In fact, it is very difficult to imagine that the Vietnamese communists could have carried out their revolutionary plans in the South without Chinese assistance, particularly prior to the U.S. escalation in 1964–65 and the Soviet assistance that followed it. In fact, a CCP Party history argues that basically all of the Vietnamese communist materiel, except weapons captured from the enemy, were supplied by China in the first half of the 1960s.[97]

Also of great importance was Beijing's direct and indirect encouragement of Vietnamese aggression in 1963–64. Several months before the alleged Gulf of Tonkin Incident and U.S. escalation of the war, Mao pushed the Vietnamese communists to increase military activities in the South, telling them not to fear U.S. intervention.[98] China offered its own territory as a strategic "rear area" if the United States were to invade the North, thus reducing the downside risks of Vietnamese communist aggression.[99] Even before the Gulf of Tonkin Resolution that authorized the expansion of U.S. military operations in Vietnam, Mao announced to the North Vietnamese chief of staff his intention to send "volun-

[95] Yang "Changes in Mao Zedong's Attitude toward the Indochina War," 23.

[96] Chen, "China's Involvement in the Vietnam War," 363. In this article, Chen emphasizes important domestic political reasons for Mao to support external revolution.

[97] Guo, ZhongYue Guanxi Yanbian Sishinian, 69

[98] Yang, "Changes in Mao Zedong's Attitude toward the Indochina War," 28–29.

[99] Chen, "China's Involvement in the Vietnam War," 360.

teers" into Vietnam.[100] In January, June, and July 1964, Mao promised his Viet-namese comrades that if their actions led to an attack on North Vietnam by the United States, China stood ready to enter the war as it had in North Korea.[101] On July 10, 1964, just one month before the Gulf of Tonkin Resolution, Zhou Enlai bolstered Vietnamese communist spirits by making a clear commitment to assist North Vietnam if the United States invaded the country. He said, "If the United States is resolved to expand this war, by invading the Democratic Republic of Vietnam, or directly sending in forces, bringing the flames of war to China's side, we cannot just stand idly by (*zuo shi bu guan*). That is to say, if they want to fight a Korean-style war, we will prepare for it."[102] Chinese insurance regard-ing backup of the North Vietnamese if their support for revolution in the South led to U.S. escalation involving a land invasion of North Vietnam could have served only to embolden the North Vietnamese in their revisionist activities in the South. The North Vietnamese leaders were not the only ones to consider the prospect of Chinese intervention in Vietnam. In a classic study, Allen Whiting, then an intelligence analyst at the State Department, notes that Washington took the possibility of large-scale Chinese intervention seriously and limited military activities accordingly.[103]

CHINA'S FOREIGN POLICY ACTIVISM:
INTERNATIONAL STRATEGY OR DOMESTIC POSTURING?

China's policy toward Southeast Asia in the early 1960s cannot be seen as fully consistent with the expectations of realpolitik, but that does not necessarily mean that ideological purity in foreign policy in competition with the Soviet Union was driving China's strategy. One could argue that all of this was simply a byproduct of domestic politics in China and not part of an ideologically driven grand strategy of sincere competition with the Soviets. In other words, CCP elites could have simply been putting on a public image of anti-Soviet ideology for domestic purposes, but might not have truly been engaged in an interna-

---

[100] Li, "ZhongSu Guanxi yu Zhongguo de YuanYue KangMei," 7.

[101] Yang, "Changes in Mao Zedong's Attitude toward the Indochina War," 28–29.

[102] July 10, 1964, entry in *Zhou Enlai Nianpu*, 2:655. In an April 1965 meeting, Zhou also asked the Pakistanis to send warning to the United States that "if the United States expands the war [to include] the Chinese people, the Chinese people will resist to the end, there is no other way out [for them to take]. If it [the United States] bombs from the air, on the ground we can adopt activities all over the place using other methods. If the United States carries out comprehensive bombing of China, that means war, and war has no limits (*jiexian*)." April 2, 1965, entry in *Zhou Enlai Nianpu*, 2:723.

[103] Allen S. Whiting, *The Chinese Calculus of Deterrence* (Ann Arbor: University of Michigan, 1975).

tional struggle with the Soviets. The evidence suggests that, at the broadest level, domestic politics were important in this period—after all, Mao's domestic ideology was clearly a driving force behind the Great Leap Forward in the late 1950s, the Socialist Education Campaign of 1963, and eventually the Cultural Revolution of the late 1960s, all of which damaged Sino-Soviet relations. But the evidence also suggests that the competition with the Soviet Union for leadership of the movement in the 1960s was quite sincere, as Mao and the CCP elites would take actions that were clearly not designed simply for domestic consumption.

There are certain things that we should expect to see if Chinese foreign policy in the early 1960s was driven in large part by Mao's desire to compete with the Soviet leadership for the hearts and minds of revolutionaries around the world but that would be relatively unimportant if Mao was simply posturing for domestic audiences by appearing to compete with the "revisionist" Soviet Union. For example, we should expect Chinese communist officials to actively lobby third-country communists by lauding their own efforts and criticizing the Soviet Union in public international forums and in private meetings with those foreign communists. We should also find serious internal assessments of that global competition and we should expect concerted marketing efforts with third parties lauding Chinese revolutionary achievements and criticizing Soviet laxity. This is exactly what the record shows.

In order to compete with the Soviets, China clearly had three different strategies in the early 1960s. With communists in Asia, Cuba, and Albania, China argued for communist internationalism and revolutionary solidarity, implicitly criticizing Khrushchev's credentials on this score. With noncommunist Third World leaders, Beijing attempted to appear moderate, self-absorbed, and nonbelligerent. With Eastern Bloc communists, Beijing tried to play up party independence and nationalism, probably in the hope of weakening the Soviet grip on these parties.[104]

[104] See for example the discussion with the DPRK representatives over economic aid in October 5, 1960, entry in *Zhou Enlai Nianpu*, 2:355. Here the Chinese Foreign ministry promises support for revolutionary movements like those in Cuba and Algeria, as well as for communist countries like the DPRK, but complains about hardships caused by the Soviets in China. In January 17, 1961, China would promise assistance to Albania on favorable terms, stating that if they had hardships they need not pay back the loans. See ibid, 2:385. On April 20, 1961, Chinese officials would discuss the nature of the Soviet-Albanian tensions with a visiting entourage from the new communist regime in Cuba. See ibid, 406. By early 1964, Albania and China would form a principled pact against "revisionism." See ibid, 2:607. In a November 14, 1961, meeting with Ho Chi Minh, a major topic of discussion was the "mobilization of communist internationalism." See ibid, 2:442. On Aug 4–10, 1963, Zhou Enlai discussed the nature of the Sino-Soviet split and China's attitudes about the unfairness to non-nuclear countries of a nuclear test-ban treaty with the Somali leader, Shemaarke. Ibid., 2:570. Finally, in a very interesting tactic, China started to appeal to nationalism and Party autonomy with some members of the Soviet bloc in Eastern Europe. Zhou Enlai told the Hungarian ambassador that every socialist country must develop independently and that "nationalism and internationalism" are

Ideological attacks on the Soviets began in earnest in spring 1960, several months after Khrushchev's and Mao's extremely contentious October 1959 summit. In April 1960 the *People's Daily* published three articles on the ninetieth anniversary of Lenin's birth. The articles took up the CCP's various problems with the notions of peaceful coexistence with the West and peaceful transition to socialism. They explicitly focused on Yugoslavian "revisionism," but the true target was obviously the Soviet Union. In late May 1960 Mao removed any remaining subtlety. In discussions with North Korean and Danish Communist Party leaders Mao criticized peaceful coexistence and peaceful transformation, accused the USSR and Eastern European states of abandoning class conflict, and attacked Khrushchev by name for his statements regarding the "Spirit of Camp David" and détente with the United States following the summit in September 1959. Mao reportedly said in a threatening fashion that he would have to "settle accounts" (*suan suan zhang*) with Khrushchev in the future.[105]

According to Chinese scholar Yang Kuisong, who performed extensive research in Chinese government archives, from late 1959 to 1961, Mao would argue privately to visiting communists from Australia to Venezuela that the notion that the communists could peacefully coexist with the capitalists over the long term was "utter nonsense."[106] In June 1960 Chinese leaders expressed important ideological differences with the Soviets before and during an international labor (*Lian Gong*) gathering in Beijing.[107] In the same month the Soviet and Chinese representatives at the international congress in Bucharest (Third Romanian Party Congress) would openly criticize one another over ideological issues.[108] In 1961 following Soviet criticisms of the CCP at the 22nd Congress of the CPSU, Beijing and Moscow exchanged polemical attacks, although they focused on proxy third targets (Albania representing China in Soviet attacks and Yugoslavia representing the Soviets in Chinese ones).[109]

In September 1963, Mao presented the Sino-Soviet ideological dispute to the world public by running a series of commentary articles (*pinglun*) in the *People's Daily*. Before launching the campaign, Mao first consulted with Kim Il-sung and Ho Chi Minh.[110] This type of consultation demonstrated the importance of his

---

integrated. See November 20, 1963, entry in ibid., 2:597. With East Germany, China recognized the big ideological battle taking place within the communist camp but called for a spirit of unity and cooperation with the DDR. April 30, 1964, entry in ibid., 639.

[105] Shen, "Sulian Dui 'Da Yue Jin' He Renmin Gongshe de Taidu ji Jieguo" 137–38.

[106] Yang "Changes in Mao Zedong's Attitude toward the Indochina War," 19.

[107] Shen, "Sulian Dui 'Da Yue Jin' He Renmin Gongshe de Taidu ji Jieguo," 137–38. The meeting was the Congress of the World Federation of Trade Unions.

[108] For coverage of the Bucharest conference, see Luthi, *The Sino-Soviet Split*, 168–74.

[109] MacFarquhar, *The Origins of the Cultural Revolution*, 3:271.

[110] Ye Yonglie, *Chen Boda Zhuan* [Biography of Chen Boda] (Beijing, Zuojia Chubanshe, November 1996), 248.

international goals in this "ideology war," rather than just domestic politics. In the early 1960s Mao spent a great deal of time meeting with Korean, Vietnamese, and Japanese communist parties, among others, to discuss Sino-Soviet ideological differences and China's plan to take the ideological dispute public. In May 1963 he would discuss Sino-Soviet ideological disputes with Kim Il-sung, and do so again in June with a visiting Vietnamese entourage headed up by Le Duan.[111] In the latter meetings, the Chairman mocked the Soviets' "so-called" assistance to its efforts during the Korean War, for which Beijing was required to repay Moscow. Mao accused the Soviets of "great power chauvinism" (*daguo shawenzhuyi*) and pointed out that China never asked for repayment for its military assistance to Vietnam in its struggle against the French and the Americans.[112] In meetings with the Vietnamese and third-country communists, Mao would go so far as to criticize himself for his advice to the Vietnamese communists during the Geneva Conference and in the years immediately following it.[113]

According to Yang Kuisong, the Vietnamese communists began leaning heavily toward the Chinese side in the Sino-Soviet dispute at the time of Le Duan's June 1963 visit, as they were frustrated with Khrushchev's continuing lack of assistance to their struggle in South Vietnam and Laos. Le Duan supported Mao's plan to publish his critical treatise on Soviet revisionism later in the year.[114] In February 1964, Beijing promised significant aid to the Vietnamese communists (two billion RMB). Then, rather abruptly, Chinese leaders asked the Vietnamese their position on the Sino-Soviet dispute.[115] In meetings with Kim Il-sung later in the month, Mao did more than accuse the Soviets of being less than generous toward their Chinese brethren, he accused them of violating the Chinese border in Xinjiang and undertaking subversive acts in the region. In meetings with Romanian officials, Liu Shaoqi accused Moscow of allying with the United States against China. He differentiated this behavior from the intra-alliance squabbles that the Chinese had with Stalin. In its public line, Beijing criticized Moscow for being lethargic in supporting revolution.[116]

It was clear that the Chinese communists were working actively to court the Cuban communists in particular. For example, in a February 21, 1963 meeting with a Cuban entourage, Zhou Enlai flattered his guests by saying that the CCP could learn much from Castro's revolution. He also played on Cuba's Third World

---

[111] Wu, *Shinian Lunzhan*, 2:538, 566, 570, 574–75.

[112] Ibid., 2:575–76.

[113] Yang "Changes in Mao Zedong's Attitude toward the Indo-China War," 27.

[114] Ibid., 26. Also see MacFarquhar, *The Origins of the Cultural Revolution*, 3:368.

[115] Wu, *Shinian Lunzhan*, 2:666–67.

[116] Ibid., 2:583–585, 681, 696–98. Kim Il-sung said to Mao that on international disputes in Asia the Soviets had tried to adopt and strengthen a "middle position" (*zhongjie shili*) but that there was not much of a market (*shichang*) for this strategy in Asia.

anticolonial nationalism. In the meantime, Zhou criticized the Soviets and defended China's policy toward nonsocialist regimes, like Sihanouk's government
in Cambodia, by stating that Sihanouk's statements are in many ways more advanced than those of capitalists and "revisionists" because they reflect the central importance of anti-imperialism.[117] At five international conferences in late
1962 through early 1963, the Chinese took a tough international stand on the
capitalist world and on revolution and either implicitly or explicitly criticized the
Soviets for being too weak toward the United States, a charge that carried more
punch after the recent Cuban Missile Crisis. Relations would devolve further in
early 1963 and by late March China was publishing open attacks on the Soviets
and questioning publicly whether the Soviet Union was still a socialist country.[118]
Later in the spring Khrushchev would publicly blast the Chinese as adventurists,
particularly on the issue of nuclear war.[119] The period 1963–64 was very tense.
In fall 1964, following Khrushchev's ouster in Moscow, the two sides launched a
half-hearted effort to patch up relations via a visit to Moscow by a CCP entourage.
This trip went from bad to worse when one of the Soviet entourage, apparently in
a drunken rampage, suggested to a Chinese representative that Mao be ousted just
as Khrushchev was; then the two sides could improve relations.[120]

As we would also expect if international goals as opposed to straightforward domestic politics were of primary importance in Mao's campaign against
Khrushchev, CCP elites performed internal net assessments of each communist
power's relative standing among third-nation communists. The Chinese communists assessed that the struggle against "Soviet revisionism" would be a long
one and the Soviets clearly had the upper hand in the near term vis à vis the
number of parties supporting them in the ideological battle (*lunzhan*) with the
Chinese. But the CCP elites also believed that, in the early 1960s, momentum
was on their side. At a work meeting of the Central Committee in February 1963,
Deng Xiaoping reported that not all parties around the world were comfortable
with backing Moscow against China. He argued that the number of countries
backing the Soviets during the 22nd Party Congress in October 1961, while still
a majority, was fewer than in 1960. Moreover, he pointed out that Khrushchev
appeared nervous about losing the support of fence-sitters and had to craft his

---

[117] *Zhou Enlai Waijiao Huodong Da Shiji, 1949-75*, (Beijing: Shijie Zhishi Publishers, 1993),
352. After the Cuban Missile Crisis ended with the Soviets backing down and removing the nuclear
weapons from Cuba, the attractiveness of Soviet leadership indeed declined among the Cuban
communists, to the benefit of the Chinese, especially among advocates of continuing third-world
revolution, like Che Guevara. Alexsei Fursenko and Timothy Naftali, *"One Hell of A Gamble":
Khrushchev, Castro, and Kennedy, 1958–1964: The Secret History of the Cuban Missile Crisis* (New
York: Norton, 1997), 327.

[118] Luthi, *The Sino-Soviet Split*, 242.

[119] Ibid., 255.

[120] Wu, *Shinian Lunzhan*, 2:861–62.

policies toward China and others to avoid losing their support for the Soviets. Deng pointed out that the CCP seemed to have a great deal of momentum in Asia in particular. He stated that the communist parties in North Korea, Malaysia, Indonesia, Japan, Vietnam, Burma, and Thailand, among others, either "oppose[ed] or [did] not support" Soviet attacks on China. For our purposes, it is important to note that Deng claimed that the Vietnamese communists in "most situations" (*duoshu de qingkuang xia*) remain consistent (*yi zhi de*, sometimes translated as "identical") with the Chinese position.[121] In 1962 in his own remarks at the Central Committee Conference that would launch the Socialist Education Campaign—an effort to rein in "rightist" thinking that he believed existed during the brief retrenchment from the Great Leap Forward—Mao also discussed how, by publishing its critiques of revisionism, the CCP could win converts over time. He boasted that many communists from capitalist countries had made contact with CCP representatives in Switzerland in order to read relevant CCP articles.[122]

The Soviets, for their part, correctly assessed what Mao was attempting to do and did so very early in the CCP's campaign. They did not see the Chinese ideological campaign as primarily aimed at domestic audiences. In September 1960 the Soviet ambassador to the PRC reported to Moscow that the CCP had increased its contacts with communist parties of all countries since 1959. The CCP was attempting to "elbow out ("*paiji*" in the Chinese-language version) the Communist Party of the Soviet Union and grab the leadership position (*lingxiu diwei*) of the international communist movement."[123]

The CCP's international campaign regarding Soviet weakness apparently had some effect on third parties and, eventually, on the Soviets themselves. In spring 1964, the Vietnamese communists began publicly discussing the risk that "revisionism" posed to the communist camp. In July and August various parties sided with China in its negative views on international peace conferences being proposed by the Soviets. At the 15th PRC National Day, foreign communists criticized the Soviet Unions' lack of revolutionary fervor. It is quite clear from the large number of parties that continued to line up with the Soviet Union in the international communist movement that Chinese gains at Soviet expense in a struggle for international leadership remained very limited at best.[124] But Mao

---

[121] Wu, *Shinian Lunzhan*, 2: 532–534. Wu Lengxi was head of Xinhua News Agency and was privy to the struggle with the Soviets for the hearts and minds of third-party communists.

[122] Wu, *Shinian Lunzhan*, 2:538.

[123] This important document is cited and quoted in Shen, "Sulian Dui 'Da Yue Jin,'" 138, note 1.

[124] This point is made forcefully by Lorenz Luthi in chapters 5 and 7 of *The Sino-Soviet Split* (esp. pp. 168–72). Luthi points out that the vast majority of parties continued to line up with the Soviets.

and his colleagues apparently believed that long-term trends favored them in the struggle, especially in Asia and elsewhere in the developing world.[125] Moscow could not have been overly confident that Mao was wrong and, as various experts on Soviet foreign policy have reported, Khrushchev's perceived need to compete with the Chinese communists for the hearts and minds of third countries had an important effect on Soviet foreign policy in Asia, Berlin, and Cuba in the early 1960s.[126]

THE EMPIRE STRIKES BACK: THE CATALYTIC EFFECT
OF THE SINO-SOVIET SPLIT ON SOVIET FOREIGN POLICY

Gaiduk writes, "The Soviet-Chinese split had a powerful impact on Moscow's considerations about Southeast Asia."[127] In his account, what drove Soviet concerns over Vietnam in the early 1960s was not a desire to counter the United States' efforts there or to spread revolution for its own sake. In fact, Khrushchev wanted détente with the West, especially after the Cuban Missile Crisis. Vietnam was a headache for him in this regard. To the degree that they were willing to support Vietnam at all, the Soviets were motivated by the fear that Chinese communist influence was growing in the region at the expense of Soviet influence. The timing of key Soviet decisions supports this argument.

Playing catchup with the Chinese, who had formally recognized the NLF the month before, Khrushchev dropped his language about peaceful coexistence and peaceful transformation in January 1961, deeming inevitable the armed struggles of national liberation against colonialism. This speech apparently made a deep impression on the incoming Kennedy administration about Soviet activism in the Third World. After the May 1961 Chinese offer of economic aid to Vietnam, in June the Soviets answered with a military package of its own of

---

But Luthi himself in ch. 9, p. 278, offers counts of pro-Chinese positions at the time of international conferences that suggest that China's influence was rising at least in Asia.

[125] For CCP elites' recognition that the majority of parties continued to line up with the Soviets in the early 1960s, see Wu, *Shinian Lunzhan*, 2:533. For the CCP elites' sense that they were gaining ground in the early 1960s, see Wu, *Shinian Lunzhan*, 2:532–34, 778, 794–99, 812–16, 823, 885–88.

[126] Houpu Halisen [Hope Harrison], "Zhongguo yu 1958–1961 Nian de Bolin Weiji" [China and the 1958–1961 Berlin Crisis] a Chinese language version of a talk and a paper presented at a conference in Beijing http://www.coldwarchina.com/wjyj/blwj/001801.html; also see Hope M. Harrison, "Driving the Soviets Up the Wall: A Super-Ally, a Superpower, and the Building of the Berlin War, 1958–1961," *Cold War History* 1, no. 1 (autumn 2000): 53–74; Richter, *Khrushchev's Double Bind*; Gaiduk, *The Soviet Union and the Vietnam War*; and Gaiduk, *Confronting Vietnam*; Fursenko and Naftali, "*One Hell of a Gamble.*"

[127] Gaiduk, *The Soviet Union and the Vietnam War*, 6.

unknown value.[128] In December 1961 the Soviet air force began airlifting supplies to the Pathet Lao. According to historians Aleksander Fursenko and Timothy Naftali, Khrushchev adopted this policy largely to maintain leverage there in competition with the Chinese.[129]

The effects of budding Sino-Soviet discord at this time were not only felt in Asia, but apparently also in Eastern Europe, Africa, the Middle East, and Latin America. Competition with the Chinese communists for prestige catalyzed the Soviets' move away from peaceful transformation and "the democratic national state" and toward political support and military assistance to irredentists and armed rebels around the world. In an excellent multi-case study of Soviet military assistance in Africa and the Middle East, an expert on Soviet security policy, Bruce Porter, wrote the following:

> [The PRC] has been a troublesome and disturbing factor in Soviet foreign policy since 1949, increasingly so since the Sino-Soviet split became serious in the late 1950s and early 1960s. The repercussions of that split have affected Soviet policy even in remote parts of the globe. Peking, particularly in the 1960s, charged the Soviet leadership with revisionism—with not pursuing a genuinely Leninist policy of fostering world revolution, supporting national liberation movements, and standing up to imperialism in local conflicts. While engaging in counter polemics against Maoism, the Kremlin leaders felt compelled to pursue a more aggressive policy in the Third World so as to refute the Chinese accusations, win the support of local communist parties, and cultivate favor with nationalist regimes ... *China plays some role in almost all of the case studies under consideration* [emphasis added].[130]

In 1964, Khrushchev's final year in office, Chinese attacks on the Soviets for not supporting armed revolution in Vietnam, the Congo, and elsewhere took their toll on Khrushchev and his policies slowly began to harden in that year.[131] When détente began to collapse and U.S.-Soviet relations worsened, as scholar James Richter acknowledges, "the rivalry with China pushed him [Khrushchev] to take an even harsher stance against the United States to prove his revolutionary credentials."[132]

In Latin America, competition with the Chinese apparently made Khrushchev extremely worried about Moscow's standing in the eyes of Cuban revolutionaries, a movement whose success did not fit with his concepts of peaceful transformation to socialism and its logical offshoot the "national democratic

---

[128] Ang, *The Vietnamese Communists' Relations with China*, 180–81, 196–98.

[129] Alexsei Fursenko and Timothy Naftali, *"One Hell of A Gamble,"* 102–3.

[130] Bruce D. Porter, *The USSR in Third World Conflicts: Soviet Arms and Diplomacy in Local Wars, 1945–1980* (Cambridge: Cambridge University Press, 1984), 64–65.

[131] Richter, *Khrushchev's Double Bind*, 166.

[132] Ibid., 138.

state." The Cuban revolution did, however match the model of rural Maoist rebellion led by a vanguard from the national bourgeoisie.[133] According to research on Soviet archives, his concern about potential Cuban leanings toward the PRC may have helped convince Khrushchev to offer Castro's revolution a nuclear umbrella against direct U.S. attack and may have influenced his decision to transfer nuclear weapons to the island. Fursenko and Naftali write, "For four years before coming right out in the open in spring 1960, the Chinese had subtly criticized Khrushchev for endorsing the doctrine of peaceful coexistence. Now that Cuba was on the verge of becoming the first socialist state in the third world since North Vietnam, Khrushchev decided he had to acknowledge the ideological promise of armed national-liberation movements."[134] Fursenko and Naftali also portray Khrushchev's main concern in Cuba in early 1962 not as fear of a loss of communist control but of a move by Cuban communists toward the Chinese position in the international communist struggle. After falling out with Castro's party, one leading Cuban communist reported in Moscow that there was a large pro-China streak in Cuban communism. Naftali and Fursenko argue that Khrushchev's desire to maintain the mantle of international communist leadership "played a part" in Moscow's May 1962 decision to transfer nuclear-capable missiles to Cuba, a decision that nearly precipitated World War III.[135]

Hope Harrison reports that East German General Secretary Walter Ulbricht referred positively to the Chinese attack on Quemoy and Matsu in memoranda to the Soviets in late 1958, drawing direct analogies between the offshore islands of Quemoy and Matsu and the "island" of West Berlin, itself surrounded on all sides by East Germany.[136] James Richter writes that Ulbricht in 1958–61 exploited "the deteriorating international conditions, as well as the pressure from China, to strengthen its [the DDR's] leverage over Moscow." In this process, East Germany was able to coax Khrushchev into taking a tougher position on West Berlin and to issue the first ultimatum that induced the Berlin crisis.[137] Harrison similarly emphasizes Khrushchev's concerns about East Germany's leanings toward the Chinese in 1961 as a major factor in Soviet support for Ulbricht's

---

[133] For an analysis of the ideological challenges that the Cuban revolution provided for Khrushchev during the budding Sino-Soviet dispute, see Jacques Levesque, "The USSR and the Cuban Revolution" *Soviet Ideological and Strategical [sic] Perspectives,* translated from the French by Deanna Drendel Leboeuf (New York: Praeger Publishers, 1978), esp. ch. 2.

[134] Fursenko and Naftali, *"One Hell of A Gamble,"* 49–52.

[135] Ibid., 167–70, 182–83.

[136] Hope M. Harrison, *Driving the Soviets Up the Wall, Soviet-East German Relations, 1953–61* (Princeton: Princeton University Press, 2003), 53–74, 104, 134.

[137] James G. Richter, *Khrushchev's Double Bind: International Pressures and Domestic Coalitions* (Baltimore: Johns Hopkins University Press, 1994), 115–16.

closing of the border with West Berlin.[138] Khrushchev's advisors recalled that he canceled the May 1960 Paris summit following the famous U-2 Incident, the shoot down of Gary Powers' spy plane over Soviet territory, in part because he feared criticism from hard-liners at home and in China if he were to appear soft toward the West at that time.[139] From the U.S. Cold War perspective, the Sino-Soviet split made the communist movement worse than a monolith not just in Asia, but also in the Western hemisphere, and it contributed to some of the most dangerous crises of the Cold War era.

---

[138] Harrison, "Driving the Soviets Up the Wall," 53–74.

[139] Luthi, *The Sino-Soviet Split*, ch. 5, p. 165. Mao in fact was geared up to criticize Khrushchev in any case because the U-2 Incident seemed to prove that Khrushchev was naïve about the United States. Deng Xiaoping opined that Khrushchev would need to respond in a tough fashion toward the United States because China had criticized him in the April articles on the occasion of Lenin's birthday, cited above. See Wu, *Shinian Lunzhan*, 1:266–67.

Chapter 6

FROM ESCALATION IN VIETNAM TO SINO-AMERICAN RAPPROCHEMENT, 1964–72

Sino-Soviet competition for the loyalties of the Vietnamese communists would begin in earnest following U.S. escalation in the war from late 1964 to early 1965. Ho Chi Minh was able to exploit Chinese and Soviet jealousies of one another to gain maximum support for his revolutionary goals in South Vietnam. From 1965 until early 1968 the rivalry between Beijing and Moscow also served to scuttle multiple Soviet-inspired proposals for peace talks between the Vietnamese communists and the United States. After the Sino-Soviet rivalry moved from intramural competition to confrontation in the period 1968–69, Sino-Soviet tensions finally served U.S. regional and global interests and facilitated rapprochement between Washington and Beijing. That rapprochement process itself, however, would be complicated by Chinese concerns about U.S. alliances with Taiwan and Japan.

HO CHI MINH AS BENEFICIARY OF SINO-SOVIET RIVALRY IN THE LATE 1960s

Despite some efforts to keep Moscow's foot in the door in Southeast Asia in the early 1960s, Soviet attention would not turn firmly toward Southeast Asia until after the August 1964 Gulf of Tonkin Resolution and the U.S. escalation that followed. In February 1964 Le Duan criticized the Soviets for being too passive in supporting revolution. The Soviets' initial response was to limit further material aid and to supply merely rhetorical support for Hanoi.[1] But Moscow would discover that this strategy only weakened Soviet influence in Vietnam, especially when compared to the relationship that Beijing enjoyed with Hanoi. After the Gulf of Tonkin Incident, Soviet advisors in Vietnam found Chinese influence strong there. Soviet military advisors complained to Polish officials stationed in Hanoi that "[f]riendly relations between the DRV and China are currently almost absolute, mainly as a result of pressure from China. At present they are conducting an internal party campaign accusing the Soviet Union of

---

[1] Zhai, *China and the Vietnam Wars*, 128.

insincere relations [with Vietnam]."[2] According to the Polish diplomats, their comrades in Moscow were so sure that the Chinese wanted to use conflict in Vietnam to improve Beijing's position in the Sino-Soviet competition that the Soviets believed that Beijing had instigated the Gulf of Tonkin Incident.[3] In late 1964, Soviet advisors reported back to Moscow the Vietnamese communists' indifference toward the USSR's help and lack of respect for its advice. For example, in November 1964 the Soviet military attaché was told that Soviet military advice was inappropriate to Vietnamese conditions and that Soviet advisors could go home.[4] In December, the Soviet Embassy Counselor in East Germany complained that the Chinese had assembled a meeting of ten countries in Beijing at which both imperialism and "modern revisionism" were criticized by DRV Foreign Minister Xuan Thuy.[5] Such reports were considered particularly irksome in Moscow. As Gaiduk argues, "Moscow now faced a problem of reconciling its Communist prestige with its geopolitical interests. The former dictated that Moscow become deeply entangled in the war while the latter demanded complete disengagement—and thus a serious risk to Soviet influence in the socialist camp and the world Communist movement. A Soviet refusal to provide assistance to the DRV would open Moscow to sharp criticism from China, which the Soviets would prefer to avoid."[6]

Once the Soviets became more heavily involved in Vietnam in 1965, Sino-Soviet rivalry would affect the war in two important ways. First, there was clearly a competition among Beijing and Moscow to provide weaponry to the Vietnamese communists, thus increasing the military power of Ho Chi Minh's movement. Second, because of Chinese pressure on the Vietnamese and criticism of the Soviets in Hanoi and around the world, the Soviets were first unable and then unwilling to push Hanoi into negotiations with Washington before 1968. If the Soviets had had their way they clearly would have preferred using Vietnam as a bargaining chip with Washington on global issues more important to Moscow. Because China rejected and undercut Soviet peace efforts, in 1965–68 the Vietnam War was much harder for the Americans to manage through the use of

---

[2] "Note on a Conversation by Tarka, Jurgas, and Milc at the Soviet Embassy in Hanoi, September 10, 1964," in *Cold War International History Project Bulletin (CWIHPB)* no. 16 (fall 2007/winter 2008): 371 in collection of documents compiled by Lorenz Luthi.

[3] Ibid.

[4] Gaiduk, *The Soviet Union and the Vietnam War* 17; for another discussion of the Vietnamese lean toward the Chinese side before 1968, see Stephen J. Morris, "The Soviet-Chinese-Vietnamese Triangle in the 1970s: The View from Moscow," Cold War International History Project (CWIHP) Working Paper No. 25 (April 1999): 5.

[5] Note No. 131/64 on a conversation between the Soviet Embassy Counselor, Comrade Privalov, and Comrade Bibow on 11/23/64 in the GDR Embassy from 10:30AM-12:45 p.m., December 10, 1964," in *Cold War International History Project Bulletin (CWIHPB)* no. 16(fall 2007/winter 2008): 372.

[6] Gaiduk, *The Soviet Union and the Vietnam War*, 17–18.

coercive diplomacy than it would have been if the movement had been firmly under Moscow's control. Chinese criticism of the Soviets meant that when the Soviets did push for negotiations, the Vietnamese were pressured by Beijing to refuse. Of course, many Vietnamese communists were more than happy to take the Chinese side in this dispute. After all, the consistent interest of Ho Chi Minh's movement from 1954 to 1965 lay in the unification of its country under its control. What had changed from the Geneva Conference in 1954 was that the Chinese and Soviets were no longer on the same page.

Foreign Arms Assistance to the Vietnamese Communists after Tonkin

As discussed in chapter 5, Mao's support for the Vietnamese communists was based both in ideological goals that Mao had set for himself in his competition with the Soviets and in national security concerns not shared by the more distant Soviets. Arguably, before U.S. escalation in the war in late 1964 and early 1965, the former was more important than the latter. Mao did discuss the U.S. threat and the need to prepare for it earlier in 1964, but in the near term he did not expect direct U.S. attacks on North Vietnam, let alone China.[7] He did worry over the longer run about air strikes, including nuclear strikes, from the United States and began conceiving of a way to make China less vulnerable to them (the so-called "third front" rural industrialization program), but it is unclear what role South Vietnam could play for the United States in such a bombing scenario in any case. Moreover, Mao believed that the United States was spread too thin and would have to rely on intermediate allies, such as Japan and West Germany, to take the fight on land to communist countries other than the Soviet Union. He believed the U.S. strategy was to be the "last to join an international war." As for U.S. policy toward China, Mao believed that Washington's strategy at the time was based more in "peaceful evolution" than military conquest.[8]

Despite such a moderate near-term threat assessment, Mao offered China's "unconditional support" to a Vietnamese General Staff entourage visiting from Hanoi in June 1964.[9] North Vietnam was a bordering state, and helping it bolster its defense might seem to be simple prudence, but Mao knew that North Vietnam was deeply involved not just in defensive preparations but in spreading revolution in the South and that Beijing's unconditional support could only encourage Vietnamese communist belligerence.

The U.S. escalation in Vietnam from August 1964 through spring 1965 would change Beijing's perceptions. On August 6, 1964, the day after U.S. Navy aviators

---

[7] Zhai, *China and the Vietnam Wars*, 140–41.

[8] Li Danhui, "Sino-Soviet Relations and the Aid Vietnam, Resist America War," in *Dangshi Yanjiu Ziliao* no. 251 (June 1998): 3–5.

[9] For the June 1964 quote, see Qu, "*Zhongguo Zhiyuan Budui*," 40.

bombed the North, the Central Committee resolved that "America's infringe-ment (*qinfan*) against the Democratic Republic of Vietnam is an infringement against China." The CCP leadership pledged to make "assistance to Vietnam our top priority. We must handle all Vietnam's requests with the utmost seriousness, conscientiousness, and activism."[10] Beijing decided that China must facilitate the provision of supplies to communists in South Vietnam as well.

It is important to note, however, that even after U.S. escalation in Vietnam, national security strategies consistent with realpolitik expectations did not dominate Beijing's approach to the Vietnam War. If they had, one would have expected Beijing to welcome any assistance to the Vietnamese communists' war effort, especially from the most powerful actor in the communist world, the Soviet Union. As it was, China poured a great deal of material and human resources into the war effort in Vietnam, but consistently opposed the introduc-tion of significant Soviet aid into the struggle. It would be wrong to say that winning the Vietnam War was not a major security priority for Mao's China, but even in the dangerous times following U.S. escalation, it apparently was not the top priority. Mao still viewed Vietnam largely as an opportunity for him and his party to gain prestige in the international communist movement at the expense of the Soviet Union. As Li Danhui, a leading Chinese diplomatic historian, puts it in an internal Communist Party journal: "From the beginning of 1965, as the Americans launched a large-scale war of aggression against Vietnam, until the late 1960s, conditions in Sino-Soviet relations would come to produce a decisive effect on the principle of aiding Vietnam and resisting America and on Chinese-Vietnamese relations."[11]

Mao greatly increased military assistance to the Vietnamese Communists in 1965. From 1965 to 1971, the Vietnamese theater would witness the rotation of 320,000 Chinese troops which in 1967–68 would reach a one-time peak of 170,000 forces. Among the Chinese forces, arguably the most important were 150,000 air defense forces. According to an internal Chinese communist history source, in the late 1960s these Chinese gunners were responsible for 38 percent of downed U.S. aircraft. Other histories discuss the military training of 6,000 of-ficers, provision of 1.28 billion bullets, 37,000 cannons, 140 boats, 500 tanks, and 16,000 motor vehicles. Moreover, China contributed an average of $10 million per year in hard currency to Ho Chi Minh's international propaganda campaign. Beijing claims it supplied $20 billion (USD) worth of assistance from 1950 to 1978, of which 90 percent took the form of outright grants. China outfitted two million Vietnamese troops, laid hundreds of kilometers of railroad, and provided

---

[10] August 6, 1964, entry in *Zhou Enlai Nianpu*, 2:663.
[11] Li, "Sino-Soviet Relations," 3.

300 million yards of cloth, 5 million tons of food, and 3,000 kms of oil pipeline. China suffered thousands of casualties in Vietnam: by one estimate there were 1,100 killed, 4,200 seriously wounded and many more wounded less seriously; another asserts that there were 1,070 killed and 5,270 seriously wounded; and a third estimate places casualties, including both dead and wounded, "in the tens of thousands" (*shu yi wan ji*) from 1954 to 1978.[12]

Beijing's strongest leverage in its competition with the Soviets for the hearts and minds of the Vietnamese communists was provided by its promise to deploy Chinese ground troops in Vietnam if the United States escalated its intervention either by invading the North with ground troops or bombing China. The promise of a large number of Chinese personnel directly involved in the war, which the Soviets could not or would not match, would be made several times to the Vietnamese and others in 1965. Even without such U.S. escalation, in April 1965, Liu Shaoqi promised Le Duan that the Chinese would "do our best to support" the Vietnamese communists in their request for "some volunteer pilots, volunteer soldiers...and other volunteers, including road and bridge engineering units."[13] In an October 1965 meeting with North Vietnamese Prime Minister Pham Van Dong, Zhou pointed out that China was arming friendly and neutral governments and forces in Laos and Cambodia (including 28,000 guns to Sihanouk) and was preparing itself for a large-scale ground war in Southeast Asia.[14]

---

[12] Guo, ed., *ZhongYue Guanxi Sishi Nian*, 68–72. For the higher casualty figures and the high percentage of the $20 billion worth of aid that was in the form of outright grants, see the back cover of Wang Xiangen, *ZhongGuo Mimi Da Fabing: Yuan Yue KangMei Shilu* [China's Secret Large Dispatch of Troops: The Real Record of the War to Assist Vietnam and Resist America] (Jinan: Jinan Publishers, November 1992). According to another source, this meant that Vietnam was China's biggest target for foreign aid, consuming 41 percent of the foreign aid budget from 1950 to 1978. See Yang Gongsu, *Zhonghua Renmin Gongheguo Waijiao Lilun yu Shixian* [The Theory and Practice of PRC Diplomacy], (internally circulated, limited edition Beijing University textbook, 1996 version), 341. For the detailed air battle, kill and POW numbers, see "Jianguo Hou Wo Dui Wai Junshi Yuanzhu Qingkuang," 32. For additional statistics, including those on the rate of increase in 1965, see Chen, "China's Involvement in the Vietnam War;" and Qu, "*Zhongguo Zhiyuan Budui*," 42. For casualty and assistance figures, also see Gong Li, *Mao Zedong yu Meiguo*," 171. It is true that China would not meet every request of the Vietnamese communists, apparently electing not to send Chinese fighter planes in large numbers into the Vietnamese fight and keeping Chinese anti-air artillery further north in Vietnam than the Vietnamese would have preferred. But Chinese assistance was robust all the same. Chen, "China's Involvement in the Vietnam War," 368–69.

[13] Liu Shaoqi and Le Duan, Beijing April 8, 1965, in Westad et al., eds., "77 Conversations," no. 7, p. 83. Also see Zhou Enlai and Nguyen Van Hieu, Nguyen Van Binh, Beijing (The Great Hall of the People) May 16, 1965, in Westad et al., eds., "77 Conversations,"no. 8, pp. 83–84. Le Duan wanted this assistance in order "1) to restrict U.S. bombing to areas south of the 20th or 19th parallels; 2) to defend the safety of Hanoi; 3) to defend several main transportation lines; and 4) to raise the morale of the Vietnamese people."

[14] Zhou Enlai and Pham Van Dong, Beijing, 4:00 p.m., October 9, 1965, in Westad et al., eds., "77 Conversations," no. 14, pp. 87–88.

Just weeks before the Vietnamese communists launched the 1968 Tet offensive, Mao promised again that China would be a "reliable rear area" should Hanoi fall into greater trouble.[15]

Although Beijing's actions were rooted in both national security concerns and communist internationalism, they were clearly not just defensive. Beijing continued to assist not just in North Vietnam's defense but in Hanoi's efforts to "liberate" the South. In May 1965 Mao supported Ho's plan to use Chinese road engineers in the north so as to free up North Vietnamese military personnel to deploy to the south. Mao also supported building roads through Laos. In preparation for escalation of the war, he even told Ho that "because we will fight large-scale battles in the future, it would be good if we also build roads to Thailand."[16] In August 1966 Zhou Enlai decided to send 100 specially trained forces into South Vietnam to serve in "command staffs, logistics, chemistry, engineering, [and] political training."[17]

The Vietnamese communists did not fail to appreciate the unique support provided by the Chinese. In November 1968, a South Vietnamese communist, Muoi Cuc, said to Mao, "We hold that the spiritual support offered by China is most important. Even in the most difficult situations, we have the great rear area of China supporting us, which allows us to fight for as long as it takes. Material assistance is also important. That we force the American troops to underground shelters [is] also because of pieces of artillery that China gave us."[18] In 1970 Le Duan would refer to China again as communist Vietnam's "great rear."[19]

Despite Vietnamese expressions of appreciation to China, Mao paid careful and jealous attention to increasing Soviet military assistance to Hanoi following U.S. escalation in 1964–65. This demonstrates, again, that Mao had much more at stake in this fight than simply the defense of North Vietnam against U.S. attack. If defense had been the only concern, Mao should have welcomed assistance to North Vietnam from any quarter.

---

[15] "Zhujia Yuenan Nanfang Minzu Jiefang Lianxian Chengli Qi Zhounian de Dianbao" [Telegram Celebrating the Seventh Anniversary of the Establishment of the National Liberation Front of South Vietnam], December 19, 1967, in *Jianguo Yilai Mao Zedong Wengao*, 12:458–59.

[16] Mao Zedong and Ho Chi Minh. Changsha (Hunan), May 16, 1965. The promise to send Chinese forces into the north to free up Vietnamese to fight into the South was apparently raised by Liu Shaoqi one month earlier as well. Concrete plans were laid out for the phasing of Chinese escalation in response to different scenarios for U.S. escalation. See Chen, "China's Involvement in the Vietnam War," 367–69.

[17] Zhou Enlai and Pham Van Dong, Hoang Tang Beijing, August 23, 1966, in Westad et al., eds., "77 Conversations," no. 20, pp. 96–97.

[18] Mao Zedong and Pham Van Dong, November 17, 1968, in Westad et al., eds., "77 Conversations," no. 39, pp.137–53. He also pointed out that China was giving more assistance to Cambodia than the U.S. and that this had helped keep supplies flowing to the South.

[19] Mao Zedong and Le Duan, May 11, 1970, 6:45–8:15 p.m., Beijing, the Great Hall of the People, in Westad et al., eds., "77 Conversations," no. 47, pp. 161–167.

According to an internal CCP study, Soviet assistance began flowing more heavily after the important February 1965 visit to Hanoi of Soviet Premier Kosygin, and it included weapons not produced in China, such as advanced air defenses. From April 1965 through 1973 Beijing calculates that the Soviets gave 1 billion rubles in economic assistance and more than $2 billion (USD) dollars in military assistance. As early as 1965, 50 percent of Soviet aid to communist countries was going to Vietnam, and 60 percent of this was military aid.[20] According to Russian figures, in 1968 the value of Soviet aid significantly surpassed that of Chinese aid, and constituted two thirds of all military aid given to Hanoi.[21] Growing alongside this meteoric rise in Soviet aid to Vietnam was Mao's feeling of envy toward growing Soviet influence there.[22]

On March 1, 1965, Zhou Enlai complained to Ho Chi Minh about the new levels of Soviet assistance, stating, "We oppose [the Soviet] military activities. ... Soviet experts have withdrawn, so what are their purposes [when they] wish to come back? We have had experience in the past when there were subversive activities in China, Korea, and Cuba." With a not so subtle threat to Ho, he continued, "We, therefore, should keep an eye on their activities, namely their transportation of weapons and military training. Otherwise, the relations between our two countries may turn from good to bad, thus affecting cooperation between our two countries."[23] In that month he also complained to Albanian leader Enver Hoxha that Soviet gratis military aid to Vietnam was catalyzed by China's earlier example but that it was being used to "get the Vietnamese under their control."[24] In May 1965, Zhou Enlai and Deng Xiaoping warned Ho Chi Minh about the danger that the "Soviet revisionists" were using aid to the Vietnamese communists to trick the North into talks with the United States.[25] Later in the year, Zhou Enlai would state baldly to Pham Van Dong, "During the time Khrushchev was in power, the Soviets could not divide us because Khrushchev

[20] Li, "Sino-Soviet Relations," 6–7.

[21] Gaiduk, *The Soviet Union and the Vietnam War*, 59. In 1968 alone, the Soviets gave nearly 400 million dollars in military aid. In addition they supplied military training in country and in the Soviet Union and also provided nonmilitary economic assistance.

[22] In fact, as the United States continued to escalate in Vietnam, Chinese threat perceptions turned north, as well as south, demonstrating the importance of the Sino-Soviet rivalry in Mao's strategic calculations. Li, "Sino-Soviet Relations," 6–7. For this reason, Li argues that from 1965 on "the gist of China's 'Help Vietnam, Resist America' [effort] was no longer just to carry forward the spirit of communist internationalism, but to contend with the Soviet Union for status in Vietnam."

[23] Zhou Enlai and Ho Chi Minh, Hanoi, March 1, 1965, in Westad et al., eds., "77 Conversations," no. 4, pp. 75–76.

[24] Minutes of Meeting between Zhou Enlai and Enver Hoxha," March 28, 1965, 4:00 p.m. in *Cold War International History Project Bulletin* (*CWIHPB*) no. 16 (fall 2007/winter 2008): see section on Albanian archives on the Sino-Albanian Summits, edited by Ana Lalaj, Christian Ostermann, and Ryan Gage, p. 287.

[25] Zhou Enlai, Deng Xiaoping, and Ho Chi Minh, Beijing, May 17, 1965, in Westad et al., eds., "77 Conversations," no. 10, p. 85.

did not help you much. The Soviets are now assisting you. But their help is not sincere. The US likes this very much. I want to tell you my opinion. It will be better without the Soviet aid."[26] In March 1966, in a conversation with Le Duan, Zhou Enlai said of the recent Kosygin and Shelepin visits to Vietnam, "the Soviets used their support to Vietnam to win your trust in a deceitful way. Their purpose is to cast a shadow over the relationship between Vietnam and China, to split Vietnam and China with a view to further controlling Vietnam to improve [their] relations with the U.S. and obstructing the struggle and revolution of the Vietnamese people." He even suggested that Soviet volunteer pilots may "disclose secrets to the enemy."[27]

In April 1966, Deng Xiaoping and Zhou Enlai demonstrated Chinese sensitivity by castigating Vietnamese representatives for mentioning Chinese and Soviet assistance to Vietnam in the same breath (*xiangti binglun*). Deng also complained bitterly to Ho Chi Minh that some in Vietnam were discussing a "threat from the North." Deng said that if the Vietnamese comrades felt this way, China could remove its large number of forces (130,000 by year's end) from Vietnam and place them in the north to counter the Soviet Union's present threat to China.[28] Deng complained that the Vietnamese communists were rejecting Chinese criticism and advice but that they were more compliant with the Soviets, because they did not want to upset them. He said bluntly, "Why are you afraid of displeasing the Soviets, and what about China?"[29]

In Spring 1967 Zhou Enlai warned a visiting Vietnamese delegation that, as the Vietnamese communists approached victory, "you will try your utmost for the final victory and we will encourage the world's people to support you. But the Soviet Union will give up."[30] For their part, the Soviets sought to undercut the Chinese position in Hanoi. Moscow complained that the Chinese were refusing to allow transport of Soviet weapons to Vietnam through China in a timely fashion.[31]

---

[26] Zhou Enlai and Pham Van Dong, Beijing, 4:00 pm, October 9, 1965, in Westad et al., eds., "77 Conversations," no. 14, pp. 87–88.

[27] Zhou Enlai and Le Duan, Beijing, March 23, 1966, in Westad et al., eds., "77 Conversations," no. 18, p. 91.

[28] Li, "Sino-Soviet Relations," 11–12; Soviet estimates placed Chinese troops in Vietnam as high as 100,000 in 1967. See Gaiduk, *The Soviet Union and the Vietnam War*, 65.

[29] Zhou Enlai, Deng Xiaoping, Kang Sheng, Nguyen Duy Trinh, Beijing April 13, 1966, in Westad et al., eds., "77 Conversations," no. 19, p. 9.

[30] Vietnamese and Chinese Delegations, Beijing, April 11, 1967, in Westad et al., eds., "77 Conversations," no. 27, pp. 105–12; also see April 7, 10, 11, 1967, entries in *Zhou Enlai Nianpu*, 3:143.

[31] "Note by the GDR Embassy in Hanoi on a Conversation with Ambassadors of the Other Socialist States in the Soviet Embassy on April 2, 1965, and April 25, 1965, in *CWIHPB* no. 16 (fall 2007/winter 2008): 376–77.

Vietnamese Communists Exploit the Sino-Soviet Rivalry: Ho Chi Minh as "Tertius Gaudens"

Rather than simply agreeing with the Chinese or the Soviets, the Vietnamese communists became skilled at playing one side off the other to maximize the support flowing to their forces. In April 1966 Le Duan combined ego-stroking with requests for future assistance by pointing out how much more sincere and meaningful Chinese support for Vietnam had been than Soviet support. He stated: "there are more than a hundred thousand Chinese military men in Vietnam, but we think that whenever there is something serious happening, there should be more than 500,000 needed. This is assistance from a fraternal country." He continued, "[H]ad the Chinese revolution not succeeded, the Vietnamese revolution could hardly have been successful. We need the assistance from all socialist countries. But we hold that Chinese assistance is the most direct and extensive." In a sort of intramural communist version of a "peaceful evolution" or "engagement" strategy, Le Duan encouraged Chinese and Vietnamese cooperation with the Soviets while congratulating China's truer revolutionary spirit. He said, "I think that there should be some revolutionary countries like China to deal with the reformist countries [according to Le Duan, this category included the USSR and Eastern European states], criticizing them, and at the same time, cooperating with them, thus leading them to the revolutionary path. ... So my personal opinion is that China, while upholding the revolutionary banner, should cooperate with reformist countries to help them make revolution."[32]

The Soviets believed that Beijing was threatening to withdraw support for the DRV war effort, including engineering troops, if Hanoi decided to pursue negotiations with the United States or invite significant forces from third countries to join their fight.[33] Whatever the case might have been, Hanoi was clearly not deterred by implicit Chinese threats to withdraw such support if the Vietnamese grew too close with the Soviets. Instead Ho effectively manipulated jealousies in Moscow and Beijing to maximize Hanoi's assistance from the two parties. In late 1964 the Vietnamese communists encouraged the Soviets to pay increasing attention and devote larger resources to their cause. Sensing Soviet concerns about lack of influence they let the Soviets know, in Gaiduk's words, that "Moscow's position in the region would depend on the scope of aid provided

---

[32] Chen, "China's Involvement in the Vietnam War, 1964–69," 382.

[33] "Note on a Talk with the Soviet Ambassador, Comrade [Ilya] Scherbakov, on 28 October 1966 in the Soviet Embassy in Hanoi," November 10, 1966, *Cold War International History Project Bulletin (CWIHPB)* no. 16 (fall 2007/winter 2009): 392–93. The document refers to Chinese pressure dating back to "last summer."

for the struggle."[34] As did the Cuban communists in 1961–62, the Vietnamese communists portrayed their party as highly factionalized between pro-Soviet nationalists and pro-Beijing ideologues. However sincere they might have been in this description, Hanoi's leaders were able to persuade the Soviets to increase assistance to Vietnam, so as to bolster the pro-Soviet factions.[35] The Vietnamese communists were relatively successful in 1966 in playing on both sides of the ideological fence. A leading scholar of the Vietnam War, Robert Brigham, reports that in 1965–66 "Hanoi hoped to distance itself from China in Soviet eyes but simultaneously to use the NLF's diplomatic corps to court China. As one Vietnamese interlocutor reported to Brigham, 'There was a little something for everybody.... We would bend toward Moscow and then send the NLF to China to sing its praises.'"[36] The Vietnamese would play on Soviet jealousies by emphasizing the degree of support from China, something not appreciated by their Soviet interlocutors any more than when the shoe was on the Chinese foot.[37]

There were mutual accusations in Beijing and Moscow that the other was actively hampering its own efforts to help the Vietnamese communists. In particular, the Soviets complained about Chinese obstruction or delay in allowing trans-shipment of Soviet war materiel to Vietnam via Chinese territory.[38] But

---

[34] Gaiduk, *The Soviet Union and the Vietnam War*, 27.

[35] Gaiduk, *The Soviet Union and the Vietnam War*, 27, 66–67.

[36] Brigham, *Guerrilla Diplomacy*, 61. In December 1966 the NLF's representative in Beijing, Tran Van Thanh, would hit all of the right buttons in Beijing. He portrayed the Vietnam War as only one important battle in the international struggle against U.S. imperialism, which could be defeated on the battlefield. He claimed that Third World people should reject "the revisionist policies of those who help the imperialists through their naive talk of peaceful co-existence." Brigham, *Guerrilla Diplomacy*, 62.

[37] "Letter from GDR Foreign Minister Otto Winzer to [SED Politburo Members] Comrade Walter Ulbricht, Comrade Willi Stolph, Comrade Erich Honecker, and Comrade Hermann Axen [Excerpts]" March 8, 1966 in *Cold War International History Project Bulletin* (*CWIHPB*) no. 16 (fall 2007/winter 2008): 388–91 at 390.

[38] Understandably, Beijing felt that it had sacrificed more than the Soviets, had done so earlier, and was the only country that could realistically save North Vietnam should the Americans choose to cross the 17th parallel as they had crossed the 38th parallel in Korea. For their part, Soviet complaints included allegations that Hanoi often dragged its feet in fulfilling its agreement to transfer captured U.S. weapons to Moscow and that Soviet ships were either prevented from entering ports promptly or placed near sensitive sites, such as air defense emplacements, and were thereby left more vulnerable to American attack. Gaiduk, *The Soviet Union and the Vietnam War*, 70–71.There are isolated incidents of the Sino-Soviet rivalry preventing certain goods and services being provided to Hanoi and the NLF. For example, Beijing is reported to have politely refused a Vietnamese request for additional air defense assistance near roads and railroads in August 1966, but this seems a relatively isolated incident. Also, allegedly because Mao feared Soviet espionage activities, China refused to supply the Soviets an air corridor through China for shipping materiel to Vietnam. Also, China subjected Soviet rail shipments to domestic regulations that the Soviets found nettlesome. Eventually the Soviets were able to convince the Vietnamese communists about Chinese foot-dragging, and this would complicate relations between Beijing and Hanoi by the late 1960s. This combined with Vietnam's frustration over Beijing's arguments about the moral equivalence of

when one considers these relatively minor affairs it seems fair to say that in the late 1960s Sino-Soviet rivalry in Vietnam led to more foreign assistance to the Vietnamese communists and a tougher diplomatic stand by the communists than we would have expected in the absence of rivalry.

THE SINO-SOVIET RIVALRY SCUTTLES PEACE TALKS BEFORE 1968

Fredrik Logevall's important and well-researched book on President Lyndon B. Johnson's decisions to escalate the war in early 1965 portrays this period as a lost chance for a negotiated settlement. Logevall posits that the communist camp seemed prepared for serious negotiations in early 1965, but U.S. policy precluded such an outcome.[39] On this one score, I disagree with Logevall's thesis. The Soviets indeed would have been willing to push for a negotiated settlement in early 1965, and Moscow was disappointed in Johnson's tenacity in Vietnam. But the Chinese were pulling in the opposite direction and, in effect, making it impossible for the Soviets to broker a peace settlement. Based on "hints of moderation" by the Chinese in January and February, Logevall argues Beijing might also have been willing to support a negotiated peace. This is not just an academic debate, as the assumptions behind it likely drove U.S. escalation policy in 1964. A 1963 Policy Planning Staff design for using coercive bombing against North Vietnam assumed that after initial U.S. bombing runs were carried out against the North, both the Soviets and Chinese would encourage Hanoi to negotiate a settlement.[40] As it turned out, this rather rough analogy to the Geneva Accords was invalid, at least for the Chinese. Beijing was pushing the Vietnamese communists to fight hard. Moreover, in 1965 the CCP would criticize the Soviets and Hanoi for straying from the socialist road, and badger both for even considering negotiations. Finally, it appears that a major reason that Moscow would increase its own support for Hanoi's belligerence is not only the U.S. escalation, but the competition with the Chinese for Hanoi's loyalties.

In Hanoi in February 1965, Soviet Premier Kosygin had urged the Vietnamese communists to consider an international conference on Indochina and talks

---

imperialism and Soviet revisionism and Beijing's jealousy over Soviet courting of Hanoi to cause major fissures in the Sino-Vietnamese alliance by 1969. Li, "Sino-Soviet Relations," 8; "Chinese Deputy Foreign Minister Qiao Guanhua and Vietnamese Ambassador Ngo Minh Loan," Beijing, May 13, 1967, in Westad et al., "77 Conversations," no. 29, pp. 119–20. On the belief in Moscow and Hanoi that Beijing thwarted Soviet assistance, see Gaiduk, *The Soviet Union and Vietnam War*, 16.

[39] Fredrick Logevall, *Choosing War: The Last Chance for Peace and the Escalation in Vietnam* (Berkeley: University of California Press, 1999), 364–65.

[40] For discussion of Walt Rostow's influential 1963 report, see Harold P. Ford, *CIA and Vietnam Policymakers: Three Episodes* (Langley, Va.: History Staff, Center for the Study of Intelligence, Central Intelligence Agency, 1998), 48.

with the United States to avoid further escalation.[41] During his stop in Beijing just after his trip to Hanoi, Kosygin also apparently urged the Chinese to assist in creating conditions in Vietnam that would encourage voluntary U.S. withdrawal.[42] Toward that end, he offered Soviet mediation services with the United States.[43] It is unclear exactly why the Soviets proposed this. Perhaps they simply wanted détente with the United States and did not see Vietnam as an important enough issue to justify risking World War III. The Soviets may have wanted to gain the political benefits on both sides of the conflict by playing the role of indispensable mediator between two governments that lacked direct contacts. Since Washington and Beijing also lacked diplomatic relations, Moscow had a unique advantage in the communist camp by virtue of its full diplomatic relations with Washington.

Increased Soviet military assistance to the Vietnamese communists in 1964–65 did not signal a change in Moscow's desire to see accommodation between Washington and Hanoi. For example, the Soviets seemed surprised and frustrated by the early February 1965 communist attack at Pleiku, which became the occasion for the dispatch of large numbers of U.S. ground troops and major air reprisals against the North. The Soviets were trying to have their cake and eat it too by purchasing Vietnamese communist loyalty with military assistance while attempting to prevent escalation of the Vietnam War from spoiling its relations with the United States and the Western alliance. According to Gaiduk, when the United States escalated in response to Pleiku, this put the Soviets in a difficult position because anything but full-scale support for the Vietnamese war effort would open Moscow up to criticism by Beijing.[44]

From the outset, Beijing rejected Kosygin's plan for negotiations and Soviet mediation as appeasement or even as a plot to assist the United States.[45] According to Russian Foreign Ministry archives, on February 27, PRC Vice Foreign Minister Liu Xiao criticized the idea of a peace conference, telling the Soviet Charge d'Affaires in Beijing that "it is evident that the proposal, promoted by your side, on conveying an international conference would mean a manifestation of weakness in front of American imperialism." Before any talks could take place, he asserted, the United States would need to withdraw all forces and South Vietnam would need to be represented by anti-government forces only.[46]

---

[41] Li, "Sino-Soviet Relations," 9.

[42] Zhai, "Beijing and the Vietnamese Peace Talks," 6.

[43] Yang, *Zhong Hua Renmin Gonghe Guo Waijiao*, 344.

[44] Gaiduk, *The Soviet Union and the Vietnam War*, 28, 33.

[45] Yang, *Zhong Hua Renmin Gonghe Guo Waijiao*, 344; Zhai, *China and the Vietnam Wars*, 157–58.

[46] "Oral Statement to the PRC Government, Transmitted by PRC Vice Foreign Minister Liu Xiao to the Charge d'Affaires of the USSR in the PRC, Cde. F.V. Mochulskii," February 27, 1965, in *Cold War International History Project Bulletin* (*CWIHPB*) no. 16 (fall 2007/winter 2008): see section on Soviet Bloc Documents and the Sino-Soviet Split, 1964-66, annotation and introduction by Lorenz Luthi.

Consistent with the thesis that China saw the Vietnam War effort as an exhibition hall for the superiority of its brand of communism over that of the Soviet Union, China advertised with third-country communists and Third World anti-colonialists its own revolutionary posture toward the Vietnam War. Zhou Enlai pointed out to Albania's Enver Hoxha in March 1965 that Kosygin's goal was to "enter into bargains" with the Americans by rallying multiple countries in the communist bloc under Soviet leadership to make such a deal. He boasted that the Chinese and the North Vietnamese rejected such "kneeling before" the United States.[47] Zhou stated that "China agrees with Vietnam's decision," which is why the United States and the Soviet Union are "not able to execute their plan" for peace talks. To Hoxha, Zhou would boast of China's lack of fear of escalation and its willingness to throw its army on the side of Vietnam if the United States were to attack Chinese bases. He went on to assert that the Vietnam struggle was worth great costs and risks because defeat there would severely harm the United States and encourage revolutionaries elsewhere.[48] In many high-level meetings and summits during this period, Chinese leaders emphasized everything from Soviet sellouts to the United States in urging Ho to negotiate, to Soviet ideological mistakes, to China's more sincere support to world anticolonial and communist revolution via Beijing's assistance in tying down the Americans in Vietnam.[49]

Especially with fellow communists, no effort was spared in Beijing to discuss esoteric points regarding true Leninism and Soviet revisionism with potentially sympathetic elites from around the world. Just several weeks after the escalation of the Vietnam War following the Gulf of Tonkin Resolution, Zhou Enlai blasted former Soviet leader Khrushchev to a visiting Romanian delegation for "selling out" (*chu mai*) the revolution, among other "adventurist" crimes.[50] He also

[47] Minutes of Meeting between Zhou Enlai and Enver Hoxha," March 28, 1965, 9:30 a.m., *Cold War International History Project Bulletin* (*CWIHPB*) no. 16 (fall 2007/winter 2008): 272-273; section on Albanian archives on the Sino-Albanian Summits, edited by Ana Lalaj, Christian Ostermann, and Ryan Gage.

[48] Minutes of Meeting between Zhou Enlai and Enver Hoxha," March 28, 1965, 9:30 a.m., *Cold War International History Project Bulletin* (*CWIHPB*) no. 16 (fall 2007/winter 2008): 276–280; quotation is on p. 276.

[49] See Zhou Enlai and the Indonesian First Vice Prime Minister Subandrio, May 28, 1965, in Westad et al., eds., "77 Conversations," no. 11, pp. 85–86, where Zhou tells Subandrio that China is prepared to fight a massive land war in Vietnam if the United States escalates. He points to Korea as the precedent; also see Zhou Enlai and Tanzanian President Nyerere, Dar Es Salaam, June 4, 1965, where Zhou links the struggle in Vietnam with anti-Western struggles in Congo. In Westad et al, eds., "77 Conversations". Also see *Zhou Enlai Waijiao Huodong*, 460.

[50] Turning the common Soviet critique of Mao's China against the Soviets, Zhou criticized Khrushchev's "adventurism" (*maoxian xing*) which included pushing around fraternal parties and failing to support revolution. September 29, 1964, entry in *Zhou Enlai Nianpu*, 2:671; In October 1966 Zhou Enlai would state to the Romanians that the Chinese and Soviet positions on peace talks were "absolutely opposite" (*juedui xiangfan*) and that China viewed the United States and the Soviet Union as mobilizing together to put pressure on Vietnam to negotiate. October 5, 1966, entry in *Zhou Enlai Nianpu*, 3:74.

criticized "modern revisionism" alongside "imperialism" with Laotian elites visiting Beijing.[51] With East Germans he accused the new Soviet leadership under Brezhnev of advocating "Khrushchevism without Khrushchev."[52] Beijing tried to impress Japanese Communist Party delegations visiting Beijing in March 1968 with China's relative revolutionary fervor.[53] Beijing also actively employed the dispute over peace negotiations with sympathetic noncommunist, anticolonial elites like the Algerians and Pakistanis.[54] In March 1965 Mao told foreign audiences that he was so concerned about Soviet-American collusion that he was preparing for the possibility of a joint U.S.-Soviet assault on China.[55]

In a diplomatic communiqué to the Soviets in June 1965, a PRC Foreign Ministry official rejected greater coordination between Moscow, Beijing, and Hanoi in the Vietnam War effort because the "Soviet Union stubbornly insists on its mistaken policy of Soviet-American cooperation for the solution of international problems, and tries to bargain with the United States on the Vietnam question."[56] In fact, according to one recent Chinese history, Beijing was more consistently in favor of belligerence toward the Americans and more consistently opposed to a negotiated truce than the Vietnamese communists themselves, who were often internally divided on the matter in 1965.[57] According to East German archival documents, in late 1964 and the first half of 1965 Soviet observers apparently agreed with this assessment, claiming that certain elements in the DRV did not want to overplay their hand by spreading revolution in the South too aggressively via military methods and were interested in negotiations with the United States, but the Chinese were pushing them toward "adventurism."[58] A

---

[51] October 10, 1964 entry in *Zhou Enlai Nianpu*, 2:675.

[52] November 10, 1964, entry in *Zhou Enlai Nianpu*, 2:687; Wu, *Shinian Lunzhan*, 2:880. In Chinese the phrase is *meiyou Heluxiaofu de Heluxiaofuzhuyi*.

[53] Zhai, "Beijing and the Vietnamese Peace Talks," 13–14.

[54] March 30, 1965 meeting in *Zhou Enlai Waijiao Huodong* , 444; Jin Chongji, ed., *Zhou Enlai Zhuan 1949–76*, Xia [Biography of Zhou Enlai 1949–76, vol 2] (Beijing: Chinese Central Documents Publishers, February 1998), 837–38; and April 2, 1965, entry in *Zhou Enlai Nianpu*, 2:687.

[55] See Zhang Baijia, "Mao Zedong yu ZhongSu Tongmeng He Zhong Su Fenlie" [Mao Zedong and the Sino-Soviet Alliance and the Sino-Soviet Split], unpublished manuscript presented to the Chinese Communist Party Central Party History Research Office's International Scholars Research Forum, Beijing, China, October 1997, p. 7.

[56] Oral Statement by the Head of Department of the USSR and the Countries of Eastern Europe of MFA PRC, Yu Zhan, Transmitted to the [USSR] Embassy on June 8, 1965, in *Cold War International History Project Bulletin* (CWIHPB) no. 16 (fall 2007/winter 2008): 380–81.

[57] Qu, "Zhongyue Zai Yinzhi Zhan Wenti Shang de Zhanlüe Yi Zhi yi Celüe Chayi," 313–15.

[58] Note No. 4 on Conversations with Comrade Scherbakov about the Developmental Tendencies in the Democratic Republic of Vietnam, on December 22 and 28, 1964 and January 6, 1965, *Cold War International History Project Bulletin* (CWIHPB) no. 16 (fall 2007/winter 2008): 373–74; also see "Note by the GDR Embassy in Hanoi on a Joint Conversation with the Ambassadors from other Socialist Countries in the Hungarian Embassy on May 4, 1965," in *Cold War International History Project Bulletin* (CWIHPB) no. 16 (fall 2007/winter 2008): 378. Interestingly, in this document the Soviet observers view pro-Chinese elements for Soviet reputational problems in Vietnam.

document from the Soviet archives quotes an official in Moscow's International Department stating, "In general, one has to acknowledge that the US, the Soviet Union, also the Vietnamese themselves would move forward toward negotiations, even if [they have] different positions and different approaches. The Soviet Union advocates a political solution … The Chinese leaders resist a political solution with full vigor."[59] Similarly, in December 1965 an East German diplomatic observer in Hanoi opined that the Vietnamese communist party was split between officials who would prefer a negotiated settlement in the South and a "pro-Chinese group" who "obstruct every initiative on negotiations, and even gathered troops on the 17th parallel without the approval of the Politburo. This group plays the Chinese declaration on the enlargement of the anti-American front in South East Asia as a major trump card in order to justify their position."[60] In January 1966, when the DRV rejected a Polish mediation effort, a leading Polish diplomat reported that the Chinese were only interested in a negotiation after military victory, that most weapons being used in Southern Vietnam were Chinese, not Soviet, and that "it is felt that some people in the DRV do not agree with the Chinese, however the influence of the PRC on the Vietnamese friends remains still great. This can be shown by the negative answer of the DRV to the Polish mediation."[61]

Even to the degree that they believed in negotiations as a useful tool in principle, elites in Beijing thought that the Vietnamese first needed to deal major blows against the United States and the South before negotiations should begin, a strategy that would preclude constructive reciprocity from the United States.[62] The Chinese communists were not only discouraging Vietnamese communist accommodation and rejecting Soviet mediation, they were encouraging Vietnamese belligerence by painting rosy scenarios of a relatively quick success. In January 1965, Zhou Enlai suggested to a Vietnamese military delegation that if Vietnam could kill enough enemy troops, reduce enemy fighting power, and "smash strategic hamlets" (*cuihui zhanlüe cun*) in the South, "then it is possible that victory can be yours even faster than we had expected."[63]

[59] "Note on a Conversation with an Unnamed Representative of the CPSU CC on the Situation in Vietnam," July 9, 1965, *Cold War International History Project Bulletin* (*CWIHPB*) no. 16 (fall 2007/winter 2008): 384.

[60] "Excerpts from a "Note by GDR Ambassador to the DRV Kohrt on the Current Policy of the Chinese Leadership," December 11, 1965, *Cold War International History Project Bulletin* (*CWIHPB*) no. 16 (fall 2007/winter 2008): 385.

[61] "Reception by Soviet Foreign Minister V.V. Kuznetsov for the general Director of the PRP FMA, Cde. Jerzy Michalowski, January 24, 1966," *Cold War International History Project Bulletin* (*CWIHPB*) no. 16 (fall 2007/winter 2008): 388.

[62] Qu Xing, "Zhongyue Zai Yinzhi Zhan Wenti Shang de Zhanlüe Yi Zhi yi Celüe Chayi," 313–15.

[63] January 22, 1965, entry in *Zhou Enlai Nianpu*, 2: 703.

In their discussions with the Vietnamese communist leadership, PRC elites blasted the Soviets in the most direct ways possible, warning their allies not to fall into the trap of joint hegemony by the American imperialists and the Soviet revisionists. They argued that the Soviet "revisionists" were dupes or even agents of the United States.[64] On March 1, 1965, a high-level Chinese delegation would visit Hanoi just weeks after the Kosygin visit there. The Chinese entourage said they had studied the new Soviet leadership carefully and had determined that what they were implementing was still "Khrushchevism," despite Khrushchev's ouster in the previous year.[65] In a December 19, 1965, meeting with Nguyen Duy Trinh, Zhou Enlai and Foreign Minister Chen Yi said that, while there was nothing wrong in principle with negotiations, the time was not ripe. They deemed "not tough" the sweeping conditions for peace negotiations in the Vietnamese communists' counteroffer to the Americans. Zhou warned that the United States might just agree with the conditions and this would give the impression that North and South Vietnam were separate political issues. This would have "a negative impact on our struggle and on our solidarity."[66] In spring 1967 Zhou told a Vietnamese delegation about how "even Stalin" compromised unnecessarily with the United States at Yalta at the Chinese communists' expense. By giving examples of how the CCP accepted negotiations during the Chinese Civil War only after they had clear strategic advantage, despite consistent Soviet advice to compromise, Zhou attempted to convince the Vietnamese communists that real negotiations in 1967 were a Soviet mistake at a minimum, and a Soviet-American trap at a maximum.[67]

Chinese pressure on the Vietnamese communists and encouragement of Hanoi's generally aggressive motivations took its toll on the prospects for peace in the mid-1960s. On at least one occasion in spring 1966, the Soviets simply refused to be placed in the role as intermediary, even in the capacity of sending messages between Washington and Hanoi.[68] Several top U.S. officials at the time seemed to appreciate the Soviet dilemma. In analyzing the Soviets' apparent reluctance to discuss resolution of the Vietnam conflict in an official manner,

[64] See Zhou Enlai, Deng Xiaoping, and Ho Chi Minh, May 17, 1965, in Westad et al., eds., "77 Conversations," no. 10, p. 85; Zhou Enlai and Pham Van Dong, Beijing, 4:00 p.m., October 9. 1965, in Westad et al., eds., "77 Conversations," no. 14, pp. 87–88; and Li Danhui. "Sino-Soviet Relations," 9.

[65] March 1, 1965, entry in *Zhou Enlai Nianpu*, 2:715; *Zhou Enlai Waijiao Huodong*, 438.

[66] Zhou Enlai and Nguyen Duy Trinh, Beijing, 10:30 a.m., December 19, 1965, in Westad et al., eds., "77 Conversations," no. 17, p. 90.

[67] Vietnamese and Chinese Delegations, Beijing 11:00 a.m., April 11, 1967, in Westad et al, eds., "77 Conversations," no. 27, pp. 105–12; also see April 7, 10, 11, 1967, entry in *Zhou Enlai Nianpu*, 3:143. Zhou also offered a very odd version of the history of Soviet and U.S. actions in spring 1949 to show that the Soviets were overly afraid of war with the United States and offered bad advice to the CCP.

[68] Gaiduk, *The Soviet Union and the Vietnam War*, 41–43, 52–55.

those U.S. officials argued that Moscow did not want Vietnam to become an occasion for the total collapse of superpower relations, but that the Kremlin also had to protect its status in comparison to Beijing in the rest of the communist world. Others understood that if the Vietnamese played their hand well, Hanoi could maximize its foreign assistance by encouraging a rivalry in Vietnam between Moscow and Beijing. This point was driven home to Averill Harriman in a meeting with Tito, who informed him that U.S. bombing of the North had put Moscow in a tough spot since "[t]he Soviet Union cannot fail in its stand of solidarity with Hanoi since it would expose itself to the danger of isolating itself in Southeast Asia and Communist parties elsewhere."[69]

Despite its difficult position and its reluctance to be seen as soft, Moscow and its Eastern European allies in the end did eventually play the role of messenger between Hanoi and Washington, a role that Beijing simply could not and would not play. Moscow also encouraged third-nation Communist Parties to urge Hanoi to accept negotiations. For example, in summer 1965, a Polish diplomat visited Moscow and then traveled to Hanoi to urge restraint and negotiation. In late 1966 early 1967, it appeared at times that Hanoi was considering Soviet advice and was willing to negotiate, and to use Moscow as a broker.[70] But before 1968, even when the Soviets braved Chinese criticism and proposed peace talks, the Vietnamese felt either empowered or obliged to reject these overtures by reference to the anger it would cause in Beijing. For example, in August 1966 General Nguyen Van Vinh rebuffed one Soviet effort to encourage talks, saying, "It would mean losing everything, and, first of all, friendship with China which is utterly opposed to negotiations."[71] Chinese pressure helped scuttle international peace initiatives raised not just by the Soviets, but by Ghana, Poland, Britain, and India.[72] China also congratulated Cambodia's Sihanouk for refusing invitations to play mediator in Vietnam.[73]

---

[69] Ibid., 49–56, 68. The Tito quotation is on pp. 55–56.

[70] The Polish diplomat stopped in Beijing as well and his ideas about peace talks were attacked with a ferocity that startled him. Poles would also be critical players in 1966 attempts to broker a peace under the name Operation Marigold. Gaiduk, *The Soviet Union and the Vietnam War*, 75, 83–84, 90–91, 100.

[71] Ibid., 80.

[72] Qiang Zhai, "Beijing and the Vietnam Peace Talks, 1965–68: New Evidence From Chinese Sources," Working Paper No.18, CWIHP, Woodrow Wilson Center for International Scholars, June 1997. For Beijing's discussion of the Ghana proposal with the Vietnamese communists, see February 16, 1966 entry in *Zhou Enlai Nianpu*, 3:13. In 1965 Beijing also tried to explain to third party noncommunists like Pakistan and Zambia why it opposed negotiations and why Beijing elites felt that even offers of "unconditional negotiations" by the United States were a conspiracy (*yinmou*). See April 2, 1965, entry in *Zhou Enlai Nianpu*, 2:723, and August 21, 1965, entry in ibid., 2:750; and *Zhou Enlai Waijiao Huodong*, 474.

[73] Zhai, *China and the Vietnam Wars*, 163.

The pernicious nature of Chinese influence on the prospects for peace negotiations even occurring, let alone succeeding, are demonstrated well in recently
available archival material about one famous late 1966 affair—Operation Marigold—an attempt by Poland to arrange for talks between the North Vietnamese
and U.S. governments in Warsaw. Debates have raged over who was to fault for
the failure of the talks to take place—with the Johnson administration blaming
the Poles for dishonesty and the Vietnamese for duplicity, and the Vietnamese
blaming U.S. intransigence and the intense December 1966 bombing campaign
against North Vietnam for scuttling any chance of the talks. In a careful study
of available relevant documents and Polish and Vietnamese officials' recollections, James Hershberg concludes that the data is too complex to draw any final
conclusions as to the mystery of "who murdered Marigold?"[74] But one thing is
absolutely clear throughout Hershberg's narrative, the Chinese were, at a minimum, an important enabling accomplice to the murderer(s). It is fairly clear
that the Soviet leadership fully supported the launching of peace talks, whether
or not they had simply ordered the Poles to play this role. The Chinese, on the
other hand, were applying pressure directly on the top leadership in Hanoi to
eschew talks and also manipulating internal Vietnamese politics by empowering
hawks among the Vietnamese communists who opposed peace talks in 1966–
67. Beijing did this by playing on divisions within the leadership of the ruling
Vietnamese Workers' Party (VWP) and between that party and the more radical
National Liberation Front (NLF), which was carrying out insurgency and revolution in the South. To the degree the Vietnamese communist top leaders were
sincerely willing to engage in discussions in Warsaw, they apparently demanded
tremendously high levels of secrecy so as to keep the initiative off China's radar
screen as long as possible and to avoid mobilizing China's hawkish allies within
the VWP and NLF. Just as the United States launched an intensified bombing
campaign against North Vietnam, the North Vietnamese were making indirect
contacts with U.S. State Department officials in Warsaw. The Poles and Soviets complained bitterly about the U.S. bombing campaign but often did so by
underscoring the manner in which that bombing allowed China and its more
hawkish Vietnamese political allies to pressure moderates in the VWP to reject
talks. One key Polish official vented his frustration early in this process stating
simply, "God damn those Chinese."[75]

Although they are frequently criticized by scholars for being blind to the
local realities in Asia, many in the Johnson administration were aware at the
time that Sino-Soviet intramural competition in the communist camp was not

[74] James G. Hershberg, "Who Murdered Marigold: New Evidence on the Mysterious Failure of
Poland's Secret Initiative to Start U.S.-North Vietnamese Peace Talks, 1966," Cold War International
History Project (CWIHP) Working Paper No. 27 (April 2000).

[75] Hershberg, "Who Murdered Marigold," 11.

only a catalyst for Soviet and Chinese assistance to Vietnam, but that it deterred Moscow from cooperating very vigorously with the United States in coaxing Hanoi into peace negotiations. In considering the possible implications of an escalation in U.S. bombing of North Vietnamese military sites for the prospect for peace negotiations, a Special National Intelligence Estimate of June 2, 1965, discusses the Soviets' "somewhat inconsistent objectives—to compete with Peiping for influence over the DRV, to contribute to deterring U.S. pressures against Hanoi, and at the same time to avoid becoming overly involved in the present US-DRV confrontation."[76] This estimate clearly and correctly envisioned Chinese competition with the Soviet Union as catalyzing the alliance's belligerence, stating that the Chinese "would feel a strong need to do something more to help the North Vietnamese, and to prevent Moscow from gaining more influence in Hanoi and in the Vietnamese situation in general. They would also seek to exploit the situation to further weaken Moscow's influence in the international communist movement and in the under-developed world."[77] The estimate entertained the notion that the Soviets might split entirely with the DRV if Moscow, fearing further escalation, were to recommend peace negotiations and Hanoi then demurred; but the authors continued, "[W]e doubt that the Soviets would pay the political price of backing down and thereby handing China a major political victory and weakening its own prestige and influence throughout the world." It went on to say, "Thus in the aftermath of the US attacks, we think the Soviets would probably conclude that they had little choice but to increase aid to the DRV."[78] Along similar lines, a January 1966 State Department document entitled "Why is Hanoi Reluctant to Negotiate?" reads as follows:

> The DRV is no doubt capable of deciding unilaterally whether or not it wants to begin negotiations; but does it want to assert its independence so bluntly[?]... [T]he Vietnamese have shown that ideologically (one can say almost spiritually) they are closest to the Chinese....The essential aid that defeated the French came from China. And Peking with its massive armies which, far more than it threatens, ultimately guarantees the survival of the North Vietnamese regime against even the most total ground invasion. Compared to China the USSR is remote indeed.[79]

[76] "Probable Communist Reactions to Certain U.S. Actions," Special National Intelligence Estimate, June 2, 1965, in *Estimative Products on Vietnam, 1948–75* (Washington, D.C.: National Intelligence Council, April 2005), 248.

[77] Ibid., 247.

[78] Ibid., 250–51.

[79] "Research Memorandum from the Deputy Director of the Bureau of Intelligence and Research (Denney) to Secretary of State Rusk," RFE-3, January 26, 1966, *Foreign Relations of the United States, 1964–68, 1966*, vol. 4, Vietnam 1966, document no. 44.

Various other documents suggest that top intelligence officials and advisors to President Johnson understood how the Soviets felt unable to push Hanoi in a more moderate direction because of pressure from China.[80]

It is ultimately unclear how much Chinese resistance to peace as opposed to Vietnamese desire to solve the issue on the battlefield actually influenced the Vietnamese decision to forego meaningful negotiations in 1965–68. Both factors pulled in the same direction. However we do know that at Geneva the Vietnamese communists were reluctant to negotiate a peace treaty that divided their country, but did so when the Chinese and the Soviets were in lockstep in pushing Hanoi to do so. Moreover, according to the Vietnam War scholar Robert Brigham, Vietnamese communist thinking on the prospect of negotiations with the United States in the mid-1960s was split between hardliners and doves. The hardliners apparently lambasted the doves with reference to the sellout of South Vietnam at Geneva.[81] Chinese encouragement of Vietnamese belligerence, combined with Beijing's apologies about the 1954 Geneva Accords and references to them as a mistake, might have been a pivotal factor in tipping the balance against the doves inside the Vietnamese deliberations.

Tet, U.S.-Vietnam Peace Negotiations, and the Last Blast of PRC Radicalism

Chinese pique with Hanoi was at its height when, after the Tet Offensive, Hanoi finally proposed peace talks with the Johnson administration against China's will. Rather than encouraging talks, just after Tet Mao encouraged Ho Chi Minh to launch a war of annihilation in the South.[82] China's position was that Vietnam should not have accepted talks especially without an unconditional halt in U.S. bombing of the North. Moreover, Hanoi should not even suggest acquiescence

---

[80] "Memorandum from the President's Special Assistant for National Security Affairs (Bundy) to President Johnson; Subject: The Peace Offensive—where we are today," January 3, 1966, *Foreign Relations of the United States*, 1864–68, vol. 2, Vietnam 1966, document 2; "Telegram from the Embassy in Vietnam to the Department of State," (Lodge) January 12, 1966, *Foreign Relations of the United States*, 1964–68, vol. 2, Vietnam 1966. See National Intelligence Estimate, "November 12, 1966: The Outlook for Sino-Soviet Relations," *Foreign Relations of the United States*, 1964–68, vol. 30, China, document no. 223; "Memorandum from the President's Special Assistant (Rostow) to President Johnson," November 30, 1966 , *Foreign Relations of the United States*, 1964–68, vol. 2, Vietnam 1966, document 319.

[81] For the argument that the Vietnamese communists were split and that a combination of U.S. bungling and domestic political in-fighting led to the failure of negotiations in 1965, see Robert K. Brigham, "Vietnamese-American Peace Negotiations: The Failed 1965 Initiatives," *Journal of East Asian Studies* 4, no. 4 (2005): 377–97.

[82] Guo Ming, *ZhongYue Guanxi Sishi Nian*, 68. Ho Chi Minh was on a secret trip to Beijing for health treatment.

to U.S. demands that it "gradually deescalate" conflict with the United States, because this would have a negative impact on the struggle of communists in the South.[83] Even as Hanoi was preparing for the beginning of talks, Beijing put pressure on Hanoi and secured a "solemn promise" from Pham Van Dong and Vo Nguyen Giap to continue prosecuting the war. The Vietnamese carried out that promise by traveling to Moscow and rejecting Soviet pressure for concessions in peace negotiations.[84] The Chinese historian Li Danhui emphasizes that Beijing's purpose in April was to interfere in the negotiating process by discouraging what the CCP elites feared would be "surrender and compromise" by their Vietnamese comrades.[85] In a meeting with the Vietnamese foreign minister in May, Zhou said that there was not much in common between the negotiations over a Korean War armistice in 1953 and what was being proposed in Vietnam.[86] The May talks immediately deadlocked.

The Tet Offensive may have created Sino-Vietnamese tensions over military tactics as well. Some believe that the Chinese Communists criticized the Vietnamese for abandoning Maoist guerrilla tactics for full-scale assaults on cities.[87] The available evidence is mixed and suggests that the biggest critique of Tet had to do with the increased Soviet assistance and influence that accompanied it, which in the Beijing elites' analysis, manifested itself more disturbingly in Hanoi's decision to negotiate than in its fighting tactics. At one point in June 1968, Zhou argued that the Soviets were suggesting large-scale assault as a military

[83] For the CCP elites' testy criticism of the Vietnamese communists on the deescalation issue, see Zhou Enlai and Pham Van Dong, Beijing, April 13, 1968, in Westad et al., eds., "77 Conversations," no. 31, pp. 121–23; and Li Danhui, "Sino-Soviet Relations," 10. Beijing also worried unnecessarily that Hanoi's agreement to talk would bolster President Johnson's chances for reelection.

[84] Zhou Enlai and Pham Van Dong, Beijing, April 13, 1968. Zhou's statement to Pham Van Dong is a rambling communist theory about the factors that strain capitalism and the relief that Vietnamese accommodation had provided to the capitalists. Also see Zhou Enlai, Chen Yi, and Xuan Thuy, Beijing (Great Hall of the People), 9:45 a.m., May 7, 1968, in Westad et al., eds., no. 35, pp. 134–35 (for partial minutes in Chinese, see *Zhou Enlai Waijiao Huodong*, 524); Chen Yi and Le Duc Tho, Beijing, October 17, 1968; and Gaiduk, *The Soviet Union and the Vietnam War*, 108–10.

[85] Li Danhui, "Vietnam and Chinese Policy toward the United States," in William C. Kirby, Robert S. Ross, and Gong Li, *Normalization of U.S.-China Relations: An International History* (Cambridge: Harvard University Asia Center, 2005), 177.

[86] May 7, 1968 entry in *Zhou Enlai Nianpu*, 3:233.

[87] For an excellent critical review of that argument, see John Garver, "The Tet Offensive and Sino-Vietnamese Relations," in Marc Jason Gilbert and William Head, eds., *The Tet Offensive* (Westport, Conn.: Praeger, 1996), 45–61. Consistent with his earlier pattern, Mao was trying to export his own distinct style of revolution not only to Vietnam but also to other Asian areas, such as Japan. In 1967 he would continue to call on Japanese communists to adopt tactics consistent with his 1938 treatise on using the "countryside to surround the cities." (Xiangcun baowei changshi). See Mao's comments on December 1, 1967, in *Jianguo Yilai Mao Zedong Wengao*, 12:273, January 1966–December 1968. But in Vietnam his biggest concern seemed to be that Tet was accompanied by increased Soviet assistance and influence and was followed by negotiations.

tactic in order to set a trap for the Vietnamese communists. This, he opined, would favor U.S. counteroffensives, decimate Vietnamese communist forces, and increase the likelihood that the Vietnamese communists would be seduced by Soviet efforts to convince them to negotiate a peace settlement.[88]

1968: THE CRITICAL TURNING POINT IN THE CHINA-SOVIET-VIETNAM TRIANGLE

As I have argued, increasing intramural competition within the revisionist communist alliance made the movement more aggressive than it otherwise would have been right up to the point at which that intramural competition turned into direct hostilities and the breakdown of the alliance. To the great benefit of the United States, open Sino-Soviet conflict would break the expansionist fever that accompanied the earlier period of intramural communist competition. The benefits for the United States of Sino-Soviet hostility came fairly quickly in the form of reduced foreign support for the Vietnamese war effort, the withdrawal of Chinese forces from the theater, and Beijing's recommendations for a less aggressive military strategy by the Vietnamese communists.

In 1968 relations between Moscow and Beijing reached a boiling point. As discussed in the previous chapter, Mao began considering the Soviets as a potential military adversary in summer 1964. With the invasion of Czechoslovakia and the announcement of the Brezhnev Doctrine in 1968, the Sino-Soviet rivalry within the communist camp morphed into Sino-Soviet confrontation. In addition to the provocative nature of Soviet policy in 1968, Chinese elites were extremely concerned about the ongoing buildup of Soviet forces along the long Chinese border with both the Soviet Union and the Soviets' close ally, Mongolia. Tensions between the two sides eventually escalated into the first major Sino-Soviet border clashes in March 1969.[89]

---

[88] Zhou Enlai and Pham Hung, June 19, 1968 in Westad et al., eds., "77 Conversations," no, 36, pp. 133–35. When the Tet Offensive proved politically successful, Mao began praising such attacks, urging more as a prelude to negotiations. See Mao Zedong and Pham Van Dong, Novermber 17, 1968, in Westad et al., eds., "77 Conversations," no. 39, pp.137–53.

[89] For a Chinese military analysis of the threat posed by Soviet forward deployed troops in Siberia and the Mongolian border in the late 1960s, see Wang Zhongchun, "The Soviet Factor in Sino-American Normalization," 150–153. Wang, a Senior Colonel at the PLA's National Defense University, offers an authoritative account of China's military threat perception in the late 1960s. His account is a fine-grained version of a widely held thesis among Chinese diplomatic and military historians and non-Chinese scholars alike that China's threat perception toward the Soviet Union was a major driver in Mao's decision to seek rapprochement with the United States. His thesis therefore differs in important ways from the controversial thesis offered by Lyle Goldstein, who argues that China did not feel threatened by Soviet deployments in 1969 and ambushed Soviet forces in March 1969 to create a diversionary conflict for domestic purposes. See Lyle Goldstein, "Research

The implications for Vietnam were dramatic. Chinese troop withdrawal from the theater began in 1968 and by mid-1970 all Chinese forces (well over 100,000 forces) were withdrawn from Vietnam, including air defense forces.[90] In the 1968–1970 period, Beijing's military and economic aid would also slow down as Beijing began fulfilling earlier promises of support more slowly and began urging more self-reliance on the Vietnamese comrades.[91] Within the reduced aid portfolio China would also place greater emphasis on economic aid over military aid, thereby reducing significantly the number of arms transferred to Vietnam between 1968 and 1969 by about 50 percent with further reductions in 1970. Moreover, Beijing would encourage patience and moderation in the Vietnamese communists' military campaign in the South, a sharp contrast to the offensive and aggressive strategies recommended by Mao and his colleagues over the past three years.[92]

For our purposes, of equal importance was China's clear reversal on the issue of negotiations between the Vietnamese and the Americans. Beginning in late 1968 Beijing began to tone down criticism of Vietnamese communist negotiations with the United States, not only directly, but with third-party communists.[93] Li Danhui posits that, as early as November 1968, in internal meetings Mao reversed course and began praising the Vietnamese for negotiating with the Americans.[94] By 1969 China would also tone down its criticism with third parties of North Vietnam's position on peace talks. For example, in stark contrast to the fiery rhetoric in opposition to negotiation earlier in the war, in a September 1969 meeting with a Romanian entourage in the days following Ho Chi Minh's funeral, Zhou seemed matter of fact about the Vietnamese decision to go ahead with the Paris Peace Talks, saying, "We exchanged ideas with them on this," but that it was a matter for the Vietnamese to decide. Interestingly, he seemed to downplay the chance for escalation in Vietnam by stating that the big problem

---

Report: Return to Zhenbao Island: Who Started Shooting and Why Does It Matter," *China Quarterly* no. 168 (2001): 985–97; for a critique of Goldstein's account and a supporting argument for a threat-based analysis for China's actions, see Thomas J. Christensen, "Windows and War: Trend Analysis and the PRC's Use of Force," in Robert Ross and Alastair Iain Johnston, eds., *New Directions in the Study of Chinese Foreign Policy* (Stanford: Stanford University Press, 2006), 70.

[90] For a brief review, see Garver, "The Tet Offensive," and Qu, "*Zhongguo Zhiyuan Budui*," 43. As Qu points out the PLA had a limited role later in the war assisting with minesweeping from May 1972 through August 1973, for example. Also see Li, "Vietnam and Chinese Policy toward the United States," 183.

[91] Li, "ZhongSu Guanxi," 12; for discussion of what she calls "subtle" (*weimiao*) pressure on the Vietnamese in the aid arena to punish them for siding too closely with the Soviets beginning in 1968, see Li Danhui, "ZhongSu Zai YuanYue KangMei Wenti Shang de Maodun yu Chongtu," 388–89.

[92] Li, "Vietnam and Chinese Policy toward the United States," 183. Li estimates that Chinese military aid in 1969 was cut by 50 percent from the previous year.

[93] Zhai, *China and the Vietnam Wars*, 173–74.

[94] Li, "Vietnam and Chinese Policy toward the United States," 178.

in Sino-U.S. relations was not Vietnam but Taiwan and UN representation. Perhaps this was an early olive branch being offered to the United States as part of the process of rapprochement.[95]

By fall 1970 China was already preparing for a high-level U.S. delegation to visit Beijing. Mao apparently tried to sell this prospect to Hanoi by portraying the Americans as on the ropes, afraid of war, and bad at diplomacy, and thereby downplaying the risks of talks with them. After assuaging fears of U.S. escalation, Mao had particularly condescending words to say about Henry Kissinger in his September 23 conversation with Pham Van Dong. He said, "Kissinger is a stinking scholar. I have read the report about the meeting between comrade Xuan Thuy and Kissinger. The last part of it is quite funny. Kissinger is a university professor who does not know anything about diplomacy."[96] He then promised significant aid to Vietnam, just as he was preparing for rapprochement with the United States.

Beijing's 1968 decisions to reduce significantly Chinese support to Vietnam followed on the heels of the Tet Offensive and Hanoi's decision to open talks with Washington in May. Vietnamese accounts make a direct connection between the issues of Vietnamese willingness to negotiate and a reduction of Chinese aid and point out that the issues were explicitly linked by Chinese authorities in bilateral discussions.[97] Stephen Morris argues that in 1968 the Vietnamese communists began leaning harder toward the Soviet side in the Sino-Soviet dispute.[98] Despite these Vietnamese perceptions, direct Sino-Vietnamese differences over war and diplomacy were much less important in Chinese decisions to moderate its positions on the war in Vietnam than were the increasing tensions in Sino-Soviet relations and Beijing's related preparations to approach the United States for rapprochement. As the leading Chinese historian on PRC-Vietnam-Soviet triangular relations, Li Danhui, points out, the main driver for the 1968–1970 adjustment in PRC policy toward Vietnam was the budding Sino-Soviet conflict and the resulting desire by Mao to improve relations with the United States. Li argues that Mao had to curb somewhat his support for revolution in Vietnam in order to reach out more effectively to the United States and create a balance against the Soviets.[99] Even under these circumstances, Mao saw the need to compete for influence with Soviets in Vietnam and eventually he would even offer Vietnam large-scale military aid once again in the early 1970s. But in the near

---

[95] September 7, 1969, entry in *Zhou Enlai Nianpu*, 3:319–20; and *Zhou Enlai Waijiao Huodong*, 538–39.

[96] Mao Zedong and Pham Van Dong, September 23, 1970, in Westad et al., eds., "77 Conversations," no. 54, p. 178.

[97] Garver, "The Tet Offensive," 52–55.

[98] Morris, "The Soviet-Chinese-Vietnamese Triangle," 10 and passim.

[99] Li, "Vietnam and Chinese Policy toward the United States," 175–76.

term (1968–1970) the CCP would compete with the Soviets in a much more restricted manner than before: by criticizing Soviet motivations verbally with the Vietnamese colleagues rather than by either encouraging Vietnamese belligerence or by transferring large-scale economic or military aid.[100]

From 1971 to 1975 Chinese material assistance to Vietnam, including military assistance, would increase again.[101] Mao became convinced in the early 1970s that the United States was not going to escalate the war, and so giving material assistance to Vietnam carried few of the downside risks of China's earlier aid to the Vietnamese communists. Moreover, although it did not play such a role in the end, such continued Chinese aid potentially still carried benefits for Beijing's efforts to prevent Vietnam from leaning even more heavily toward the Soviets in the Sino-Soviet dispute. As Mao put it, the lack of an even verbal U.S. response to China's clear activity in the war showed that it was fighting with "one hand behind its back [*liu zi yi shou de*]." But Mao still wanted to compete in Vietnam, if possible, with the Soviets.[102] To do so he needed to cushion somewhat the blow of Sino-American rapprochement for his Vietnamese comrades.[103]

It must be recognized that the increase in Chinese military aid to Vietnam from 1971 to 1975 diverges from the expectation of the argument here. But when one considers the overall change in China's military and diplomatic strategy toward Vietnam during 1968–72—including the withdrawal of Chinese forces and the support for peace talks—the general thesis here holds up quite well. Chinese revolutionary activism in Vietnam reduced sharply from the time the Chinese and the Soviets moved from rivalry to open conflict. Moreover, as Li argues, even Mao's last binge of material support for Vietnam was rooted in Beijing's continuing struggle with the Soviet Union, a notion fully consistent with the thesis here.[104]

Once the Sino-Soviet rivalry turned to Sino-Soviet conflict, China's overall support for an uncompromising and belligerent Vietnamese strategy would never again reach pre-1968 levels. The last sizable Chinese military units left Vietnam by July 1970.[105] Moreover, as Sino-Soviet confrontation and U.S-PRC

---

[100]In 1968–70, the Chinese continued to blast the Soviets in conversations with the Vietnamese communists by accusing Moscow of evil, imperialist intent in their aid programs worldwide. In such a conversation Kang Sheng began questioning whether Vietnam needed to have 6,000 fighting-age students in China. Zhou Enlai, Kang Sheng and Pham Van Dong, Hoang Van Thai, Pham Fung and Others in the COSVN Delegation, Beijing April 20 and April 21, 1969, in Westad et al., eds., "77 Conversations," no. 41, pp. 158–59.

[101] Li, "Vietnam and Chinese Policy toward the United States," 184.

[102] Li, "Sino-Soviet Relations," 11–13.

[103] Zhai, *China and the Vietnam Wars*, 195.

[104] Li Danhui, "ZhongSu Zai YuanYue KangMei Wenti Shang de Maodun yu Chongtu (1965-72)," 372–414.

[105] Chen, "China's Involvement in the Vietnam War, 1964–69," 384–85.

rapprochement became realities, on the diplomatic front China's revolutionary zeal dissipated relatively quickly. In one telling example of the change in approach, Zhou Enlai, in a March 1971 conversation, simply refused to pick up the mantle of anti-imperialism offered by Le Duan and Pham Van Dong. The Vietnamese entourage wanted China to take the lead in opposing the 1969 Nixon Doctrine and increases in U.S.-Japan security cooperation to which it might lead. Le Duan attempted to flatter his Chinese comrades, saying, "Japan has a plan for Southeast Asia. It wants to control the region. We want to smash the US-Japan alliance as well as the alliance between the US, Japan, and the regional bourgeois class. We have to establish a world front that will be built first by some core countries and later enlarged to include African and Latin American countries . . . The world's people oppose the "Nixon Doctrine" which also means opposing the US-Japan alliance. The questions, therefore, are how [do] we establish this front, who is capable of doing this. Only China and no one else."[106] In the days before the Brezhnev Doctrine and the Sino-Soviet border conflict of 1969, it is hard to imagine that such an offer and portrayal of the world would not have met with anything but the warmest reception in Beijing. But after the Soviets and Chinese were no longer competing within the communist movement but were in direct conflict, China's priorities changed. In order to patch up relations with the United States, China would need to avoid unnecessary provocations relating to the U.S.-Japan alliance. So, Zhou responded to Le Duan cryptically, "This is a new issue. East Asia is part of the world. The people in Asia ... are suffering from American and Japanese reactionaries' policies. Yet, on establishing a People's Front to oppose them, we need more time to think. Sometimes, you are in a more advantageous position than we."[107]

China's views on a negotiated peace and a coalition government in the South changed nearly 180 degrees from 1968 to 1972. Just after the Kissinger visit to China in July 1971, Zhou Enlai reportedly made a secret visit to Hanoi to urge the Vietnamese brethren to negotiate a peace agreement with the Americans.[108] Consistent with U.S. wishes, Zhou would ask the Vietnamese communists to push only for the removal of U.S. forces from South Vietnam and not to insist on the removal of Nguyen Van Thieu's regime in the South as a precondition for peace.[109] Apparently, this met with an angry response in Hanoi.[110] In the weeks

---

[106] Zhou Enlai with Le Duan and Pham Van Dong, March 7, 1971, in Westad et al., eds., "77 Conversations," no. 55, pp.176–77.

[107] For Zhou's statement see Zhou Enlai with Le Duan and Pham Van Dong, March 7, 1971, in Westad et al., eds., "77 Conversations," no. 55, pp.176–77.

[108] James Mann, *About Face: A History of America's Curious Relationship with China, From Nixon to Clinton* (New York: Knopf, 1998), 34.

[109] Li, "Vietnam and Chinese Policy toward the United States," 193.

[110] Mann, *About Face*, 34.

and months following the February 1972 Nixon visit, Zhou urged the Vietnamese communists to consider allowing a coalition government in the South that included Thieu's government, arguing that this would surprise the Americans and that, in negotiations, "surprise is necessary." He also argued that North Vietnam could use a period of peace to consolidate its power. Finally, in another reversal, Zhou spoke positively about the 1954 Geneva conference and how it had been wise to negotiate with the French government.[111] At the end of the year, Zhou would again urge flexibility in negotiations on the Vietnamese, saying that they should give Nixon some "face" so as to ease the negotiation process.[112] By December 1972, Mao was even blasting the "so-called communists" who were advising the Vietnamese to break off the Paris Peace Talks. He cynically blamed the then-deceased Lin Biao for such excesses in China in the past![113] In June 1973 Zhou urged a high-level Vietnamese delegation to "relax and build forces" for five to ten years and to "build peace, independence, and neutrality," in South Vietnam, Cambodia, and Laos.[114] This was a huge change from 1965–68, when at the height of Sino-Soviet rivalry short of outright conflict, the Chinese adopted their most aggressive policies toward the war in Vietnam.

Despite the renewed flow of Chinese military assistance to Vietnam in the final four years of the Vietnam War, Beijing's change of overall strategy toward the Vietnam conflict put a major strain on Sino-Vietnamese relations.[115] The PRC's general policy toward the conflict in Vietnam became much closer to the more moderate stance that Beijing had adopted at Geneva in 1954 once Sino-Soviet intramural competition had finally turned into Sino-Soviet confrontation. This affected not only Sino-Soviet, but Sino-American, and Sino-Vietnamese relations.

In terms of U.S. equities in Vietnam, this outcome would have a limited effect. The U.S. war effort was already considered lost there in Beijing, Hanoi, Moscow, and even among many in Washington. The question was whether or not China would play some role in helping the United States leave Vietnam under the cover of a negotiated settlement that few in Asia expected to hold. All indications suggest that China did try to help on this score.

---

[111] For Zhou's comments, see Zhou Enlai and Nguyen Tien, Beijing, April 12, 1972, in Westad et al., eds., "77 Conversations," no. 59, p.179; Zhou Enlai and Xuan Thuy, Ly Ban, Beijing 5:40 pm, July 7, 1972, in ibid., no. 60, p. 179; and Zhou Enlai and Le Duc Tho, July 12, 1972, in ibid., no. 61, pp. 179–82.

[112] Qu Xing, "ZhongYue Zai Yinzhi Zhan Wenti Shang de Zhanlüe Yi Zhi yi Celüe Chayi," 317.

[113] Mao Zedong and Nguyen Thi Binh, Beijing, December 29, 1972. Mao blamed the "bad guys" under Lin Biao for radical acts during the Cultural Revolution such as burning the British embassy in Beijing; also see Zhou Enlai and Truong Chinh, Beijing, 8:40 p.m., December 31, 1972, in Westad et al., eds.,"77 Conversations," no. 62.

[114] Zhou Enlai and Le Duan, Pham Van Dong and Le Thanh Nghi, Beijing, June 5, 1973, in Westad et al., eds., "77 Conversations," no. 68, pp. 184–87.

[115] Li, "Vietnam and Chinese Policy toward the United States," 198.

From a global perspective, however, the transformation of Sino-Soviet rivalry into Sino-Soviet conflict had a hugely positive benefit for the United States' position within the Cold War. The Sino-Soviet Cold War would tie down a huge number of Soviet troops along the Sino-Soviet and Sino-Mongolian border, more than were deployed to Eastern Europe. China would also assist in intelligence-gathering and in the very effective efforts to supply anti-Soviet forces in Afghanistan. Finally, the Sino-Vietnamese conflict that arose after the Vietnam War ended helped limit the expansion of Vietnamese and, thereby, Soviet influence in Southeast Asia.

U.S. ALLIANCES AND THE SLOWING OF U.S.-PRC RAPPROCHEMENT

The weakening of the U.S. position in Asia with the slowly progressing defeat in Vietnam and the growth of Soviet regional power provided the opportunity for the U.S.-PRC rapprochement under the Nixon administration. Uncertainties in Beijing about aspects of U.S. relations with its allies in the region made that rapprochement quite complicated to achieve.

It is widely understood in the existing literature that the United States under Richard Nixon and Henry Kissinger and the People's Republic of China under Mao and Zhou Enlai needed to come up with some formula in which the two sides could "agree to disagree" on the U.S. alliance relationship with Taiwan in order to move U.S.-PRC relations in the direction of eventual normalization. This feat was not easy to accomplish. Despite the initial claims in Henry Kissinger's memoir that Taiwan was not an issue of "primary concern" to the PRC, the subsequently declassified documents reveal that Taiwan was front and center in the bilateral negotiations of 1971–72.[116] When the United States and the PRC were exchanging feelers via Romania in 1969–70, Zhou Enlai made it clear to the Romanian deputy premier, "Between China and the U.S., there is one main issue—the issue of Taiwan."[117] In a May 29, 1971, message to President Richard Nixon, Zhou Enlai stated that the "crucial issue" to be addressed first "is the question of the concrete way of the withdrawal of all the U.S. Armed Forces from Taiwan

[116] Henry Kissinger, *White House Years* (Boston: Little Brown), 705. Xia, *Negotiating with the Enemy*, 151–52 states that for Zhou Enlai, progress on Taiwan was the "sole purpose" of having direct talks with the Americans.

[117] "Memorandum of Conversation between Romanian Deputy Premier Gheorghe Radulescu and Zhou Enlai," December 12, 1970, *Cold War International History Project Bulletin* (*CWIHPB*), no. 16 (fall 2007/winter 2008): 439. This theme runs through the documents uncovered in the Romanian archives by Mircea Munteanu in this volume. See, for example, "Telegram from Romanian Ambassador in Beijing Aurel Duma to Foreign Minister Corneliu Manescu Regarding Conversations with Representatives of the PRC Ministries of Trade, Foreign Affairs, and Defense," May 13, 1969, in ibid., 411.

and Taiwan Strait Area."[118] Taiwan would then figure prominently in the discussions between Kissinger and Zhou in July 1971. During the former's secret trip to Beijing, Zhou would once again call Taiwan "the crucial issue" in U.S.-PRC relations.[119] In an October 1971 meeting in Beijing, Kissinger stated plainly to Zhou that "we understand and recognize that the most important issue for you is the issue of Taiwan."[120] In those meetings, Zhou Enlai consistently expressed concerns about a Taiwan independence movement and U.S. support for that movement or a "two-Chinas" solution. Kissinger responded that the United States would not support either and offered in the near term a U.S. removal of forces in Taiwan related to the war in Southeast Asia, and, over time, other U.S. forces on the island as relations between the United States and the PRC improved.[121]

Given the distance between what the PRC wanted from Washington and what the United States was willing to offer on America's posture toward cross-Strait relations, the Shanghai Communiqué of February 1972 was a masterfully drafted document. It hails the common interests in opposing Soviet "hegemony" and allows both sides to express their quite different views on Taiwan's sovereignty. For its part, the PRC made unilateral assertions about China's sovereignty over Taiwan while the United States maintained for the time being its alliance with the Republic of China on Taiwan and public ambiguity over Taiwan's sovereign status in relation to mainland China. Because it has been very ably covered elsewhere, there is no need to rehash here the way in which the two sides used split statements and word-smithing in the Shanghai Communiqué, which allowed the United States to "acknowledge" "without dispute" the claim of Chinese on both sides of the Strait that Taiwan is part of China, without actually recognizing or acceding to that position in a positive manner.[122] Although Nixon and Kissinger promised Mao and Zhou to avoid saying so in public, and thereby unnecessarily poking Beijing in the eye, the U.S. legal position on Taiwan remained the same: the island's sovereign status remained "undetermined," with the United States taking no position one way or the other about Taiwan's ultimate disposition,

---

[118] "Message from Zhou to Nixon," May 29, 1971, in William Burr, ed., "The Beijing-Washington Back Channel and Henry Kissinger's Secret Trip to China," *National Security Archive Electronic Briefing Book* No. 66, February 27, 2002, document no. 26, p. 1.

[119] See, for example, "Memorandum for Henry A. Kissinger from Winston Lord. Subject: Memcon of Your Conversation with Chou En-lai," July 29, 1971, in ibid., 13.

[120] "Memcon, Kissinger and Zhou: Opening Statements, Agenda, and President's Visit," October 20, 1971, 4:40–7:10 p.m. in William Burr, Sharon Chamberlain, Gao Bei, and Zhao Han, eds., "Negotiating U.S.-Chinese Rapprochement," *National Security Archive Electronic Briefing Book* No. 70, National Security Archives, May 22, 2002, document no. 10, p. 7.

[121] Ibid., 12–13.

[122] For classic and authoritative coverage of the negotiations in 1971–72 to produce the Shanghai Communiqué , see Harding, *A Fragile Relationship*, chs. 1–2; Ross, *Negotiating Cooperation*, chs. 1–3; and Romberg, *Rein In at the Brink of the Precipice*, chs. 2–3.

as long as differences across the Strait were managed and settled peacefully.[123] Unfortunately from the point of view of consistency and clarity in U.S. policy, President Nixon, like many top leaders in the United States, had a tendency to leave the script and simplify to a fault, suggesting that the United States affirms the PRC position that Taiwan is part of China. So, on February 22, 1972, he said to Zhou more baldly than is stated in actual U.S. policy,

> Dr Kissinger when he was here stated our agreement to five principles. I completely endorse these principles, and the Prime Minister can count on that no matter what we say on other subjects:
>
> > Principle one. There is one China, and Taiwan is a part of China. There will be no more statements made—if I can control our bureaucracy—to the effect that the status of Taiwan is undetermined.[124]

As scholar and former State Department official Alan Romberg argues in an authoritative study, it was one thing to say that we would keep the bureaucracy from stating that Taiwan's status was undetermined in U.S. legal analysis and quite another to state that Taiwan is factually "part of China" as an American legal position. Such an affirmative statement was carefully avoided in the U.S. statement in the Shanghai Communiqué and has not been subsequently stated publicly in any official document.[125] In fact, in later private meetings on the same trip, Nixon pointed out that if he were to be successful in achieving his ultimate goal of normalization of U.S.-PRC relations, he could not allow U.S. opponents of his approach to China to criticize him for repudiating a "commitment to the government on Taiwan" and for making "secret deals" on Taiwan.[126]

---

[123] The best treatment in the literature of this issue is Romberg, *Rein In at the Brink of the Precipice*, ch. 2. For an extended back and forth on this issue between Kissinger and Zhou, see "Memcon, Kissinger and Zhou: President's Visit, Taiwan, and Japan" October 21, 1971, 10:30 am-1:45 pm, in Burr et al., eds., "Negotiating U.S.-Chinese Rapprochement," document no. 11, pp. 21–28. Kissinger had to finesse the issue by saying "let me separate what we can say and what our policy is." Not only does he promise that U.S. officials will not continue to call Taiwan's status "undetermined," but he goes a bit further saying, "We do not challenge the fact that all Chinese maintain that there's only one China and that Taiwan is part of China. And therefore we do not maintain that the status *in that respect* is undetermined. How this can be expressed is a difficult matter, but we would certainly be prepared in a communiqué that might be issued to take note of the fact that all Chinese maintain that there is only one China. So that is the policy of this government." Ibid., 26. In subsequent years the United States has made it clear that it does not support Taiwan independence and that it opposes unilateral changes to the status quo by either side of the Taiwan Strait.

[124] Memorandum of Conversation, Tuesday, February 22, 1972, 2:10 p.m.–6:00 p.m., Great Hall of the People, Peking, in William Burr et al., eds., "Negotiating U.S.-Chinese Rapprochement," Nixon's Trip to China, document no. 1, p. 5.

[125] For an excellent review of Nixon's logic and the ways that he and Kissinger pulled back finally from abandoning U.S. ambiguity on Taiwan's legal status, see Romberg, *Rein In at the Brink of the Precipice*, ch. 2.

[126] Memcon, February 24, 1972, 5:15 p.m.–8:05 p.m., Peking, in Burr et al., ed.s, "Negotiating U.S.-Chinese Rapprochement," Nixon's Trip to China. document no. 3, p. 10.

What is less understood than the importance of the issue of cross-Strait relations in U.S.-PRC rapprochement are the ways in which new trends in the U.S.-Japan alliance concerned Beijing. In particular Mao and his colleagues worried about Japanese independence of action in the security sphere and, especially, Japan's return to intervention in the Taiwan Strait.[127] The term "Nixon Shocks" generally refer to Japanese astonishment at two key Nixon administration decisions about which Tokyo had not been forewarned: the July 1971 announcement that Nixon would visit the PRC and Nixon's August 1971 decision to end the Bretton Woods system under which American dollars were freely convertible into gold, and to level across-the-board 10 percent tariffs on imports. But from the perspective of Beijing and to a significant degree Tokyo itself, there was another shock that preceded and set the stage for these two. Two years before Kissinger's secret journey to Beijing, President Nixon announced in Guam his new strategic doctrine of relying more on Asian allies to provide for their own defense in light of America's failing efforts in Vietnam and the high cost of that war for the U.S. economy. Nixon's move, labeled the "Guam Doctrine" or "Nixon Doctrine," was part of his plan for increased responsibility among Asian allies for checking communism. Followed in December of the same year by a joint communiqué between President Nixon and Prime Minister Sato Eisaku of Japan on the alliance, the Nixon Doctrine created a fresh debate in the latter's country about the meaning of Article IX of its "Peace Constitution," and the proper role of Japanese defense policy within the U.S.-Japan alliance structure. In order to entice Japan to take on greater burden-sharing roles as part of the alliance, Nixon held out Japanese recovery of Okinawa (and, to China's dismay, the Diaoyudao/Senkaku islands) as a prize.[128] In addition to material burden-sharing demands by the United States, there were additional political sacrifices requested of Sato in order to guarantee that Japan did not stray far from American policy in East Asia. Sato was urged during the December 1969 Washington Summit to declare that both South Korea and Taiwan were critical to Japanese security. The Nixon-Sato Communiqué did just that and thereby rankled both Japanese pacifists and those in Japan opposed to Tokyo's pro-Taiwan stance.[129]

What might have appeared like a tightening of the U.S.-Japan alliance in late 1969 actually signified serious weaknesses. Not only was there concern in Japan about U.S. willingness or ability to stand up to the Soviets in the region in the

---

[127] Zhou described Japan to Kissinger in October 1971 as "an issue which has a far-reaching influence on the relaxation of tension in the Far East." "Opening Statements, Agenda, and President's Visit," October 20, 1971, in Burr et al., eds., "Negotiating U.S.-Chinese Rapprochement,"document no. 10, p. 16.

[128] Yoshihide Soeya, "U.S.-Japan-China Relations and the opening to China: The 1970s," Working Paper No. 5, U.S.-Japan Project, National Security Archive, p. 4.

[129] For Beijing's reaction, see Doak Barnett, *China and the Major Powers in East Asia* (Washington, D.C.: Brookings Institution, 1977).

future, Sato resented American pressure on new Japanese military roles and on increased restrictions on Japanese exports to America. In the short term, Tokyo's unwillingness to consider some added military responsibilities as part of the alliance exacerbated American suspicions about Japan's reliability as an ally. According to Assistant Secretary of State for East Asia Marshall Green, President Nixon and Kissinger viewed Japanese leaders, unlike Mao and Zhou, as obsessed with economics rather than strategic interests.[130] John Holdridge, then of the NSC, argues that there was a general fear that Sato was untrustworthy and that he would have raced to beat America to the punch on China policy and thereby reaped the domestic political benefits of being in the vanguard if the Nixon administration had hinted to him about Kissinger's plan to visit Beijing.[131] Nixon's frustration with Japan in the economic sphere and Sato's claim that he could not implement his agreements because of domestic resistance probably helped shape the administration's dim view of Tokyo politics. The failure of Sato to carry through on his original commitment to limit Japanese textile exports to the United States upset the American leadership and almost certainly affected Nixon's decision to leave Japan out of the loop on the Kissinger visit to Beijing in July 1971. Perhaps this is why Nixon announced the second Nixon Shock, the end of Bretton Woods and a 10 percent surcharge on imports, on the anniversary of V-J Day.[132]

From the Chinese perspective, rather than alliance challenges, the Nixon-Sato communiqué of 1969 and the alliance politics that flowed from it signaled a period of increased strategic cooperation between the United States and Japan to include greater Japanese room to maneuver in the security realm. Most alarming to PRC elites was an explicit, reinvigorated interest in Tokyo in Taiwan's security. Requests for Japan to do more as part of the U.S.-Japan alliance initially triggered deep anxieties in Beijing about Japan's restoration as a major military power and, more specifically, the prospect of Japan's replacing the United States as the major military ally of Chiang Kai-shek's Taiwan.[133] Nothing would push the buttons of Chinese nationalism like the vision of revived Japa-

---

[130] Marshall Green, "The State Department Perspective," in Marshall Green, John H. Holdridge, and William N. Stokes, eds., *War and Peace with China: First-Hand Experiences in the Foreign Service of the United States* (Washington, D.C.: Dacor Press, 1994), 180–81.

[131] John Holdridge, *Crossing the Divide: An Insider's Account of the Normalization of U.S.-China Relations* (Lanham, Md.: Rowman & Littlefield, 1997).

[132] LaFeber, *The Clash*, 354.

[133] On Zhou Enlai's reaction to the Nixon-Sato Comminiqué, see Kissinger, *White House Years*, 334. In 1972 Zhou himself cited the communiqué to Nixon as a source of his earlier concerns about future Japanese aggressiveness. Memcon, February 23, 1972, 2:00 p.m.–6:00 p.m., Peking in Burr et al., eds., "Negotiating U.S.-Chinese Rapprochement," Nixon's Trip to China, document no. 2, p. 18. Also see, Memcon , July 9, 1971, Peking, in Burr, ed., "Secret U.S. Chinese Diplomacy Behind Nixon's Trip," *National Security Archive Electronic Briefing Book* No 145, document no. 9, p. 29.

nese imperialism in Taiwan, and for all of its ideological pretenses and follies, Mao's CCP still portrayed itself as a nationalist regime that had been forged in the struggle against Japan.[134]

By 1970, Beijing's fears would be further fueled by the drafting of Defense Minister Nakasone's "Fourth Defense Build-up Plan," even though that plan would be paralyzed by domestic resistance for two more years.[135] As a result, in 1970 concerns about Japanese militarism (*junguozhuyi*) reached a post-Korean War peak in China as the Sato government, along with other domestic conservative forces in Japan, appeared to be gaining the upper hand domestically. From Beijing's perspective this portended the revival of a militarily aggressive Japan. In a joint communiqué with Kim Il-sung in April 1970, Zhou Enlai claimed that "under the active protection of the American imperialists, Japanese militarism has already been revived."[136] In his discussions with the Romanians, who served as a channel between Beijing and Washington, Zhou Enlai stated that "the economic expansion of Japan will be the source of military expansion. That is why the U.S.-Japan security treaty has been extended indefinitely. The Japanese economic expansion requires the expansion of its rearming plans. ... Thus their position as a great economic power will mean their expansion as a military power. In this context, it must be stated that the Sato-Nixon declaration clearly states that South Korea and Taiwan are indispensible to the security of Japan."[137] In his initial meetings with Kissinger in July 1971, Zhou Enlai summed up his view of the U.S. strategy as follows:

> After 25 years it's no longer possible for the U.S. to exercise a position of hegemony. Japan has become strengthened, and if you will now withdraw all foreign troops from the Far East, it's your purpose to strengthen Japan so it can serve as your vanguard in the Far East in controlling Asian countries.[138]

---

[134] For *People's Daily* coverage of the U.S.-Japan alliance in 1970 along these lines, see *Zhanhou ZhongRi Guanxi Wenxian Ji: 1945–70*, [Collected Documents on Post-War Sino-Japanese Relations] (Beijing: Zhongguo Shehui Kexue Chubanshe, 1996), 909–40.

[135] Momoi, "Basic Trends in Japanese Security Policies," 348. The Fourth Plan, slated to be twice as expensive as the Third Plan, ran into trouble at home and abroad. It would not be adopted until Prime Minister Tanaka returned from Beijing in September 1972. For Zhou Enlai's expression of concern about the Plan see, Memcon, July 9, 1971, Peking, in Burr, ed., "Secret U.S. Chinese Diplomacy Behind Nixon's Trip," document nol 9, p. 29.

[136] See Wang Shu and Xiao Xiangqian, et al., *Bu Xunchang de Tanpan* [Unusual Negotiations] (Jiangsu People's Press, 1996), 243.

[137] "Memorandum of Conversation between Radulescu and Zhou," December 12, 1970, *CWIHPB*, no. 16 (fall 2007/winter 2008): 440.

[138] "Memorandum for Henry A. Kissinger from Winston Lord. Subject: Memcon of Your Conversations with Chou En-lai," July 29, 1971, in William Burr, ed., "The Beijing Washington Back-Channel and Henry Kissinger's Secret Trip to China, September 1970-July 1971," document no. 34, p. 37.

Even though Beijing recognized some of the sources of tension in the U.S.-Japan relationship, these were viewed as factors that could only strain the relationship in the near term, producing a break only in the future. Presaging a similar view by their compatriots in the early 1990s, which is outlined in the next chapter, Beijing analysts believed that a U.S.-Japan break would occur only after a period in which Japan, under America's protective wing, would become militarily stronger, domestically more militaristic, and diplomatically more aggressive toward China and others.[139]

Beijing's concerns about the U.S.-Japan relationship were at their height just as the slow process of U.S.-China rapprochement began in 1969. Kissinger notes that it would take a long time for him in 1971–72 to convince Zhou Enlai that a strengthened U.S.-Japan alliance was the best way to achieve the dual goals of countering growing Soviet power and limiting the prospect of a highly nationalistic and more independent Japanese security policy.[140] In his 1997 memoir, Nixon NSC official John Holdridge mentioned the "hot and heavy" discussions that Kissinger's entourage had in July 1971 about the prospect of a remilitarized Japan and the need for the United States to restrain Japanese military assertiveness.[141] In a July summation of Kissinger's discussion with Zhou, Kissinger's Special Assistant at the NSC, Winston Lord, noted the latter's "preoccupation with Japan."[142] Declassified U.S. government documents regarding the U.S.-PRC rapprochement process reveal that, if anything, Holdridge and Lord understated the Chinese worries about these issues. Beijing appeared highly concerned not only about the United States relationship with Taiwan after rapprochement, as is commonly known, but about Japan's future regional role, particularly toward Taiwan. In fact, in one of his meetings with President Nixon, Zhou stated that from the Chinese perspective, the statement in the Shanghai Communiqué opposing "hegemony" was aimed at Japan as well as the Soviet Union.[143] Zhou believed that there was a conspiracy in Japan to absorb Taiwan into the Japanese

---

[139] For Chinese reactions to the Nixon-Sato communique, see Kurt Werner Radtke, *China's Relations with Japan, 1945–83: The Role of Liao Chengzhi* (Manchester: Manchester University Press, 1990), 183; Barnett, *China and the Major Powers*, 111; Wang Shu and Xiao Xiangqian, et al., *Bu Xunchang de Tanpan* [Unusual Negotiations], 244; and Wu Xuewen, Lin Liande, and Xu Zhixian, *Dangdai ZhongRi Guanxi* [Contemporary Sino-Japanese Relations] (Beijing: Shishi Chubanshe, 1995), 140–42.

[140] Kissinger, *White House Years*, 334.

[141] Holdridge, *Crossing the Divide*, 60.

[142] "Memorandum for Henry A. Kissinger from Winston Lord. Subject: Memcon of Your Conversation with Chou En-lai," July 29, 1971, in document no. 34, p 1.

[143] Memorandum of Conversation, February 25, 1972, 5:45 p.m. to 6:45 p.m., Peking, in "Record of Historic Richard Nixon-Zhou Enlai Talks in February 1972 Now Declassified," *National Security Archive Electronic Briefing Book*, available at http://www.gwu.edu/~nsarchiv/nsa/publications/DOC_readers/kissinger/nixzhou/.

island chain. He feared that Taiwan's resistance to both Japanese troops on the island and a significantly improved political relationship with Japan was mainly driven by Chiang Kai-shek's personal aversion to such outcomes. But Zhou noted nervously that Chiang was 85 years old.[144] Consistent with his assessment that the Guam Doctrine would lead to a restoration of Japanese military assertiveness, Zhou expressed similar concern that, if the United States were to leave South Korea, Japanese forces would replace them.[145] In fact, Beijing's fear about a resurgent Japan inserting itself in cross-Strait relations was so great that in October 1971, Zhou Enlai admitted privately to Kissinger that he preferred some U.S. forces to remain on the island for the foreseeable future lest they be replaced by Japanese forces! He stated that "objectively speaking for you to keep forces on Taiwan to prevent Japan from sending its forces into Taiwan is beneficial to the relaxation of tension in the Far East. If Japan puts forces in to bring about a so-called independent Taiwan, that will be the beginning of the end for peace in the Far East."[146] Zhou pointed out that, as a sovereign state, the PRC could not state this publicly. In summing up the meetings for President Nixon, Kissinger would list as China's "primary concern(s)," "Japanese intentions and the possibility of Taiwan independence."[147]

It might have been possible that the concerns expressed by Chinese leaders about Japanese militarism and Taiwan were insincere and designed only for U.S. consumption. But we know from an internal report, which Zhou had prepared for the Chinese Politburo in spring 1971 in anticipation of direct meetings with the Americans, that the Chinese elites' concerns about Japan were heartfelt. As revealed in an excellent account by historian Yafeng Xia, the second point in the eight-point report states that "Taiwan is China's territory, and the liberation of Taiwan is China's internal affair. No foreign intervention should be allowed. Vigilance toward the activities of Japanese militarism in Taiwan must be pursued."[148] In October 1971 Zhou demanded of Kissinger a commitment to

[144] "Memcon, Kissinger and Zhou: President's Visit, Taiwan, and Japan," document no. 11, pp. 31–33.

[145] "Memcon, Kissinger and Zhou. "Korea, Japan, South Asia, Soviet Union, Arms Control," October 22, 1971, in Burr et al., eds., "Negotiating U.S.-Chinese Rapprochement," document no. 13, pp. 10–11.

[146] "Memcon, Kissinger and Zhou, "Communiqué," October 25, 1971, 5:30–8:10 p.m., *Negotiating U.S.-Chinese Rapprochement: National Security Archive Electronic Briefing Book* No. 70, document no. 18, pp. 13–14. About U.S. forces on the island, Zhou also stated, "I think there are some points of advantage in leaving some so as to not let the Japanese send in their forces ... in actual practice we are not opposed to your acting like that because so long as you still have forces on Taiwan you will have the responsibility of not letting Japanese armed forces go in." Ibid., 13.

[147] "Kissinger to Nixon," November 11, 1971 in Burr et al., ed., "Negotiating U.S.-Chinese Rapprochement," *National Security Archive Electronic Briefing Book* No. 70, document no. 20, p. 14.

[148] Xia, *Negotiating with the Enemy*, 156.

oppose Japanese armed forces moving to the island. Kissinger did so, and added that "to the extent that we have influence in Japan we will oppose an attempt by Japan to support the Taiwan Independence Movement."[149] President Nixon himself would list as his third principle to Zhou, U.S. good faith efforts to "discourage" Japan from supporting Taiwan independence, or moving into Taiwan as U.S. presence there diminishes, and he would reiterate this principle during his stay.[150] Within a few days of Nixon's arrival, Zhou seemed very much comfortable with the arguments for U.S. forces in both Japan and Taiwan, as long as the United States refused to support Taiwan independence or encourage Japan to do so militarily or politically.[151]

Kissinger and his Nixon administration colleagues deftly turned the Chinese concern about Japan to their advantage, convincing Beijing not only that growing Soviet power in Asia warranted a sustained and strengthened U.S.-Japan alliance but also that a stronger conventional role for Japan within the U.S.-Japan alliance, as advocated in the Nixon Doctrine, should be greatly preferred by both Washington and Beijing over a Japan on its own, with a very strong military and, perhaps, nuclear weapons, and with no clear check on Tokyo's behavior by Washington.[152] Even before Kissinger departed for Beijing in July 1971, President Nixon urged him to explain more clearly to the Chinese "the threat of Japan's future orientation," and the danger of a U.S. withdrawal. This was ranked as a top priority alongside emphasizing the severity of the Soviet threat and the dangers of a prolonged war in Vietnam.[153] In an October 1971 meeting with Zhou, Kissinger would argue,

> A Japan which defends itself with its own resources will be an objective danger to all countries around it because it will be so much more powerful. Therefore, I believe that its present relationship with the U.S. is actually a restraint on Japan. If we wanted to pursue a cynical policy, we would cut Japan loose and encourage it to stand on its

[149] "Memcon, Kissinger and Zhou: President's Visit, Taiwan, and Japan," document no. 11, pp. 20, 31–33.

[150] Memorandum of Conversation February 22, 1971," Nixon's Trip to China, document no. 5, p. 5.

[151] Memcon, February 24, 1972," Nixon's Trip to China", document no. 3, pp. 12–13.

[152] Holdridge, *Crossing the Divide* 60; "Memorandum for Henry A. Kissinger from Winston Lord. Subject: Memcon of Your Conversation with Chou En-lai," July 29, 1971, document no. 34, p. 42. Kissinger reassured Zhou that the United States did not want to see Japan "heavily re-armed," because this could lead to the expulsion of U.S. forces from Japan, something not in the interests of the PRC or the United States.

[153] "Memorandum for the President's Files, "Meeting Between President, Dr. Kissinger, and General Haig," July 1, 1971" in William Burr, ed., "The Beijing-Washington Back Channel and Henry Kissinger's Secret Trip to China," *National Security Archive Electronic Briefing* Book No. 66, February 27, 2002, document no. 33, pp. 1–2. On these points, also see Xia, *Negotiating with the Enemy*, 164.

own feet, because this would cause so much tension between Japan and China, and we could come between them. That would be very shortsighted. Either you or we would be the victim."[154]

He went on to state that the United States would oppose nuclear weapons development in Japan, encourage defensive conventional military development, and discourage Japan from the "extension of military power to Taiwan, Korea, or elsewhere." Kissinger would emphasize that Japan could easily decide to develop nuclear weapons without a strong U.S.-Japan alliance.[155]

Kissinger's impression, reflected in a memo to the president upon his return, was that his arguments were effective in making Zhou begin to rethink the advantages to China of a revitalized U.S.-Japan alliance. But he expressed concern that the U.S.-Japan-PRC triangle could become "one of our most difficult problems."[156] In briefing President Nixon for his trip to China in February 1972, Kissinger stated, "They don't want us out of Taiwan precipitously only to see the Japanese move in. They don't want us to withdraw from Asia generally and leave the field for Moscow."[157]

In Beijing in February, Nixon picked up this theme in conversations with Zhou, insisting that "the interests of China are just as great as those of the U.S." in a continued U.S. military presence in Japan and Europe, and sustained U.S. naval forces in the Pacific.[158] Playing directly on Chinese fears of a more aggressive Japan, he added later that "where Taiwan is concerned and where Korea is concerned, the U.S. policy is opposed to Japan moving in as the U.S. moves out, but we cannot guarantee that. And if we had no defense arrangement with Japan, we would have no influence where that is concerned."[159] Appealing to the CCP's Marxist and realist tendencies relating to economic determinism, Nixon went further, claiming that the alliance needed to be strengthened because "if they [the Japanese] are left alone as an economic giant and a military pygmy the inevitable result, I think, will be at this point to make them more susceptible to the demands of the militarists."[160] During his visit, Nixon recognized the Chinese and American differences over Japan early in his conversations with Mao. He

[154] Memcon, Kissinger and Zhou. "Korea, Japan, South Asia, Soviet Union, Arms Control," document no. 13, p. 24

[155] Ibid., document no. 13, pp. 24–25.

[156] "Kissinger to Nixon," November 11, 1971, document no. 20, pp. 4 and 24, quotation on p. 4.

[157] "Kissinger to Nixon," "Your Encounter with the Chinese," February 5, 1972, Burr et al., eds., "Negotiating U.S.-Chinese Rapprochement," *National Security Archive Electronic Briefing* Book No. 70, no. 27, p. 2.

[158] Memorandum of Conversation, Tuesday, February 22, 1972, Nixon's Trip to China, document no. 1, p. 9.

[159] Ibid., Nixon's Trip to China, document no. 1, p. 12.

[160] Memcon, February 23, 1972, 2:00-6:00 pm, Nixon's Trip to China, document no. 2, p. 19.

attempted, however, to explain the benefits of America's Japan policy for China within the context of a Soviet threat to China that outstripped in certain ways the Soviet threat to Europe. He argued that a neutralist Japan would create a dangerous vacuum that someone (i.e., the Soviets) could fill.[161]

Henry Kissinger and Richard Nixon both argued that they convinced Beijing that the U.S.-Japan alliance was in China's national interest.[162] Though it is common for leaders to exaggerate their own historical influence, in this case the available evidence supports their claim. After their visits, China apparently did think in more realpolitik and less emotional terms about both the United States and Japan within the context of the apparent shift occurring in the US-Soviet balance of power, particularly in Asia. Beijing adjusted its attitudes about the U.S.-Japan alliance in two significant ways. Fearing the implications of a potential U.S. military drawdown in the region and the Nixon Doctrine, Beijing became more comfortable with a continued U.S. presence and active leadership in the U.S.-Japan alliance. The second, and arguably more dramatic, change was Beijing's grudging acceptance of Japanese military strengthening as part of a burden-sharing role with the United States. Four factors contributed to these important shifts in Beijing's perceptions: following the border clashes in 1969 and subsequent Soviet buildups in the region, China feared the Soviet Union much more than it feared the United States; because of reassuring military and political gestures from the United States, including a halt of naval patrols in the Taiwan Strait in November 1969, China worried less about the U.S. fleet based in Japan being deployed to harm China's interests in maintaining stable cross-Strait relations; China became increasingly worried about the prospects of a thawing of relations between Moscow and Tokyo, especially if the latter's relations with the United States were to become more tenuous; and, finally, China feared that a Japan less tethered to the United States would remilitarize very quickly and without restraint on nuclear weapons and power projection capabilities even if a more independent Japan continued to oppose the Soviets on its own.[163] Given the twin fears of Soviet hegemony and American atrophy, the American presence in Japan appeared preferable to Japanese neutralism, a Japanese-Soviet condominium, or a nuclearized, militarily muscular, and diplomatically independent Japan. A Japan that built up its military capabilities more modestly as part of a strengthened U.S.-Japan alliance was preferable in Beijing to these alternative scenarios.

[161] Wang Yulin. *DaGuo de Duikang: Ershi Shiji Zhongmei Guanxi* [Great Power Confrontation] (Beijing: Shishi Chubanshe, 1997), 913.

[162] Kissinger, *White House Years*, 1089. Richard M. Nixon, *RN: The Memoirs of Richard Nixon* (New York: Grosset & Dunlap, 1978), 567.

[163] In January 1972, Mao told the visiting Alexander Haig that he was concerned that the Gromyko visit to Tokyo was part of a Soviet strategy to encircle (*baowei*) China. See Wang, ed., *DaGuo de Duikang*, 3:901.

A shift in attitudes toward Japanese military strengthening became evident in Chinese public statements in spring 1972, following the Nixon visit and the February 1972 passage of Japan's "Fourth Defense Build-up Plan."[164] In early 1973, Zhou Enlai stated that "inasmuch as Japan constitutes a nation, weapons are essential to it."[165] In late 1973, top Japan policy advisor Liao Chengzhi and Foreign Minister Ji Pengfei would similarly make positive statements about the U.S.-Japan alliance.[166] By 1975, internally circulated CCP educational materials were practically lauding Japanese military efforts as necessary counters to the Soviet threat in the Western Pacific. The degree of empathy for Japanese security concerns in these materials shows a tremendous intellectual journey from the time of the Nixon-Sato communiqué.[167] It is also notable that retrospective internal PLA writings on those Japanese buildups do not suggest any opportunism on the part of Japanese military planners in the late 1970s and early 1980s.[168]

This second shift in Chinese attitudes is more significant because the notion of Japanese military power plays not only on China's realist, great power concerns, but also on historically rooted, emotional fears about the aggressiveness of "Japanese militarism," a phenomenon believed to be lurking beneath the surface of even outwardly pacific Japanese governments. Japanese leaders in the 1970s were very sensitive to this Chinese view. In fact, Makato Momoi argues that the full budget for Defense Minister Nakasone's "Fourth Defense Build-up Plan," first drafted in 1970, was accepted only in late 1972 after Prime Minister Tanaka returned from Beijing, where he learned to his surprise that the Japanese buildup would not be overly upsetting to Beijing.[169] One major reason for the marked improvement in Sino-Japanese relations on this score was the implicit understanding of the United States and China regarding Taiwan. A second reason was that the newly elected Japanese Prime Minister Tanaka was much more eager than his predecessor, Sato, to improve relations with China and seemed much less likely to take provocative actions in Taiwan policy. So the more worrisome clauses of the 1969 Nixon-Sato communiqué referring to

---

[164] Allen S. Whiting, *China Eyes Japan* (Berkeley: University of California Press, 1989), 2. Akira Iriye, "Chinese-Japanese Relations," in Christopher Howe, ed., *China and Japan: History, Trends, Prospects* (Oxford: Clarendon Press, 1996), 50.

[165] Barnett, *China and the Major Powers*, 115.

[166] Ibid., 115–16.

[167] "Riben Fandui Chaoji Daguo Kongzhi, Weixie he Qifu de Douzheng," [The Japanese Struggle to Oppose Superpower Control, Threats, and Bullying"] *Neibu Jiaoxue Cankao Ziliao* [Internal Educational Reference Materials], Beijing University, International Politics Department, 1975.

[168] Pan Sifeng, *Riben Junshi Sixiang Yanjiu* [*Japanese Military Thought*], ch. 14.

[169] Momoi, "Basic Trends in Japanese Security Policies," 348. The budget for the Fourth Plan was eventually struck down in late 1973, but for reasons related to the Oil Crisis and Japan's need for fiscal and energy conservation, not because of Chinese resistance.

Taiwan and Korea no longer greatly concerned Mao and Zhou, at least in the near term.[170]

Although we do not have the detailed minutes of the Chinese leaders' meetings on these issues, it is safe to assume that it took a very big threat to override the anti-Japanese biases of the Chinese leadership. The special comparative advantage of the Soviet Union was its ability to scare others and, in the process, help assemble a balancing coalition that eventually included not only communist China, Japan, and the United States, but other unlikely coalition-mates such as Israel and Muslim fundamentalists in Afghanistan. In fact, it seems that Gromyko's visit to Tokyo in January 1972 was so disturbing to Mao, that Nixon's tutoring efforts on the benefits of the U.S.-Japan alliance for China were made much easier.[171] Judging from internally circulated educational materials from late 1971, the CCP did not need much prodding to begin worrying about the prospect of a Japanese-Soviet condominium. Two months before the Nixon visit and one month before the Gromyko visit to Tokyo, one internal CCP report states that, given the economic frictions between the United States and Japan, the security relationship between the two could become poisoned. Moreover, the report continued, signs of Japanese-Soviet rapprochement and potential collusion to contain China were already arising from growing U.S.-Japan tensions.[172] It was in this context that Nixon and Kissinger were able to reassure Beijing on both Taiwan and Japan simultaneously while gaining Beijing's acceptance of a continued U.S. military presence in the region, the maintenance of the U.S.-ROC alliance for the time being, and an increase in Japanese military spending and capabilities as part of a continuing U.S.-Japan alliance. This process took time and the initial delay in U.S.-PRC rapprochement following the devolution of Sino-Soviet rivalry into Sino-Soviet conflict in 1968–69 can be attributed most clearly to continuing Chinese concerns about the U.S. relationship with two regional allies—the ROC and Japan—in the period 1969–72. There were, of course, other aggravating factors that delayed the process, such as the U.S. invasion of Cambodia in 1970, but the alliance-based hurdles discussed in this chapter seemed the most important for the two sides to cross.

---

[170] See Marius B. Jansen, *Japan and China: From War to Peace, 1894–1972* (Chicago: Rand McNally College Publishers, 1975), 504–5.

[171] See Wang, ed., *DaGuo de Duikang*, 901. China also criticized Japan and the Soviet Union for supporting internationalization of the Malacca Strait. See Rajendra Kumar Jain, *China and Japan, 1949–76* (Atlantic Highlands, N.J.: Humanities Press, 1977), 98.

[172] See "Riben Fandongpai Qin Mei Lian Su Fan Hua, Zheng Ba Yazhou de Waijiao Luxian," (Neibu Jiaoxue Cankao Ziliao), [The Pro-America, Align with the Soviets, Oppose China, Struggle for Asian Hegemony Diplomatic Line of the Japanese Reactionary Faction] (Beijing University, International Politics Department, December 1971); internally circulated educational reference materials.

Chapter 7

THE FALL AND REVIVAL OF COERCIVE DIPLOMACY: SECURITY PARTNERSHIPS AND SINO-AMERICAN SECURITY RELATIONS, 1972–2009

The legacies of U.S. regional alliances, especially with the Republic of China (Taiwan), continued to complicate U.S.-PRC relations from the time of rapprochement in 1971–72 to full normalization of relations in 1979, but the dynamics analyzed in this study were not prevalent in either that period or the longer one between normalization and the end of the Cold War in 1991. The United States and the PRC were for the most part aligned on the same side in the Cold War, and coercive diplomacy was not a major issue in the two states' relations with each other. Nor were trends in the U.S.-Japan alliance particularly sensitive in Beijing, as they were in the first two decades of the Cold War. Following the switch of recognition from Taipei to Beijing in 1979 and the abolition of the U.S.-ROC alliance relations, the issue of continuing security relations between the United States and Taiwan in accordance with the Taiwan Relations Act of 1979 remained very sensitive in Beijing, but this served more as an irritant in a generally improving and cooperative economic and political relationship between the United States and the PRC than a potential source of crisis and conflict between the two states.

With the collapse of the shared Soviet threat in 1991 and shifts in Taiwan politics toward assertions of Taiwan independence from the Chinese mainland in the 1990s, Chinese elites became much more sensitive about U.S. security relationships in the region, especially U.S. policy toward former and current allies in Taiwan and Japan, respectively. It is critically important to note, however, that U.S.-PRC relations from 1991 to the present were incomparably better in the first two decades of the post–Cold War era than they were in the first two decades of the Cold War, when the two countries were sworn enemies. Despite these large differences, there are some similarities between the two periods. Coercive diplomacy, particularly on issues related to Taiwan, has returned to U.S.-China relations for the first time since the early 1970s. As was the case in both the 1950s and the early 1970s, the United States is trying to revise and refine its burden-sharing relationship with Tokyo in a period of fast-paced international change. As in the earlier period, Beijing's fears about changes in the U.S.-Japan alliance in a period of flux have fed into long-term concerns about the potential for conflict across the Taiwan Strait.

In the first decade of the twenty-first century, the United States was able to reassure Beijing that it did not support Taiwan independence and would oppose provocative unilateral assertions by Taiwan in that direction, even as it continued to offer arms sales to Taiwan and to oppose the mainland's buildup across the Taiwan Strait and its refusal to eschew the use of force in cross-Strait relations. Washington thereby bolstered stability in both cross-Strait relations and U.S.-PRC relations. Especially after pro-independence parties and initiatives in Taiwan fared poorly in the 2008 elections, Beijing became more relaxed about cross-Strait relations and about trends in the U.S.-Japan alliance. Largely because of domestic politics in Japan, during the first two years of the Obama administration, there have been major new challenges in the U.S.-Japan security relationship related to burden-sharing. In a different context, these problems might have complicated U.S.-PRC-Japan triangular relations. But the dramatic improvement in relations between the mainland and Taiwan has seemingly rendered Beijing much less concerned about the most recent period of flux in the U.S.-Japan alliance than it had been in the past.

THE END OF COLD WAR COERCIVE DIPLOMACY: SINO-AMERICAN RELATIONS 1972–91

Following Sino-American rapprochement in 1972, and especially after normalization of bilateral diplomatic relations in 1979, many of the problems discussed in the first six chapters of this book dissipated, at least for the remainder of the Cold War. Despite some public pronouncements in China regarding "an independent line" between the superpowers in the early 1980s, for the remainder of the Cold War the People's Republic of China remained firmly aligned, though not allied, with the United States in an anti-Soviet coalition. Moreover, the PRC underwent a major domestic transformation beginning in the late 1970s. Deng Xiaoping consolidated his power after Mao's death in 1976 and, in late 1978 launched an ambitious and successful economic reform program that emphasized pragmatism over ideology, market incentives over communism, and foreign investment and trade over autarky. Having eschewed radical leftism even at home, Chinese leaders no longer had any incentive to compete with the Soviet Union for leadership in spreading communist revolution abroad. Moreover, since China and the United States were in the same broad anti-Soviet camp, there really was no direct coercive diplomacy between the two sides, and alterations or adjustments in U.S. security policies in East Asia were extremely unlikely to create Sino-American conflict in the period 1979–91.

Washington's transfer of formal diplomatic recognition from the ROC in Taipei to the PRC in Beijing in January 1979, together with the abolition of the U.S.-ROC Mutual Defense Treaty the following year, was over seven years in the making, dating back to Kissinger's secret visit in July 1971. During that period, U.S. relations with its ally on Taiwan remained the single most sensitive issue and biggest stumbling block on the road to formal diplomatic relations between Washington and Beijing. Tokyo and Beijing sealed such relations as early as 1972 because the Japanese government under Prime Minister Tanaka was more able to navigate the domestic politics surrounding compromise with the PRC on Taiwan policy than were Japan's American allies. The United States remained committed to a political and security relationship with Taiwan that would delay normalization with the PRC. That security relationship would be codified into law in the Taiwan Relations Act of April 1979 even after normalization was secured on January 1, 1979. The ongoing U.S. commitment to Taiwan includes the sale of defensive military equipment, as U.S. officials deem necessary, to help Taiwan maintain the wherewithal to defend against both invasion and coercion from the mainland, and a commitment to maintain sufficient U.S. capabilities in the theater to come to Taiwan's military assistance if the president were so to choose.[1]

Washington's arms sales to Taiwan would remain the most sensitive in U.S.-PRC bilateral relations for the remainder of the Cold War and the issue would come to a head in 1982, when the PRC pressured Washington to accept a third communiqué on the issue of arms sales.[2] On the surface, in that communiqué the United States appeared to commit to reducing arms sales, with the ultimate purpose of eventually halting them. As with previous communiqués, however, the August 1982 version provided enough ambiguity to allow the United States to continue to sell defensive arms to Taiwan if tensions in cross-Strait relations did not decrease or if military developments in the PRC posed an increased security threat to Taiwan.[3]

Beijing never viewed U.S. arms sales to Taiwan as legitimate, and they proved an important irritant in bilateral relations in the last decade of the Cold War. But given common U.S.-PRC geostrategic interests and the improving relations across the Taiwan Strait, U.S. arms sales decisions could not cause serious tensions, let alone conflict, in the bilateral relationship. After all, in the 1980s the United States and the PRC were increasing their cooperation in the Cold War, particularly in blunting Soviet power in Afghanistan. Moreover, bilateral economic relations were blossoming and, at least until the Tiananmen massacre of

---

[1] For classic accounts, see Harding, *A Fragile Relationship*, chs. 3–4; Ross, *Negotiating Cooperation*, chs. 3–6; and Romberg, *Rein In at the Brink of the Precipice*, chs. 4–6.

[2] Harding, *A Fragile Relationship*, ch. 6; Romberg, *Rein In at the Brink of the Precipice*, ch. 6; Ross, *Negotiating Cooperation*, ch. 6; and Mann, *About Face*, ch. 6.

[3] For a blow-by-blow history that describes how the Reagan administration never intended to leave Taiwan vulnerable to mainland coercion by agreeing to the August 1982 communiqué, see Mann, *About Face*, ch. 6. For an insider's look at China policy at this time, see James Lilley with Jeffrey Lilley, *China Hands: Nine Decades of Adventure, Espionage, and Diplomacy in Asia* (New York: Public Affairs, 2004), part 4, esp. ch. 14.

June 1989, Beijing was viewed in Washington as moving in a positive direction at home. Moreover, cross-Strait relations were improving in the 1980s under the leadership of Chiang Kai-shek's son, Chiang Ching-kuo. Since there really was not much occasion for coercive diplomacy between Washington and Beijing or between mainland China and Taiwan, we did not see the kind of security dilemma dynamics that could have been exacerbated by aspects of the U.S. relationship with either Japan or Taiwan.

The End of the Cold War and the Return of U.S.-PRC Coercive Diplomacy

This stable reality would not survive the collapse of the Soviet Union and, with it, the collapse of a common threat that helped bind the United States, Japan, and the PRC. Even though Sino-American bilateral relations were still much better in the 1990s than they were in the first half of the Cold War, from a U.S. perspective some of the core challenges of coercive diplomacy returned. The political blowback in the United States from the Tiananmen Massacre, and the democratization of Taiwan further complicated the picture. Beijing lost its reform-era luster almost overnight just as Taiwan became more attractive to the United States through the legalization of a major opposition party and the holding of competitive elections. Moreover, upon his death in 1988, Chiang Ching-kuo would be succeeded by the first Taiwan-born President of the Republic of China, Lee Teng-hui, a skilled politician who would both reflect and lead the growing sentiment on the island throughout the 1990s for a sovereign existence entirely separate from mainland China. Such sentiment, always there but repressed during the authoritarian period, would find full voice in the new competitive politics on the island.

In this geostrategic and cross-Strait context, Beijing's traditional Cold War concerns about the United States relationship with both Taiwan and Japan would return after two decades of absence. The controversy over continued U.S. arms sales to Taiwan and the prospect of U.S. intervention in a cross-Strait conflict would take on new meaning to the bilateral relationship. Similarly, any perceived or projected changes in the U.S.-Japan alliance were viewed in Beijing with much more concern than at any time since 1972. China's increased attention to Taiwan occurred at a time in which the PLA was developing for the first time conventional military capabilities designed to coerce the island over a sustained period of time. These new capabilities were also apparently designed to raise the costs of U.S. and Japanese assistance to the island during a crisis or war.[4]

---

[4] For a more complete discussion of these capabilities and the logic behind their development, see Thomas J. Christensen, "Posing Problems Without Catching Up: China's Rise and Challenges for U.S. Security Policy," *International Security* 25, no. 4 (spring 2001): 5–40.

At the same time, Beijing grew increasingly sensitive about perceived slights from Taiwan and concern grew in elite circles there about the prospect that long-term trend lines might be leading toward Taiwan independence if the PRC did not prepare to take remedial coercive action to delay, halt, or reverse those trends. There was a tradition in the PRC, demonstrated first in 1954–55 in the Taiwan Strait and repeated later, to use limited force to send a coercive message about the costs enemies would pay if they continued down their current path.[5] If this pattern were to hold in the post–Cold War world, an initially limited military conflict could occur without either a single, obvious red-line having been crossed by Taiwan (e.g., a formal, legal declaration of independence by Taiwan) or a near-term vision in Beijing of how to solve the outstanding problems in cross-Strait relations once and for all through a single military campaign (e.g., a full-scale invasion and occupation of Taiwan).

Deterring such a limited, coercive attack on Taiwan requires a delicate balance in coercive diplomacy for Washington, Taipei, and Tokyo. On the one hand, Beijing needs to be convinced and reminded that any mainland attempt to use force would fail at great cost to the PRC. Such a strategy requires not only being able to defeat PRC forces in a toe-to-toe struggle but maintaining sufficient superiority to convince Beijing that it cannot easily and cheaply raise the costs of conflict for the United States and its partners. At the same time Beijing needs to be assured that if it does not use force against the island, the PRC's key interests in preventing Taiwan's sovereign independence would not be harmed now or in the future.[6] Complicating the picture is the perceived need to adjust burden-sharing roles in the post–Cold War era and to bolster Taiwan's native defenses in the face of the mainland's military modernization and, more specifically, the military buildup across from Taiwan in the Nanjing Military District.

The United States and its security partners have been successful in finding a proper balance in coercive diplomacy under what have been generally challenging conditions: when Taiwan domestic politics were clearly moving in a pro-independence direction. The United States continued to abide by the commitments in the Taiwan Relations Act in the face of the PLA buildup and to warn against the use of force to solve cross-Strait differences. The U.S.-Japan alliance was fortified to adjust to the post–Cold War environment that included

[5] For a detailed argument about the importance of trend analysis in the history of PRC coercive diplomacy, see Thomas J. Christensen, "Windows and War: Changes in the International System and China's Decision to Use Force," in Alastair Iain Johnston and Robert Ross, eds. *New Approaches to China's Foreign Relations: Essays in Honor of Allen S. Whiting* (Stanford: Stanford University Press, 2006).

[6] For a lengthier explication of this logic, see Thomas J. Christensen, "The Contemporary Security Dilemma: Deterring a Taiwan Conflict," *Washington Quarterly* 25, no. 4 (autumn 2002): 722.

challenges from North Korea, a changing balance of power in the region and, beginning in 2001, the Global War on Terror. These alliance adjustments included specific mention of the importance of peace and stability in the Taiwan Strait in the joint statement of the U.S.-Japan alliance in 2005. Strengthening of the alliance arguably had positive implications for deterrence, as it seemed less likely that Beijing would be able to separate the two allies in a time of crisis. But there was a concurrent danger that adjustments in Washington's relations with Tokyo might undercut the critically important assurance component of coercive diplomacy. Fortunately, at key junctures, Washington and Tokyo were able to convey to Beijing that the point of the revitalized U.S.-Japan alliance was not to contain the growth of Chinese influence in the region and around the world, nor to promote a unilateral declaration of Taiwan independence.

THE U.S.-JAPAN ALLIANCE AND THE SECURITY DILEMMA IN SINO-AMERICAN RELATIONS IN THE 1990s

To international relations scholars of both the liberal and realist persuasion, East Asia appeared more dangerous in the immediate aftermath of the Cold War than did other areas of potential great power competition, such as Western Europe. The common bond that the Soviet Union had provided for the U.S.-PRC and even for the Sino-Japanese relationship had been removed. East Asia was characterized not only by potentially destabilizing shifts in the balance of power (especially between a rising China and its neighbors), but unlike Western Europe, it was also saddled with skewed distributions of economic and political power within and between countries, political and cultural heterogeneity, growing but still relatively low levels of intraregional economic interdependence, anemic security institutionalization, and widespread territorial disputes that combine natural resource issues with postcolonial nationalism.[7]

Beijing had to be concerned about two new potential scenarios that could harm its security environment: a shift in the U.S. relationship with Taiwan, which itself was undergoing significant domestic changes; and a shift in U.S. policy toward its regional allies, especially Japan. For its part, the United States needed to think about how to adjust and bolster its alliances in the post–Cold War world and to address remaining potential challenges on the Korean penin-

---

[7] Aaron L. Friedberg, "Ripe for Rivalry: Prospects for Peace in a Multipolar Asia," *International Security* 18, no. 3 (winter 1993/94): 533; Richard K. Betts, "Wealth, Power, and Instability," *International Security* 18, no. 3 (winter 1993/94): 34–77; Stephen Van Evera, "Primed for Peace: Europe after the Cold War," *International Security* 15, no. 3 (winter 1990/91): 757; and James Goldgeier and Michael McFaul, "A Tale of Two Worlds," *International Organization* 46, no. 2 (spring 1992): 467–492.

sula and in the Taiwan Strait, as North Korea seemed dedicated to developing nuclear weapons and China began modernizing its military and building up coercive capabilities across from Taiwan.

Back to the Future: China's Early Post–Cold War Fears about Japan

In the first several years after the end of the Cold War I did extensive interviews in Beijing and Shanghai about Chinese security perceptions toward Japan and the United States.[8] In the early 1990s, Chinese security analysts expressed many of the same concerns about Japan that Zhou Enlai expressed to Nixon administration officials in 1971–72. In the early 1970s the defeat in Vietnam and the Nixon administration's strategy of relying more on Asians to provide for their own security led Chinese elites to fear a resurgent Japan that, over time, would likely challenge PRC security interests, particularly in and around Taiwan. As discussed in chapter 6, only guarantees from the United States about the sustainability of the U.S.-Japan alliance and Washington's commitment to attempt to restrain Japan from a hawkish turn in the future would reassure Beijing about Japan's future role within the Cold War alliance. In the early years of the Clinton administration, it was U.S. victory in the Cold War and the initial expectation of a reduction of forward deployed U.S. forces in Asia that led many Chinese analysts to fear a similar prospect. As did Zhou two decades earlier, Chinese security analysts, particularly military officers, feared that under those conditions Japan could again assert itself as a formidable military power in the first quarter of the twenty-first century. They believed that it was only natural that Japan would want to graduate from a great economic power with some real international political clout, to a nation with the full panoply of great power accessories: economic, political, and military power.[9] Applying some of the same geopolitical economic determinism of Chinese elites in the early 1970s, more pessimistic Chinese analysts often stated that Japan's material interests have not changed much from the 1930s to the present. They believed that, because Japan is still heavily dependent on foreign trade and investment, it could again choose to develop power projection capabilities designed to protect its economic interests far from home.[10]

---

[8] For a full review of this research, see Thomas J. Christensen, "China, the U.S.-Japan Alliance and the Security Dilemma in East Asia," *International Security* 23, no. 4 (spring 1999): 49–80.

[9] Christensen, "China, the U.S.-Japan Alliance and the Security Dilemma in East Asia." Also see Banning Garrett and Bonnie Glaser, "Chinese Apprehensions about Revitalization of the U.S.-Japan Alliance," *Asian Survey* 37, no. 4 (April 1997): 383–402.

[10] Vigilant about this possibility, Chinese analysts reacted negatively to even mild new Japanese initiatives away from the home islands (such as sending peacekeepers to Cambodia or minesweepers to the Persian Gulf after the first Gulf War). This argument was made particularly forcefully in my interviews with three military officers in 1994. See also Pan, *Riben Junshi Sixiang Yanjiu*, 502–3; and Wu Peng, "Riben Weihe Jianchi Xiang Haiwai Paibing" [Why Japan Insisted on

Especially in the minds of my most pessimistic interlocutors in Beijing, the only thing that had prevented this from happening earlier was the protective umbrella supplied by the rather lopsided U.S.-Japan alliance, which allowed Japan to enjoy security at low cost and allowed moderates in Japan to keep more hawkish elements of Japanese politics at bay. But many experts in China feared that the loss of a common foe, growing economic tensions in U.S.-Japan relations at the beginning of the 1990s, and the early Clinton administration's focus on "economic security," might strain the U.S.-Japan alliance, leaving Japan more to its own devices. Alternatively, the United States might simply begin to ask Japan to carry a much bigger burden in the U.S.-Japan alliance and thereby transform Japan's attitudes about defense spending and the projection of military power. Eventually, they believed, this process would lead to a more independent and dangerous Japanese neighbor for China. Especially given the emotional history between Japan and China, the notion of increased Japanese power and independence is never welcome in Beijing. But the prospect of a more assertive Japan down the road was particularly troublesome at a time when Taiwan was led by Lee Teng-hui, the Republic of China's first Taiwan-born leader who enjoyed impeccable Japanese language skills and had friends in Japan's most staunchly pro-Taiwan and anti-PRC factions.[11]

Chinese Assessments of Japanese Military Power and Potential

In assessing Japan's current military strength in the early 1990s, Chinese analysts emphasized the advanced equipment that Japan has acquired, particularly since the late 1970s, when it began developing a navy and air force designed to help the United States contain the Soviet Union's growing Pacific fleet. Chinese military writings highlighted Japanese antisubmarine capabilities (such as the P-3C aircraft), advanced fighters (such as the F-15), the E-2 advanced warning aircraft, Patriot air defense batteries, and Aegis technology on surface ships.[12] Chinese analysts correctly pointed out that, excluding U.S. deployments in the region, these weapons systems constituted the most technologically advanced arsenal of any East Asian power. They also cited the Japanese defense budget, which, although small as a percentage of gross national product (GNP), was second only to U.S. military spending in absolute size (until Chinese spending overtook it at some point during the past ten years).[13]

Sending Forces Abroad], *Shijie Jingji yu Zhengzhi* [World Economy and Politics] no. 12 (December 1992): 46–50.

[11] Lee's father was a local official on Taiwan in the days of Japanese colonialism. Lee was schooled during the Japanese occupation in Taiwan and served briefly in a noncombat branch of the Japanese Army toward the end of the Pacific War.

[12] See, for example, Pan, ed., *Riben Junshi Sixiang Yanjiu*, 388–92; internally circulated.

[13] Multiple interviews, 1993–98. It is difficult to say exactly when China's defense spending overtook Japan's defense budget because of lack of transparency in PRC defense spending. At the

Despite their highlighting of Japan's existing defense budget and high levels of military sophistication, Chinese analysts in the early post–Cold War era understood that Japan could easily do much more militarily if it were so to choose. While they generally did not believe that Japan had the requisite combination of material capabilities, political will, and ideological mission to become a Soviet-style superpower, they did believe that Japan could easily become a great military power (such as France or Great Britain) in the first decades of the twenty-first century with power projection capabilities, strike weapons, and, perhaps, even a nuclear arsenal. On the one hand, analysts often argued that it was in Japan's economic interest to continue to rely on U.S. military protection in the near future, but on the other hand, they did not think that significantly increased military spending would strongly damage the Japanese economy.[14] They expressed deep suspicions about the massive stockpiles of high-grade nuclear fuel that was reprocessed in France and shipped back to Japan in the early 1990s. Many in China view Japan's acquisition of this plutonium as part of a strategy for the eventual development of nuclear weapons, something, they point out, Japanese scientists would have little difficulty producing.[15] Chinese security analysts also have stated that Japan could become a great military power even if it were to forgo the domestically sensitive nuclear option. In the mid-1990s Chinese military and civilian experts emphasized that nuclear weapons may not be as useful in the future as high-tech conventional weapons, and that Japan was already a leader in dual-use high technology.[16]

Chinese experts recognized that Japan practiced a great deal of self-restraint during the Cold War, in particular by eschewing weapons designed to project power far from the home islands. For example, in 1996 one military officer stated that despite the long list of current Japanese capabilities mentioned previously, Japan certainly was not yet a normal great power because it lacked the required trappings of such a power (e.g., aircraft carriers, nuclear submarines, nuclear weapons, and long-range missile systems).[17] For this officer and many of his compatriots, the question was simply a matter of if and when Japan would

---

time of this writing, the Chinese official budget is larger than Japan's, but China's actual defense spending probably outstripped Japan's in the early part of this decade at the latest.

[14] In 1992 an internally circulated analysis of Japan's military affairs points out that Japan could easily spend 4 percent of GNP on its military without doing fundamental harm to its long-term economic growth. The examples of much higher levels of spending in healthy economies in the United States and Europe during the Cold War are cited as evidence. See Pan, *Riben Junshi Sixiang Yanjiu*, 499. Similar positions were taken by active and retired military officers in 1996 and 1998.

[15] This was a particularly sensitive issue in 1993 and 1994.

[16] Multiple interviews, 1996. For written materials, see Gao Heng, "Shijie Junshi Xingshi" [The World Military Scene], *Shijie Jingji yu Zhengzhi* [World Economy and Politics], no. 2 (February 1995): 14–18. For a similar Western view on Japanese "technonationalism," see Richard J. Samuels, *Rich Nation, Strong Army: National Security and the Technological Transformation of Japan* (Ithaca: Cornell University Press, 1994).

[17] Interview, Beijing, 1996.

decide to adopt these systems. Chinese analysts therefore had reason to view as dangerous Japan's adoption of even additional defensive military roles as part of the U.S.-Japan alliance. It was not because these measures were in and of themselves particularly provocative, but because they might begin to erode the constitutional (Article 9) and nonconstitutional norms of self-restraint (e.g., 1,000-nautical-mile limit on power-projection capability, prohibitions on the military use of space, and tight arms export controls) that have prevented Japan from realizing its full military potential.[18]

Although almost all Chinese analysts in the 1990s expressed concern about the prospect of a Japanese military buildup, they differed in their assessment of the likelihood that Japan would attempt to realize its military potential over the next few decades. The more pessimistic argued that this outcome was extremely likely or even bound to happen. Even though they were aware of arguments about why Japan would continue to restrain itself—for reasons having to do with U.S.-Japan alliance politics, Japanese domestic politics, and Japan's economic interest—and did not dismiss them out of hand, some analysts viewed such obstacles to Japanese military buildups merely as delaying factors in a long-term and inevitable process. Other more conditionally pessimistic and cautiously optimistic analysts placed greater faith in the hypothetical possibility of preventing significant Japanese buildups and military assertiveness over the longer run, but expressed concern over the hardiness of the delaying factors that could theoretically prevent such buildups. The most optimistic analysts argued that these factors should remain sturdy and would prevent Japan from injuring its regional relations by pursuing a more assertive military role long into the future.[19] The vast majority of these optimists and pessimists believed that along with the domestic political and economic stability of Japan, the most important factor that might delay or prevent Japanese military buildups was the status of the U.S.-Japan relationship, particularly the security alliance.[20] The common belief in Beijing

[18] Interestingly, many Chinese analysts in the 1990s did not consider economic hard times in Japan to be particularly reassuring. On the contrary, in terms of intentions, some feared that economic recession and financial crises could improve the fortunes of relatively hawkish Japanese elites by creating a general sense of uncertainty and threat in Japanese society, by fueling Japanese nationalism more generally, and by harming relations with the United States (Japan's main provider of security). In terms of capabilities, some Chinese analysts argued that Japan's technological infrastructure, which would be critical to a modern military buildup, did not seem affected by Japan's economic woes in the past decade. This was a consistent theme in interviews from 1993 to 1998 and was repeated in 1998, during the financial crisis.

[19] The simplest versions of the most optimistic and most pessimistic forecasts about Japan's future were offered most frequently during my first three research trips from 1993 to 1995. After the Taiwan Strait crisis of 1995–96, one heard less often the most optimistic liberal argument that economic interests will trump security interests in the post–Cold War world. Following the 1995 Nye Report, one heard the simplest versions of the pessimists' scenarios less often because they were often predicated on fragility in the post–Cold War U.S.-Japan alliance.

[20] Interviews, 1993–98. See also Pan, *Riben Junshi Sixiang Yanjiu*, 501. This book states in typical fashion, of all the factors that could compel Japan's military policy to change, U.S.-Japan relations

security circles had been that by reassuring Japan and providing for Japanese security on the cheap, the United States had fostered a political climate in which the Japanese public remained opposed to military buildups and the more hawkish elements of the Japanese elite were kept at bay. If, however, the U.S.-Japan security alliance were to either become strained or undergo a transformation that would give Japan a much more prominent military role, Chinese experts believed that those ever-present hawks might find a more fertile field in which to plant the seeds for an eventually much more powerful and assertive Japan.[21]

THE CHINA-JAPAN SECURITY DILEMMA AND U.S. POLICY CHALLENGES

As discussed in the introductory chapter, problems of burden-sharing, abandonment, and entrapment in one alliance can create true security dilemmas in relations between the alliance and real or potential adversaries. During the Clinton administration, Chinese analysts expressed concern about almost any change in the U.S.-Japan alliance. In the early 1990s, the prospect of a breakdown of U.S.-Japan ties worried pessimists and optimists alike. Later in the decade, Chinese analysts of all stripes also worried to varying degrees whenever Japan adopted greater defense burden-sharing roles as part of efforts to revitalize the bilateral alliance. These dual and almost contradictory Chinese fears demonstrate what Snyder referred to as the complex security dilemma and posed major problems for Clinton administration officials. For their part, U.S. experts and officials saw a need for Japan to do more to shoulder the burden of the alliance: they often expressed concerns that the alliance was dangerously vague and out of date and was therefore unsustainable over the long run in its current form, but most still wanted the United States to maintain the reassurance role for Japan and its neighbors that was outlined in key documents such as the 1998 East Asia Pacific Strategy Report.[22] Especially before the 1996–97 Guidelines review, the

---

would be the deciding factor. See also Wang, ed., *Riben Junshi Zhanlüe Yanjiu*, 308–10; and Liu Shilong, "Dangqian Rimei Anbao Tizhi de San Ge Tedian" [Three Special Characteristics of the Current U.S.-Japan Security Structure], *Riben Yanjiu* [Japan studies], no. 4 (1996): 27. One article bases its optimism largely on the author's belief that, despite economic frictions, the U.S.-Japan alliance is stable. See He Fang, "Lengzhan Hou de Riben Duiwai Zhanlüe" [Japan's Post–Cold War International Strategy], *Waiguo Wenti Yanjiu* [Research on Foreign Problems] no. 2 (1993): 14.

[21] For an early discussion of the two very different potential paths to Japanese buildups, see Cai Zuming, ed., *Meiguo Junshi Zhanlüe Yanjiu* [Studies of American Military Strategy] (Beijing: Academy of Military Sciences Press, 1993), 218–33, internally circulated.

[22] For the logic of reassurance in official U.S. defense policy, see the Pentagon's *United States Security Strategy for the East Asia-Pacific Region 1998*, which states the following: "In addition to its deterrent function, U.S. military presence in Asia serves to shape the security environment to prevent challenges from developing at all. U.S. force presence mitigates the impact of historical regional tensions and allows the United States to anticipate problems, manage potential threats, and encourage peaceful resolution of disputes."

U.S.-Japan alliance had often been thought of in the United States as lopsided and unfair because the United States guaranteed Japanese security without clear guarantees of even rudimentary assistance from Japan if U.S. forces were to become embroiled in a regional armed conflict outside of that country.[23]

In the first years after the Cold War ended, some U.S. elites focused on economic security over military security, arguing that the alliance was overrated and that it had prevented the United States from pursuing its economic interests in the U.S.-Japan relationship. Some even argued that the United States should use the security relationship as leverage against Japan in an attempt to open Japanese trade and financial markets to American firms.[24] In their view Japan had been able to ride free for too long on the U.S. economy because of Washington's excessive concern over preserving an apparently unfair alliance relationship.

Following the North Korean nuclear crisis of 1993–94 and with the publication of the critically important February 1995 East Asia Strategy Report (also labeled "the Nye Report," after Assistant Secretary of Defense Joseph Nye), U.S. leaders began expressing very different concerns about the U.S.-Japan relationship. The Nye Report and the broader Nye Initiative of which it was a part placed new emphasis on maintaining and strengthening the security alliance and on keeping economic disputes from poisoning it. The report reaffirmed the centrality of U.S. security alliances in Asia, placed a floor on U.S. troop strength in East Asia at 100,000, and called for increased security cooperation between Japan and the United States, including greater Japanese logistical support for U.S. forces operating in the region and joint research and development of theater missile defenses (TMD).[25]

Despite the Clinton administration's decision to insulate the U.S.-Japan security relationship from economic disputes, the alliance still seemed potentially vulnerable. There had been a widely held concern that, purely on security grounds, the alliance could be dangerously weakened if Japanese roles were not

[23] This common view often ignores the clear benefits to the United States of the Cold War version of the alliance. The United States was guaranteed basing in Japan, and 70 to 80 percent of those basing costs were covered by the Japanese. Without this basing, the United States would have great difficulty maintaining its presence in the region. For a cost analysis, see Michael O'Hanlon, "Restructuring U.S. Forces and Bases in Japan," in Mike M. Mochizuki, ed., *Toward a True Alliance: Restructuring U.S.-Japan Security Relations* (Washington, D.C.: Brookings, 1997), 149–78.

[24] See Eric Heginbotham and Richard J. Samuels, "Mercantile Realism and Japanese Foreign Policy," *International Security* 22, no. 4 (spring 1998): 179.

[25] The Nye Report, named for former Assistant Secretary of Defense Joseph S. Nye, Jr., is titled "United States Security Strategy for the East Asia Pacific Region," Office of International Security Affairs, Department of Defense, February 1995. For an insider's look at concerns about how acrimonious economic disputes were harming the alliance, see David L. Asher, "U.S.-Japan Alliance for the Next Century," *Orbis* 41, no. 3 (summer 1997): 346–348.

clarified and expanded and if the two militaries were not better integrated in preparation for joint operations.[26] Japan's checkbook diplomacy in the Gulf War was considered insufficient support for U.S.-led efforts to protect a region that supplies Japan, not the United States, with the bulk of its oil. It also became clear during the 1993–94 crisis with Pyongyang over North Korea's nuclear weapons development that, under the existing Defense Guidelines, in a Korean conflict scenario Japan was not even clearly obliged to allow the U.S. military use of its civilian airstrips or ports. In fact, if the crisis had escalated, it was at least conceivable that Japan might not have provided overt, tangible support to U.S. forces operating outside of Japan. Even U.S. access to its bases in Japan for combat operations not directly tied to the defense of the Japanese home islands was seen by some as questionable.[27] Aside from the obvious military dangers inherent in such Japanese passivity, Japanese obstructionism and foot-dragging could undermine elite and popular support in the United States for the most important security relationship in East Asia. So, it appeared to many American elites that the Cold War version of the U.S.-Japan alliance could be one regional crisis away from its demise. Such concerns were a major driver behind the Nye Initiative, which was designed to clarify and strengthen Japan's commitment to support U.S.-led military operations. Fearing instability in Japanese elite and popular attitudes on defense issues, Washington also wanted to increase the number of functional links between the two militaries to tie Japan more firmly into the U.S. defense network for the long run.[28]

Chinese security analysts followed these trends in U.S.-Japan relations with great interest and concern. Before 1995 most pessimistic Chinese analysts predicted and feared Japanese military buildups largely because they sensed the potential for trouble, not strengthening, in the post–Cold War U.S.-Japan alliance. Those analysts posited that, given the lack of a common enemy and the natural clash of economic interests between Japan and the United States, politi-

[26] For discussion of these issues, see Mike M. Mochizuki, "New Bargain for a New Alliance," and "American and Japanese Strategic Debates," in Mochizuki, *Toward a True Alliance*, 5–40, 43–82, esp. pp. 35, 69–70.

[27] For the importance of the 1994 Korean crisis in officials' calculations, see Kurt M. Campbell, "The Official U.S. View," in Michael J. Green and Mike M. Mochizuki, *The U.S.-Japan Security Alliance in the Twenty-first Century* (New York: Council on Foreign Relations Study Group Papers, 1998), 85–87.

[28] For discussion of these issues, see Bruce Stokes and James Shinn, *The Tests of War and the Strains of Peace: The U.S.-Japan Security Relationship* (New York: Council on Foreign Relations Study Group Report, January 1998). For a strong critique of the weaknesses in the alliance in this century along the same lines, see Michael Finnegan, *Managing Unmet Expectations in the U.S.-Japan Alliance* National Bureau of Asian Research Special Report, November 2009. For the fear among U.S. officials in the 1990s that the Japanese public was moving away from support for the alliance in the 1990s, see Campbell, "The Official U.S. View."

cal conflict between the two allies was very likely. This conflict could eventually infect and destroy the U.S.-Japan security relationship, which in turn could lead to the withdrawal of U.S. forces and eventually Japanese military buildups. In this period some Chinese analysts also discussed how domestic factors such as U.S. neo-isolationism, rising Japanese nationalism, the inexperience and lack of security focus in the newly elected Clinton administration, and domestic instability in Japan could combine with worsening U.S.-Japan trade conflicts to speed the alliance's demise.[29]

By mid-1995 it seemed to an increasingly large group of Chinese analysts that U.S.-Japan trade conflict was being contained and that the Clinton administration was paying more attention to international security affairs and to Asia in particular.[30] Key contributors to this growing confidence in U.S. staying power were the Nye Report and the failure of the recent automobile parts dispute between Tokyo and Washington to escalate.

As one would expect when studying alliance security dilemmas, from a Chinese perspective, the news was far from all good, however. By spring 1996 the Nye Initiative led to harsh reactions in China, exposing the subtle challenges facing the United States in managing the U.S.-China-Japan triangle. China's cautious optimism about trends in the U.S.-Japan alliance turned to pessimism, as concerns about future Japanese military assertiveness grew rapidly. But the new reasons for pessimism were quite different than in the period before 1995. The fear was no longer potential discord in the U.S.-Japan relationship, but concern that the United States would encourage Japan to adopt new military roles and develop new military capabilities as part of a revitalized alliance in which Japan carried a greater share of the burden and risk.[31]

On April 17, 1996 President Clinton and Prime Minister Hashimoto issued a joint communiqué that called for revitalization of the alliance in order to better

[29] In particular, three military officers whom I interviewed in 1994 stressed these themes. For fears about Democrats and neo-isolationism, see Cai, *Meiguo Junshi Zhanlüe Yanjiu*, 223; and Liu Liping, "Jilie Zhendanzhong de Meiguo Duiwai Zhengce Sichao" [The Storm over Contending Positions on U.S. Foreign Policy], *Xiandai Guoji Guanxi* [Contemporary International Relations], no. 6 (1992): 15–18. For a similar argument made before Bill Clinton was elected president of the United States, see Li Shusheng, "Sulian de Jieti yu MeiRi zai Yatai Diqu de Zhengduo" [The Disintegration of the Soviet Union and U.S.-Japan Rivalry in the Asia Pacific], *Shijie Jingji yu Zhengzhi* [World Economy and Politics] no. 7 (July 1992): 56–58. For an article about the emphasis on trade and the lack of strategic focus in Washington, see Lu Zhongwei, "Yazhou Anquanzhong de ZhongRi Guanxi" [Sino-Japanese relations in the Asian Security Environment], *Shijie Jingji yu Zhengzhi* [World Economy and Politics] no. 3 (March 1993): 23–35, 42.

[30] Multiple interviews, 1995. For a published work arguing along these lines, see Yang Yunzhong, "Meiguo Zhengfu Jinyibu Tiaozheng dui Ri Zhengce" [Further adjustments in America's Japan Policy], *Shijie Jingji yu Zhengzhi* [World Economy and Politics] no. 7 (July 1995): 61–65.

[31] For elaborations of these arguments, see Garrett and Glaser, "Chinese Apprehension"; and Thomas J. Christensen, "Chinese Realpolitik," *Foreign Affairs* 75, no. 5 (September/October 1996): 37–52.

guarantee the security of the "Asia Pacific region." In the communiqué and in the agreements reached in the days preceding it, Japan guaranteed base access for U.S. forces and committed itself to increased logistics and rear-area support roles. In what might appear in retrospect to be a mild and rather vague commitment by Japan, the two sides also agreed to cooperate in the "ongoing study" of ballistic missile defense.[32]

However mild its wording might appear in retrospect, the communiqué created significant concerns among elite experts in Beijing at the time.[33] One major catalyst of Chinese concerns in the period between the issuing of the Nye Report in early 1995 and the Clinton-Hashimoto Communiqué in April 1996 was the notion that changes in the U.S.-Japan alliance could easily facilitate U.S. intervention in a Taiwan contingency. That contingency seemed more than just hypothetical following President Lee Teng-hui's visit to his alma mater, Cornell University, in June 1995, to give a highly political speech about Taiwan's role in the world. Perhaps more than anything else, the issuance of a visa to Lee by the United States, after apparent high-level guarantees that such a visa would not be issued, portended long-term negative trends in the cross-Strait relationship, and PRC leaders reacted with coercive missile and surface exercises in July 1995, December 1995, and, finally, March 1996. The last of these exercises sparked a Sino-American crisis, including the dispatch of two U.S. aircraft carrier battle groups to the waters near Taiwan. This crisis played out during Taiwan's first direct election for president, which Lee won handily.[34] Chinese analysts at this time believed that the United States was largely in control of the U.S.-Japan alliance's military policy, but they also worried that Japan had stronger emotional and practical reasons than the United States for opposing Taiwan's integration with the mainland and an even greater stake than the United States in issues such as sea-lane protection in Asia far from the Japanese home islands.[35]

[32] For the English language text of the April 17, 1996, Clinton-Hashimoto Joint Communiqué, see http://www.ioc.u-tokyo.ac.jp/~worldjpn/documents/texts/docs/19960417.D1E.html.

[33] Author interviews, Beijing, 1996.

[34] For authoritative coverage of the issuance of the visa to President Lee and the ensuing crisis by a former USG official, see Suettinger, *Beyond Tiananmen* (Washington, D.C.: Brookings Institution Press, 2003), ch. 6.

[35] Interviews, 1996 and 1998. Taiwan is a former Japanese colony (1895–1945). It is near international sea-lanes that are important to Japan. In addition, Chinese analysts argue that, for straightforward reasons relating to relative national power, Japan has a strategic interest in preventing Taiwan's high-technology and capital-rich economy from linking politically with the mainland. Moreover, some Chinese analysts view Taiwan as having geostrategic significance for Japan as a potential ally because of its location near the Chinese mainland. Another issue fueling mistrust of Japan is the feeling that Taiwan's president, Lee Teng-hui, who attended college in Japan and who speaks Japanese fluently, may be more pro-Japan than pro-China. For a particularly alarmist argument along these lines, see Li Yaqiang, "What Is Japan Doing Southward?" *Beijing Jianchuan Zhishi* [Naval and Merchant Ships] no. 6 (June 6, 1997): 7–8, in Foreign Broadcast Information Service Daily Report China, September 4, 1997. For a more sober analysis, see Yang Xuejun and Li

The Clinton-Hashimoto Joint Communiqué was issued one month after the most intense phase of the 1995–96 Taiwan Strait crisis. The communiqué suggested to some in China that Japan might soon begin scrapping various norms of self-restraint and begin expanding its military operations into the Taiwan area and the South China Sea. In addition to focusing on new logistics roles for Japan and the potential for future joint development of missile defenses, many Chinese observers believed (falsely) that the joint communiqué expanded the geographic scope of the alliance from the area immediately around Japan to a vaguely defined, but clearly much larger, "Asia Pacific."[36] It seemed that rather than restraining Japan, the U.S.-Japan alliance was starting to foster Japan's assertiveness. As one leading Chinese expert on Japan argued in 1998, the U.S. presence in Japan could be seen either as a "bottle cap," keeping the Japanese military genie in the bottle, or as an "egg shell," fostering the growth of Japanese military power under U.S. protection until it one day hatches onto the regional scene. Since 1996, this analyst argued, fears about the "egg shell" function of the U.S.-Japan alliance had increased markedly, while faith in the "bottle cap" function had declined.[37]

In September 1997 Chinese analysts' concerns turned to the announcement of revised Defense Guidelines for the U.S.-Japan alliance. These Guidelines put in writing many of the changes suggested in the joint communiqué. New and clarified Japanese roles in the alliance included those logistics and rear-area support roles mentioned in the joint communiqué and added "operational cooperation" missions for Japan's Self-Defense Forces in time of regional conflict, including intelligence gathering, surveillance, and minesweeping missions. Although Washington and Tokyo quickly abandoned the most vague and therefore most provocative term "Asia Pacific" following the issuance of the joint communiqué, the 1997 revised Guidelines were not entirely reassuring to Chinese ears on this score either. They stated that the scope of the alliance covered "situations in the areas surrounding Japan," but the definition of those areas would be determined by "situational" rather than "geographic" imperatives. This only confirmed fears among Beijing elites regarding the potential inclusion of Taiwan and the South China Sea in the alliance's scope.[38] Following the issuance of the

---

Hanmei, "Yingxiang Weilai Riben Dui Wai Zhanlüe he Xingwei de Zhongyao Yinsu" [Important Factors Influencing Future Japanese Foreign Strategy and Conduct], *Zhanlüe yu Guanli* [Strategy and Management] no. 1 (1998): 21.

[36] Interviews 1996. See also Liu, "Dangqian Rimei Anbao Tizhi de San Ge Tedian," 20–22; and Yang Bojiang, "Why [a] U.S.-Japan Joint Declaration on [the] Security Alliance?" *Contemporary International Relations* 6, no. 5 (May 1996): 112.

[37] Liu Jiangyong, "New Trends in Sino-U.S.-Japan Relations," *Contemporary International Relations* 8, no. 7 (July 1998): 113.

[38] See "The Guidelines for U.S.-Japan Defense Cooperation," in Green and Mochizuki, *The U.S.-Japan Security Alliance in the Twenty-first Century*, 65.

revised Guidelines, President Jiang Zemin announced that China is on "high alert" about changes in the alliance.[39]

In 1998 Chinese concerns focused on Japan's September agreement to research theater missile defense jointly with the United States. The initial proposal for joint development of TMD was made by Washington in 1993, long before the Nye Initiative had been launched. It was later folded into the initiative, but Japan still seemed reluctant to commit itself to the project.[40] After five years of U.S. coaxing and Japanese foot-dragging, Tokyo finally agreed to joint TMD research after the launch of a North Korean rocket across Japanese territory on August 31 1998. Although Chinese analysts do recognize the threat to Japan from North Korean ballistic missiles, they still believe that development of U.S.-Japan TMD is also designed to counter China's conventionally tipped missile capabilities, which the People's Liberation Army (PLA) and civilian analysts recognize as among China's most effective military assets, especially in relations with Taiwan.[41]

Taiwan, the U.S.-Japan Alliance, and the Offense-Defense Factor

As they did in the early 1950s and the early 1970s, strategic thinkers in Beijing linked their concerns about real or perceived changes in the U.S.-Japan alliance with their analysis of trends in relations across the Taiwan Strait. The importance of cross-Strait relations in Chinese calculations about TMD and the revised Guidelines cannot be overstated. Along with the brutal legacy of World War II and the ways in which Japan's historical responsibility has been avoided, insufficiently recognized, or even denied by certain politicians and in certain history texts, tensions in relations across the Taiwan Strait and changes in the U.S.-Japan alliance have constituted the most critical exacerbating factor in the China-Japan security dilemma in the post–Cold War world.[42] Also important

---

[39] Interviews, 1996 and 1998. The Jiang quotation comes from a Reuters news service report on October 18, 1997.

[40] For the earliest discussions of joint U.S.-Japan development of TMD and Tokyo's resistance to the plan, see David E. Sanger, "New Missile Defense in Japan under Discussion with U.S.," *New York Times*, September 18, 1993, p. A1. A year-and-a-half later, the language on TMD in the 1995 Nye Report belies Japan's reluctance to agree to joint research, stating that the United States "is exploring with Japan cooperative efforts" in TMD.

[41] Interviews, 1998. See also Wu Chunsi, "Tactical Missile Defense, Sino-U.S.-Japanese Relationship, and East Asian Security," *Inesap Information Bulletin* no. 16 (November 1998): 20–23. For a more complete analysis of this issue, see Thomas J. Christensen, "Theater Missile Defense and Taiwan's Security," *Orbis* 44, no.1 (winter 2000): 79–90.

[42] For an innovative and thorough treatment of the history issue in Sino-Japanese relations, see Yinan He, *The Search for Reconciliation: Sino-Japanese and German-Polish Relations since WWII* (Cambridge: Cambridge University Press, 2009). For a contrary position on the importance of apologies and reconciliation in international relations, see Jennifer Lind, *Sorry States: Apologies in International Politics* (Ithaca: Cornell University, 2008).

are continuing sovereignty disputes in the East China Sea over the Senkaku (Japanese name)/Diaoyu (Chinese name) island chain and the waters and seabed around them.[43]

The intensity of Chinese reactions to adjustments in the U.S.-Japan alliance vary in large part based on the assessment of stability or instability in cross-Strait relations and, most importantly, the positions of the United States and Japan at the time regarding support, indifference, or active opposition to attempts on the island to assert Taiwan's sovereign independence from the mainland. In the mid-1990s changes in the U.S.-Japan alliance were accompanied by what appeared to be an upgrading of Washington's relations with an increasingly assertive President Lee Teng-hui. Both during and after the 1995–96 crisis, which began in the weeks after Lee's heavily politicized June 1995 visit to Cornell, the Clinton administration sought to reassure Beijing that the United States did not support Taiwan independence and that the issuance of a visa to Lee Teng-hui was not a change in long-standing U.S. policy.[44] President Clinton would later publicly and famously reassure the Chinese leadership and public in Shanghai in 1998 that Washington did not support the following scenarios: Taiwan independence; "two Chinas" or "one China, one Taiwan"; or Taiwan's membership in international organizations for which statehood is a requirement (commonly referred to as the "3 no's"). The Clinton administration would also quickly distance itself from President Lee's assertion to a German journalist in July 1999 that the cross-Strait political situation could best be described as "special state to state relations," a statement that sent high-level envoys almost immediately from Washington to both sides of the Taiwan Strait to clarify U.S. policy.[45]

But it is not at all clear that Beijing was reassured enough during this period to view with equanimity the concurrent evolution of the U.S.-Japan alliance. Then NSC Director for Asian Affairs Robert Suettinger, in his magisterial re-

---

[43] M. Taylor Fravel, *Strong Border, Secure Nation: Cooperation and Conflict in China's Territorial Disputes* (Princeton: Princeton University Press, 2008), ch. 6. For an article addressing how Beijing has carefully managed the nationalist sentiments surrounding sovereignty disputes with Japan, see Phillip C. Saunders and Erica Strecker Downs, "Legitimacy and the Limits of Nationalism: China and the Diaoyu Islands," *International Security* 23, no. 3 (winter 1998/1999): 114–46. For a somewhat more negative view of CCP manipulation of anti-Japanese sentiment, see Jessica Chen Weiss, "Powerful Patriots: Nationalism, Diplomacy, and the Strategic Logic of Anti-Foreign Protest," Ph.D. diss., University of California, San Diego. 2008.

[44] Suettinger, *Beyond Tiananmen*, 200–263; Tucker, *Strait Talk*, 205–12; Patrick Tyler, *A Great Wall: Six Presidents and China, An Investigative History* (New York: Century Foundation Books, 1999), 320–22; Mann, *About Face*, 320–30; Romberg, *Rein In at the Brink*, 161–62.

[45] David M. Lampton, *Same Bed, Different Dreams: Managing U.S.-China Relations, 1989–2000* (Berkeley: University of California Press, 2001), 61; Romberg, *Rein In at the Brink*, 185–89; Suettinger, *Beyond Tiananmen*, 382; Richard Bush, *Untying the Knot: Making Peace in the Taiwan Strait* (Washington, D.C.: Brookings Institution, 2005), 218–19. For a transcript of President Lee Teng-hui's July 9, 1999, *Deutsche Welle* interview, see http://www.taiwanheadlines.gov.tw/state/1.htm.

count of this era, labels the period 1999–2000 "Back to the Cold," and a leading scholar of U.S.-China relations, David M. Lampton, calls the period the fourth major turning point in the bilateral relationship in the post-Tiananmen era.[46] Lingering Chinese concerns festered in the late 1990s about U.S. Taiwan policy, including arms sales notifications in 1999–2000. These concerns were amplified greatly during the Kosovo War in Spring 1999, which included not only NATO intervention against what was viewed in Beijing as a legitimate Serbian operation against regional separatism, but also the accidental bombing of the PRC Embassy in Belgrade as part of the air operations against Milosevic's government. Much attention understandably has been paid to that dramatic and tragic episode, but for China's security analysts, of equal importance was the entire Kosovo operation, which seemed to carry a negative precedent for China in terms of the willingness of the United States and its allies to carve up sovereign countries on what Chinese analysts viewed as a mere pretext of humanitarian intervention.[47]

It was in the months immediately following the Kosovo War and Lee's July presentation of the "Two State Theory," and before the Taiwan presidential election of March 2000, that Jiang Zemin is reported to have called for a significant modernization drive in the PLA with a special emphasis on the development of new options for Taiwan scenarios.[48] It was rumored in scholarly and journalistic circles in China that Jiang had grown impatient about the mainland's lack of progress in cross-Strait relations, despite the growing economic interdependence across the Strait, and that he had, therefore, created a timetable within which Taiwan must be coerced back into accepting the broad concept of "one China" and engage in serious talks on the terms for unification.[49] Beijing's "Taiwan White Paper" of early 2000 seemed at least consistent with these rumors if not fully confirmatory of them. The White Paper, which Chinese Premier Zhu Rongji said was explicitly a response to Lee's "state to state theory," added a new condition that could lead to the use of force by the PRC: Taiwan's "indefinite" (*wu xianqi*) refusal to engage in negotiations on reunification.[50]

[46] Suettinger, *Beyond Tiananmen*, ch. 9; Lampton, *Same Bed, Different Dreams*, 55–61.

[47] Author interviews with military officers, civilian government experts, and Western military attachés, spring 1999 and January 2000. For official analyses that link the Kosovo operation to American and Japanese containment strategies toward China through interference in China's internal affairs, see Gao Qiufu, ed., *Xiao Yan Weigan: Kesuowo Zhanzheng yu Shijie Geju* [The Kosovo War and the World Structure] (Beijing: Xinhua Publishers, July 1999), esp. ch. 3; internally circulated.

[48] Suettinger, *Beyond Tiananmen*, 402.

[49] For an article based on the "timetable" rumor circulating in Beijing, see Willy Wo-Lap Lam, "Dual Edge to 'Liberation' Timetable," *South China Morning Post*, March 1, 2000. For an analysis of the timeline issues, see Alan Romberg, "Cross-Strait Relations: In Search of Peace," *China Leadership Monitor* no. 23 (winter 2008), at http://media.hoover.org/documents/CLM23AR.pdf.

[50] Taiwan White Paper, February 2000, in *Beijing Review*, March 6, 2000, pp. 16–24; for Premier Zhu's statement about the origins of the White Paper, see Romberg, *Rein In at the Brink*, 192.

This was the backdrop for a rather severe security dilemma at the end of the last century and the beginning of the new one. This security dilemma was driven by mainland China's coercive reaction to trends in Taiwan politics, U.S. responses in accordance with the Taiwan Relations Act, and ongoing Chinese suspicions about the nature and direction of the U.S.-Japan alliance in the post–Cold War world. In the cross-Strait context, security dilemma dynamics are particularly difficult to manage. The nature of the cross-Strait conflict is such that the usual remedies based in distinctions between defensive and offensive weapons and doctrines apply very poorly. The basic theoretical distinction, simply stated, is that the buildup of defensive weapons and the adoption of defensive doctrines should not fuel the security dilemma and spirals of tension, because such capabilities and methods are not useful for aggression.[51] Defensive weapons are stabilizing because they shore up the territorial status quo by dissuading or physically preventing aggressors from achieving revisionist goals, while reassuring the target state that its passivity will not simply lead to aggressive attacks against its interests in the future. Offensive weapons are destabilizing as deterrent tools because they threaten that status quo and thereby undermine such assurances to the target, which are essential components of successful coercive diplomacy.[52]

What makes offense-defense theories less applicable in the case of cross-Strait relations is that Beijing's main security goal is to prevent Taiwan from declaring permanent independence from the Chinese nation, a de facto administrative and territorial condition that Taiwan already enjoys. In other words, the main threat to Beijing is a political change in cross-Strait relations that would legalize and thereby permanently freeze the territorial and political status quo. The PRC's main method of countering that threat is a combination of military coercion to deter Taiwan independence and economic levers, both positive and negative, to encourage greater integration and raise the opportunity costs of alienating the mainland.[53] In cross-Strait relations Beijing considers traditionally defensive weapons in the hands of Taiwan and any of its potential allies to be dangerous, because they may give Taiwan officials additional confidence in their

[51] Jervis, "Cooperation under the Security Dilemma."

[52] Although scholars differ on definitions of what specifically constitutes a destabilizing offense and a stabilizing defense in the current literature, they all focus on states' capacity for fighting across borders and seizing enemy-held territory as the measure of the offense-defense balance. See, for example, Van Evera, "Offense, Defense, and the Causes of War" and Charles L. Glaser and Chaim Kaufmann, "What Is the Offense-Defense Balance and Can We Measure It?" *International Security* 22, no. 4 (spring 1998): 44–82.

[53] On the use of economic levers in cross-Strait relations, see John Qunjian Tian, *Government, Business, and the Politics of Interdependence and Conflict across the Taiwan Strait* (New York: Palgrave, 2006); and Scott Kastner, *Commerce in the Shadow of Conflict: Political Conflict and Economic Interdependence across the Taiwan Strait and Beyond* (Stanford: Stanford University Press, 2009).

efforts to legalize the territorial status quo. In fact, given that the PRC seems willing to risk extreme costs to deter Taiwanese independence, and, if necessary, to compel a reversal of any such decision by the Taipei authorities, and that Taiwan has fully abandoned Chiang Kai-shek's irredentist designs on the mainland, Taiwan's ability to attack the mainland, strangely, may be only somewhat more worrisome to Beijing than Taiwan's ability to fend off the mainland's coercive attacks on Taiwan.

In this context, it is worth considering China's strong reaction in the 1990s to the plans of Tokyo and Washington to jointly develop mobile, navy-based theater ballistic missile defenses as part of the Nye Initiative and the Guidelines review process. Given the Chinese concerns over Taiwan, future U.S. and Japanese TMD, if effective, and if transferred in peacetime or put at the service of Taiwan in a crisis, could reduce Beijing's ability to threaten the island with ballistic missile attacks, the PLA's main means of coercing Taiwan. Particularly relevant here are the ship-based systems that Japan and the United States agreed to research jointly in September 1998. Analysts in Beijing worried for the same reason that many Americans supported the choice of a ship-based TMD system.[54] As one U.S. commentator applauded, ship-based systems can be moved quickly to other regions to support out-of-area conflicts.[55] Like their U.S. and Japanese counterparts, Chinese analysts in the 1990s expressed serious doubts about the likely effectiveness of such a system, particularly given the proximity of Taiwan to the mainland and the ability of the PRC to launch a large number and wide variety of missiles. Nevertheless, they still worried about the psychological and political impact the system could have on Taipei's attitudes about seeking more diplomatic space, and on U.S. and Japanese attitudes about cross-Strait relations.[56] Although Japan's agreement to help the United States develop defenses against theater missiles as part of a revitalized U.S.-Japan alliance was the most sensitive among my Chinese interlocutors in the 1990s, other generally nonoffensive roles such as mine-clearing and intelligence gathering included in the revised Guidelines also caused concern in Beijing, in part because those roles could come

[54] See "U.S., Japan Agree to Study Missile Defense," *Washington Times*, September 21, 1998, p. 1; and "Japan Makes Missile-Defense Plan High Priority," *Washington Times*, November 6, 1998, p. 12. Statements by Chinese arms control and missile experts in the United States in August 1998, and discussions with one active and one retired military officer in China in November 1998, demonstrated Chinese elite concerns about the mobility and large-area protection that could be provided to Taiwan by U.S. and Japanese navy-based theater missile defenses in the future.

[55] Richard Fisher, quoted in Rob Holzer and Barbara Opall-Rome, "U.S. Anticipates Approval from Tokyo on Joint TMD" *Defense News*, September 2127, 1998, p. 34. See also Peter Landers, Susan Lawrence, and Julian Baum, "Hard Target," *Far Eastern Economic Review*, September 24, 1998, pp. 2021. For a discussion of China's more general concerns about TMD, see Benjamin Valentino, "Small Nuclear Powers and Opponents of Ballistic Missile Defenses in the Post–Cold War Era," *Security Studies* 7, no. 2 (winter 1997/98): 229–232.

[56] Interviews with civilian analysts, November 1998.

into play in scenarios in which the mainland used coercive force against Taiwan, such as a limited and punitive blockade of the island.

The Asian Alliance Security Dilemma in the Early Twenty-first Century

In a sense the first decade of this century has demonstrated very well the importance of U.S. policy in the region in maintaining positive U.S.-China relations and a stable regional security environment. There have been many challenges that, under different diplomatic circumstances, could have created instability: the unexpected 2000 election in Taiwan of President Chen Shui-bian from the traditionally pro-independence DPP and his subsequent pursuit of policies designed to move Taiwan in that direction; the fast-paced development of significant new PLA coercive capabilities, especially since 1999, including the deployment of new naval and air assets and many hundreds of missiles across from the island; the election in the United States of George W. Bush, whose campaign had portrayed China as a "strategic competitor," and who came into office promising to support Taiwan's defense in a robust fashion; domestic politics trends in Japan that engendered nationalist gestures by Japanese elites, including visits to the Yasukuni Shrine by Prime Minister Koizumi, all of which soured Japan's relations with China and South Korea and precluded high-level summits and bilateral confidence-building measures between Beijing and Tokyo; a tighter U.S.-Japan alliance to include more active Japanese support roles far from the Japanese homeland after the terrorist attacks on September 11, 2001; the first post-Deng political transition in China from Jiang Zemin to Hu Jintao in 2003–4; and the growth of unruly anti-Japanese nationalism in Chinese society, made manifest in soccer riots in August 2004 in Beijing following the Japanese victory in the Asian Cup Soccer final in Beijing, and reflected in unruly protests filled with invective and vandalism in Shanghai and other Chinese cities in April 2005 as Japan pursued membership in the United Nations Security Council.[57]

Despite all of these challenges, we have seen some rather positive results: U.S.-China relations blossomed in the early part of this century, especially after September 11, 2001. China has not taken direct political aim at the continued U.S. alliances and military presence in the region. Sino-Japanese relations have improved as Koizumi's successors have eschewed trips to Yasukuni since 2005, thus enabling the restoration of summits and the deepening of lower-level security and economic dialogues between the two governments. For China's part, President Hu Jintao adopted a decidedly less anti-Japanese posture than his

---

[57] Robert Marquand, "Anti-Japan Protests Jar an Uneasy Asia: Demonstrations Spread from Beijing to Several Southern Cities Sunday," *Christian Science Monitor*, April 11, 2005; available at http://www.csmonitor.com/2005/0411/p01s04-woap.html. For a novel political analysis of the anti-Japanese protests in China, see Weiss, "Powerful Nationalists."

predecessor, Jiang Zemin.[58] China, the United States, and two U.S. allies—Japan and South Korea—have cooperated actively on the North Korean nuclear issue in the Six Party Talks framework. Relations across the Taiwan Strait have also improved markedly since spring of 2008, following the failure of a politically provocative referendum in Taiwan on applying to the United Nations under the name Taiwan, and the election of the KMT's Ma Ying-jeou, who has rejected the pursuit of Taiwan independence and has actively sought practical cooperation with the mainland.

This book is not primarily about contemporary affairs, and I have written elsewhere more extensively on some of the topics addressed in the remainder of this chapter.[59] But I would like to draw a few conclusions about the first decade of this century that are consistent with the themes of this book. It should not be taken for granted that in a period in which China's influence rose rapidly there would also be quite positive and deepening bilateral U.S.-China relations. A major factor in the maintenance of peace and security in the region was not only U.S. policy toward China but also the diplomacy surrounding adjustments in U.S. security policies toward its allies and security partners in East Asia. Because of careful U.S. diplomacy, the United States was able to avoid sparking excessive Chinese concern even as Washington encouraged a more active role for Japan as part of the U.S.-Japan alliance. For example, Tokyo sent engineers to Iraq and oil tankers to the Indian Ocean as part of Operation Enduring Freedom in Afghanistan.

There are several reasons for this successful outcome. First, after September 11, 2001, there was a clear reason for increased activity by Japan, Korea, Australia, and other U.S. allies that had nothing to do with countering the rise of China per se, and this message could be easily conveyed to a skeptical Beijing. More important, especially from mid-year in 2002 to the end of the Bush administration, Washington took private and public positions that reduced China's concerns about U.S. and allied activity in the East Asia region and beyond. The United States clearly rejected a hostile stance toward Beijing and sought China's cooperation in solving a range of security issues around the world, most notably the North Korean nuclear issue.[60] Perhaps more important still, both Washington and Japan handled the rising tensions in cross-Strait relations with moderation. Although the signals sent to Beijing about Taiwan from Washing-

[58] For elite Chinese analysis of the warming trends in U.S.-Japan relations, see Ma Hao-liang, "China-Japan Relations will Usher in a Period of Relative Stability," *Ta Kung Pao*, Internet Version, May 3, 2008 in FBIS T08:50:47Z.

[59] See, for example, my articles in the first fifteen volumes of the *China Leadership Monitor* at http://www.hoover.org/publications/china-leadership-monitor.

[60] Thomas J. Christensen, "Shaping the Choices of a Rising China: Some Recent Lessons for the Obama Administration," *Washington Quarterly* 32, no. 3 (July 2009): 89–104.

ton and Tokyo were not always perfectly clear, and we can only know now in hindsight that the policies adopted were effective, the positions adopted by the United States and Japan arguably contained the right mix of credible threats and assurances to prevent conflict in the Taiwan Strait during what was otherwise a politically very volatile time. As in the early 1970s, because issues related to Taiwan were handled carefully in Washington, Beijing's sensitivities about changes in the U.S.-Japan alliance were constrained. The road was not always smooth and Beijing would at times become very concerned about trends in the United States' relationship with both Taiwan and Tokyo. In general, however, the first decade of this century has demonstrated that a strong U.S. security presence in Asia and the strengthening of U.S. alliances and nonallied security relationships need not come at the expense of improved relations between the United States and the PRC.

The Bush administration inherited a lot of the bilateral U.S.-PRC tension from the second half of the 1990s and was elected to office on a platform that was characterized by a generally increased level of suspicion about China's increased power (summarized by the label of "strategic competitor" for China during the campaign). The Bush campaign emphasized the strengthening of U.S. alliances in East Asia (codified in the 2000 "Armitage Report," an influential prescriptive analysis for the alliance written during the presidential campaign in the United States by several future Bush administration officials, including future Deputy Secretary of State Richard Armitage). The Bush team was also dedicated to improving Taiwan's defensive capabilities (demonstrated less than three months after the adminsitration took office by the offering of a very large arms package to Taiwan in April 2001). Moreover, less than two years after the bombing of the PRC embassy in Belgrade in May 1999, the U.S.-China relationship would suffer a major setback again as a result of the EP-3 crisis of April 2001. The situation was exacerbated by the ham-fisted way in which Beijing handled the detaining of the aircrew of the crippled U.S. plane that had been rammed by a Chinese fighter jet intercepting it over international waters in the South China Sea, resulting in the loss of the Chinese pilot.[61] The ability to handle that crisis through bilateral dialogue arguably facilitated the beginning of a more constructive U.S.-PRC relationship, a process accelerated by the terrorist attacks on September 11, 2001. Nevertheless, it is fair to say that, in spring 2001, relations between the United States and China were worse than at any time since the 1999 embassy bombing. Following the early 2001 EP-3 crisis, the Bush administration announced the in-

---

[61] The best coverage of the EP-3 crisis is by a major player in the events at the U.S. Embassy in Beijing, John Keefe, who was special assistant to Ambassador Joseph Prueher at the time. See his monograph, *Anatomy of the EP-3 Incident, April 2001* (Alexander, Va.: Center for Naval Analysis, 2001).

tention to offer large-scale arms package to Taiwan, and President Bush stated in an interview that the United States would "do whatever it takes" to help Taiwan defend itself.[62]

In alliance security dilemmas, policies are rarely purely stabilizing or destabilizing and the Bush administration's posture toward Taiwan and its regional allies, while certainly viewed as provocative in Beijing in 2001, also had real benefits in terms of long-term coercive diplomacy. That posture made clear that the United States was committed to a peaceful settlement of cross-Strait differences and emphasized to Beijing that the United States had noted the ongoing large-scale military build-up across from Taiwan and was willing to respond to it vigorously. The biggest constraining factor in this process was Taiwan's domestic politics, which precluded the purchase of many items in the package of arms offered to Taiwan until 2007–8. During those years the Bush administration finally was able to notify Congress on an unprecedentedly large sale of arms in a one-year period (USD $3 billion in late summer of 2007 and $6.5 billion in fall of 2008). During both terms, the administration consistently criticized the massive PLA build-up across from Taiwan as a force for instability in cross-Strait relations. For its part, under Prime Minister Koizumi, Japan coordinated its policies with Washington, expressing concerns about China's regional buildup in its defense reports. In February 2005 the "2+2" statement of the U.S.-Japan Alliance's Security Consultative Committee (which includes leaders in the State Department, Defense Department, and their Japanese counterparts) highlighted a shared concern among the two allies about the preservation of peace and stability in the Taiwan Strait, signaling that neither Beijing's nor Taipei's actions toward the other would drive a wedge between the two allies.[63]

Such expressions of resolve and alliance coordination are essential elements of successful coercive diplomacy, but as discussed in chapter 1, resolve, or the "credibility of threat," is only one part of successful coercive diplomacy. What bolsters the credibility of threat can unintentionally undercut coercive diplomacy if it seems provocative and inconsistent with assurances that the key interests of the target will not be harmed now or in the future if the target foregoes the use of force in the near term. To be effective, coercive diplomacy must blend resolve with credible assurances. And while they were important parts of a long-term strategy of deterring conflict in the Taiwan Strait at a time in which China was

---

[62] President Bush made this comment in a televised interview with Charlie Gibson of ABC News on April 25, 2001.

[63] For an assessment of the U.S.-Japan joint statement, see Yuki Tatsumi, "U.S.-Japan Security Consultative Committee: An Assessment," *Pacific Forum* 10, at http://www.csis.org; for a general treatment of Asia's reaction to Japan's treatment of history, see Hugo Restall, "'Opposing the Sun': Japan Alienates Asia," *Far Eastern Economic Review*, April 2005.

building up its own coercive capacity toward Taiwan in potentially threatening ways, all of the policies just mentioned also caused concern and sometimes even harsh reactions in Beijing that could have potentially undercut U.S. coercive diplomacy over time if they were not accompanied with appropriate assurances from Washington and Tokyo.[64] In a period in which Taiwan's leadership was making fairly frequent verbal assertions of Taiwan's independent sovereignty and promoting a series of policy measures to remove historical links between Taiwan and the mainland, Beijing could have concluded that force was necessary to avoid further erosion of its long-term position in cross-Strait relations, regardless of the degree of U.S. resolve to resist such a move.

The relatively assertive policies in Washington, Tokyo, and Taipei did not, however, lead to high levels of sustained tension in the Strait or across the Pacific, let alone conflict. I would assert that this was the case in large part because the U.S. and Japanese policies described above were coupled with reassurances from Washington, Tokyo, and other allied capitals that the United States and its security partners did not support or approve of Chen Shui-bian's divisive pursuit of pro-independence initiatives in Taiwan. Moreover, especially since mid-2002 the United States pursued cooperation with Beijing in ways that undercut any notion that the purpose of the U.S. presence in the region was to contain the growth of Chinese influence. In other words, a strong U.S. presence in the region and a strong set of alliance relationships do not foster conflict across the Taiwan Strait, or harm China's basic relations with Washington or Tokyo, if Beijing does not view the primary purpose of those assets to be the containment of China or the promotion of Taiwan independence. On the contrary, especially when placed in the proper diplomatic context, a strong U.S. presence, a clear commitment to the peaceful resolution of differences across the Taiwan Strait in a manner amenable to both sides, and well-coordinated alliance relations all should serve as essential elements of stability in cross-Strait relations and regional security politics as PRC power and regional influence grow.

After the EP-3 incident, relations had begun to thaw with Secretary Colin Powell's trip to China in late July 2001. That process would accelerate after the September 11 attacks on New York City and Washington, which emphasized to leaders in both capitals that, despite the two countries' differences, the United States and China have many common interests and should not view their relations as a zero-sum game in which one actor's pain is somehow to the other's benefit. Beijing has not been a leading player in the Global War on Terror that followed the attacks but it did take constructive public and private positions very quickly after the attacks. One example of such a constructive Chinese policy

[64] Hugo Restall, "'Opposing the Sun': Japan Alienates Asia," *Far Eastern Economic Review* (April 2005).

initiative was a small but symbolically important aid program to Pakistan just after the attacks, which helped support President Musharraf's domestically controversial decision to cooperate with the U.S. military effort to take down the Taliban regime in Afghanistan, a regime that the Pakistani security and intelligence agencies had been instrumental in creating.[65]

But all was not well in U.S.-China relations at the time, in large part, I would argue, because the administration seemed very supportive of Taiwan, almost unconditionally so, at a time in which the island's leadership seemed to place identity politics over cross-Strait peace and stability. The problem went beyond the April 2001 arms notifications and Bush's "whatever it takes" comment on assistance to Taiwan's defense. In a speech at Tsinghua University during his trip to China in February 2002, the president would mention the Taiwan Relations Act exclusively as the basis of U.S. policy toward cross-Strait relations, failing to mention the three U.S.-PRC joint communiqués as guiding U.S. policy toward cross-Strait relations. The deepening of U.S. defense ties with Taiwan in early 2002 also catalyzed PRC concerns about United States' activities in the region. Perhaps the most controversial event was the invitation of Taiwan Defense Minister Tang Yao-ming to Florida for a mid-March defense industry meeting attended by Deputy Secretary of Defense Paul Wolfowitz and Assistant Secretary for East Asia James Kelley, among other top officials. Official and unofficial protests from Beijing branded the invitation of such a high-level cabinet member from Taiwan a violation of the 1979 U.S.-PRC normalization agreement.[66] Finally, the DOD's "Nuclear Posture Review," leaked to the press in March 2002, led to a backlash in China because it specified a future Taiwan scenario as one in which U.S. nuclear weapons might be useful. Moreover, this section was part of a broader document that suggested that the United States should not exclude the first use of small, bunker-busting nuclear weapons.[67]

---

[65] On October 1, 2001, the official news agency, Xinhua reported that Jiang Zemin had spoken with President Musharraf directly by phone and had promised 10 million RMB in aid. For coverage of U.S.-China relations just after 9-11, see Thomas J. Christensen, "China," in Richard Ellings and Aaron Friedberg with Michael Wills, eds., *Asian Aftershocks: Strategic Asia 2002–2003* (Seattle: National Bureau of Asian Research, 2002), 51–94.

[66] For official CCP reaction, see "China Summons U.S. Ambassador to Make Representations," *Xinhua News Agency*, March 16, 2002; "U.S.-Taiwan Secret Talks on Arms-Sales: Analysis," *People's Daily Online*, March 18, 2002, Foreign Broadcast Information Service (hearafter FBIS) FBIS CPP-2002-0118-000088. For press reports of Chinese reactions to the visit of ROC Minister of Defense Tang Yao-ming, see Murray Hiebert and Susan V. Lawrence, "Taiwan: Crossing the Red Lines," *Far Eastern Economic Review* (April 4, 2002); and Willy Lo-lap Lam, "China's Army to Prepare for Military Struggle," CNN.com, March 13, 2002.

[67] On the DOD's 2002 "Nuclear Posture Review," see William Arkin, "Secret Plan Outlines the Unthinkable," *Los Angeles Times*, May 11, 2002, p. 1. For Chinese reactions, see "China Summons U.S. Ambassador to Make Representations," *Xinhua News Agency*, March 16, 2002; "Where Lies the

In this context, other aspects of the U.S. regional presence ostensibly related to the Global War on Terror were treated with some suspicion in Beijing. While President Bush was still on his return trip from China, the *PLA Daily* criticized recent U.S. security initiatives with India as potentially destabilizing in South Asia.[68] In early 2002 the PRC press published a fairly steady flow of implicit and explicit criticisms of aspects of the U.S. war on terrorism as it applied to both the immediate region and Southwest and Central Asia. The United States and Japan were singled out in multiple articles for opportunistically exploiting September 11 to increase their military power-projection capacities in areas surrounding China. The underlying themes of these articles are that Tokyo had planned to break out of the constraints of its peacetime constitution and that the United States had planned to increase its presence in Central Asia and Southeast Asia even before September 11. A focus of this criticism was Japan's decision to send Japanese Maritime Self Defense Force ships to the Indian Ocean in support of the U.S.-led effort there. The terrorist attacks on the U.S., the argument runs, only provided a pretext for the United States and its friends and allies to carry out their geostrategic plans, which are aimed as much at hegemony and countering China's rise as they are at countering terrorism.[69]

The U.S.-China relationship would begin to stabilize and improve over the course of 2002. In spring 2002 then–Vice President Hu Jintao paid a visit to Washington and had what were, by all accounts, very useful discussions with Vice President Dick Cheney. This set the backdrop for Deputy Secretary of State Richard Armitage's important trip to Beijing in late August 2002. Three weeks

---

Mistake"; "U.S.-Taiwan Secret Talks"; and John Pomfret, "U.S.-China Relations Appear Headed for Shaky Ground," *Washington Post Foreign Service*, March 19, 2002.

[68] See, for example, Ding Zengyi, "YinMei Junshi Hezuo Yinren Guanzhu" (U.S.-India Military Cooperation Draws People's Attention), *Jiefangjun Bao*, February 24, 2002. For more criticism of improved U.S.-India ties, see Qian Feng, "India Wants to Be the International Maritime Police of the Malacca Strait," Beijing *Renminwang*, April 20, 2002, FBIS CPP-2002-0420-000026.

[69] For Chinese press reactions to U.S. basing in Central Asia, see, for example, Gao Qiufu, "U.S. Wishful Thinking on Its Military Presence in Central Asia and Real Purpose," Beijing *Liaowang*, April 29, 2002, FBIS CPP-2002-0506-000066; He Chong, "The United States Emphasizes the Purposes of the Long-term Stationing of Troops in Central Asia," Hong Kong *Tongxun She*, January 9, 2002, FBIS CPP-2002-0109-000124; and Shih Chun-yu, "United States Wants Long-term Military Deployment to Control Central Asia," Hong Kong *Ta Kung Pao*, January 11, 2002, FBIS CPP-2002-0111-000037. For descriptions of U.S.-Japanese activities during the war on terrorism, see "Two MSDF Ships Set Sail for Indian Ocean," *Kyodo News*, February 12, 2002, and "MSDF to Extend Anti-terror Tour," *Yomiuri Shimbun*, May 10, 2002. For Chinese reactions to those deployments, see "9-11 Cheng Guanjian Zhuanzhe Riben Junshi Xingdong Huoyue Wei Wushinian Zhi Zui" [September 11 was the Most Critical Turning Point in 50 years for the Invigoration of Japanese Military Activity], *Nanfang Dushi Bao*, April 16, 2002, and "Riben 'Jinjun' Dongnanya Qitu Hezai" [For What Purpose is Japan Planning to Enter Southeast Asia Militarily], *Canwang Xinwen Zhoukan*, May 3, 2002. For PRC criticism of U.S. policy toward North Korea, see Yan Guoqun, "Sunshine Policy Is Shining Again," Beijing *Jiefangjun Bao*, April 11, 2002, FBIS CPP-2002-0411-000088.

before that trip, Chen Shui-bian made a speech that asserted Taiwan's sovereignty and seemed to violate his promises to eschew the pursuit of Taiwan independence while in office, which he made in his moderate inaugural address of May 2000. On August 3, 2002, Chen described relations across the Taiwan Strait as "one country on each side," (*yi bian yi guo*) and suggested that he would pursue a popular referendum to determine Taiwan's status. Chen's formulation arguably went considerably further than Lee Teng-hui's formulation of "special state-to-state relations" (*teshu de guo yu guo guanxi*), which could be interpreted as two governments negotiating on an equal basis inside one notional country. Because Chen's speech refers to "China" and "Taiwan" separately, his statement cannot be interpreted in any way other than as an assertion of Taiwan's independence as a country. This is especially true after his government offered the official English translation of the speech with the word *guo* given as "country," not "state" (the Chinese character *guo* itself is ambiguous on this score).[70] For our purposes here, it is also notable that Chen's speech was made in a teleconference with Taiwan compatriots in Japan, a fact that could only exacerbate Chinese concerns about the link between Japan and Taiwan independence and therefore, between the U.S.-Japan alliance and Taiwan independence.

During his August 2002 Beijing visit, Deputy Secretary Armitage not only clearly distanced himself from the statements by Chen but implicitly labeled them as actions designed to promote Taiwan independence. When answering questions to the press about the statements, Armitage replied simply and firmly that the United States "does not support Taiwan independence." Also of importance, during his trip Armitage treated China as a partner in the War on Terror by publicly labeling the East Turkestan Islamic Movement (ETIM), which seeks independence for the PRC's Xinjiang region, as an international terrorist organization with links to Al Qaeda.[71]

At the Crawford Summit in October 2002 President Bush would have another chance to explain to President Jiang that his administration was not seeking to change the historical U.S. "one China policy." But it would be in December

---

[70] "Chen Stresses Urgency for Referendum Legislation for Taiwan's Future," Taipei Central News Agency, August 3, 2002, and Taipei Office of the President, "Apparent Text of Chen Shui-bian's Speech on Taiwan's Future, Referendum" (in Chinese), August 3, 2002, FBIS CPP-2002-080-3000098. For the CCP's official reaction, see "Text of Taiwan Affairs Spokesman's Remarks on Chen's Call for Referendum," Xinhua News Agency (Chinese), August 5, 2002, FBIS CPP-2002-0805-00002.

[71] Romberg, *Rein In at the Brink of the Precipice*, 207–208. For the official U.S. position on the Armitage visit, see U.S. Department of State, *Transcript of Deputy Secretary of State Richard Armitage Press Conference—Conclusion of China Visit, Beijing, China*, August 26, 2002. For very positive Chinese coverage of the meetings, see Li Xuanliang, "Zhongguo Daodan Guande Hen Yen" [China Severely Restricts Missiles], and Song Nianshen, "Meiguo Shouci Rending 'Dongtu' jiu shi Kongbuzuzhi" [For the First Time the United States Maintains that "ETIM" is a Terrorist Organization], *Huanqiu Shibao* [Global Times], August 29, 2002, pp. 1–2.

2003, in the leadup to the 2004 Taiwan presidential elections, that President Bush would state most clearly that Washington not only did not support Taiwan independence but actively opposed statements and proposals by President Chen that seemed designed to change unilaterally the status quo across the Taiwan Strait. On the campaign trail in late 2003, President Chen had suggested that he would pursue "defensive referenda" on aspects of Taiwan's relationship with the mainland during the presidential election of March 2004. He also suggested the need for constitutional reform, and made various assertions of Taiwan's sovereign independence from mainland China. Singling out Chen's recent actions and statements for comment in a public address, President Bush, with the visiting PRC Premier Wen Jiabao at his side, clearly and publicly stated that he opposed actions by either side of the Taiwan Strait to unilaterally change the status quo.[72]

President Chen won a second term in a much disputed electoral process that included an apparent eleventh-hour assassination attempt on him and Vice President Annette Lu. The result was extremely bad news in Beijing, even though the scaled back "defensive referendum" on Taiwan's pursuit of missile defense that accompanied the election failed to receive sufficient ballots for passage. Chen would return to divisive rhetoric as he campaigned for his party's candidates in the legislative elections in December 2004. China's leadership responded by having the China's National People's Congress draft and eventually pass an "anti-Secession Law," which outlined in rather vague yet quite provocative terms the conditions under which the mainland should use "nonpeaceful" means in its policy toward Taiwan.[73]

Contributing to China's concerns about regional security in this period were trends in Japanese politics and in U.S.-Japan alliance relations. In Japan, Prime Minister Koizumi had visited the Yasukuni Shrine every year since he took office in 2001, aggravating nationalist sentiments in both China and Korea. Especially since the September 11, 2001, terrorist attacks, the United States and Japan had strengthened the alliance in ways consistent with the 2000 "Armitage Report." The report urged Japan to play a larger role in the alliance and to shed some of the historical and legal constraints on collective self-defense and on Japan's ability to project power abroad. In February 2005, the aforementioned "2+2" statement of the U.S.-Japan alliance was issued. That statement

---

[72] For the very positive reaction in China to Bush's statement, see John Pomfret, "China Lauds Bush for Comments on Taiwan," *Washington Post*, December 11, 2003.

[73] For coverage of the political tensions in the leadup to the December 2004 legislative elections and the creation of the anti-secession law, see Thomas J. Christensen, "Taiwan's Legislative Yuan Elections and Cross-Strait Relations: Reduced Tensions and Remaining Challenges," *China Leadership Monitor* 13 (winter 2005), at http://media.hoover.org/documents/clm13_tc.pdf.

mentioned peaceful and stable relations across the Taiwan Strait as a mutual interest of the two allies.[74]

In a revealing 2005 article, an influential Shanghai-based scholar, Wu Xinbo, outlined China's growing pessimism about the U.S-Japan alliance at that time. The degree of concern he expressed seemed more severe than the mainstream view in China at the time, but Wu's analysis was hardly an unconditional dia-tribe against the alliance. Wu presented real concerns in China and offered a sophisticated analysis of what does and does not set off alarm bells in Beijing about the U.S.-Japan alliance. The article begins by breaking away from standard official PRC public line that opposes all U.S. basing overseas, instead noting that many analysts in China have viewed the U.S.-Japan alliance in the past as con-structive in reducing the prospect of a more militarily assertive and independent Japanese security policy. Wu claimed that the increased levels of concern about the alliance had to do with three key issues: nationalist political trends in Japan during the Koizumi years; the perception in some quarters in Japan of China's rise as a threat; and Japan's apparent growing attention and interest in Taiwan, as evidenced for Wu by the mention of the Taiwan Strait in the February 2005 "2+2" statement. Wu claimed, however, that these negative reactions had been tempered in the past and could be tempered in the future by proactive diplo-macy in Washington and Tokyo. Wu emphasized the importance of the allies coordinating with a rising China in international efforts to address common security concerns. Moreover, he said that both Washington and Tokyo should take a clear position against President Chen's efforts to unilaterally assert Taiwan independence. Finally steps should be taken in Washington and Tokyo to foster better bilateral PRC-Japan relations and trilateral dialogue between the United States, Japan, and China.[75] Although his article tends to emphasize the negative, Wu himself concedes that some efforts along these lines had already been made. For example, he notes that in 2002–2003 leaders in both Washington and To-kyo had rejected the provocative rhetoric and actions by President Chen during the elections.[76]

In 2005–2006 the United States and Japan would take actions consistent with the prescriptions in Wu's article. The Bush administration would adopt policies that clearly ran counter to any notion that Washington was trying to contain a

[74] Wang Te-chun, "Strengthening of U.S.-Japan Alliance Explicitly Aimed at China," *Ta Kung Pao*, February 20, 2005 in FBIS TOO:22:31Z. The article is an interview with influential Chinese Academy of Social Sciences Japan scholar, Jin Xide. Jin calls the then forthcoming "2+2" statement, the second major adjustment to the alliance after the Guidelines Review of the 1990s.

[75] Wu Xinbo, "The End of the Silver Lining: A Chinese View of the U.S.-Japan Alliance," *Washington Quarterly* 29, no. 1 (September 2005): 119–130.

[76] Wu, "The End of the Silver Lining," 128.

rising China. Washington engaged Beijing in high-level security dialogues at which Beijing was invited to play a larger role on the international stage, albeit for purposes that served common interests in bolstering regional and global stability in security and economic spheres. This approach took on almost doctrinal stature through Deputy Secretary Robert Zoellick's important speech at the National Committee on U.S.-China Relations gala dinner on September 21, 2005, a statement that invited China to become a "responsible stakeholder" on the international stage and outlined the philosophy behind the U.S.-PRC "Senior Dialogue" on Security and Political Affairs. The same spirit of U.S.-PRC cooperation and the same rejection of a zero-sum mentality in bilateral relations underpinned the late 2006 initiative to create the Strategic Economic Dialogue, headed by Treasury Secretary Henry Paulson, on how to improve economic cooperation between the two states.[77] Cooperation and collaboration among China, the United States, and two U.S. allies—Japan and the ROK—during the Six Party Talks on Korean denuclearization would continue and intensify, especially following the North Korean nuclear test of late 2006, and Washington would continue to engage in productive bilateral discussions with Beijing about how the United States and the PRC could better coordinate the two countries' responses to challenges around the world.[78]

Perhaps the most important instances of confidence-building between the United States, Japan, and the PRC were Washington's and Tokyo's policies toward relations across the Taiwan Strait in the lead-up to the March 2008 Taiwan Presidential elections. In 2007–2008 the United States and Japan would join many other important actors in the region and around the world in criticizing the referendum on applying to the UN under the name Taiwan. President Chen's DPP was clearly using the referendum and the prospect of significant constitutional revision as both a campaign strategy and as a launching pad for future pro-independence policies. After private diplomacy failed to dissuade Chen and the DPP from this approach, the Bush administration opposed the referendum publicly in a series of high-level statements in late summer 2007 by Deputy Secretary of State John Negroponte, Deputy National Security Advisor James Jeffrey, and the National Security Council Senior Director for Asia, Dennis Wilder. In this process, the author, then–Deputy Assistant Secretary of State for East Asian and Pacific Affairs, gave a speech to a high-level U.S.-Taiwan defense conference explaining in detail the basis of the Bush administration's op-

---

[77] For a text of the speech, see "Whither China: From Membership to Responsibility?" Robert B. Zoellick, Deputy Secretary of State, Remarks to National Committee on U.S.-China Relations, September 21, 2005, New York City, at http://www.ncuscr.org/articlesandspeeches/Zoellick.htm.

[78] For an overview of U.S.-China relations in this period, see Christensen, "Shaping the Choices of a Rising China," 89–104.

position to the referendum.[79] In December, Secretary of State Condoleezza Rice again rejected the referendum as "provocative" in a press briefing, even as she reminded Beijing of the other half of coercive diplomacy by calling for a peaceful settlement of all cross-Strait differences.[80]

For our purposes here, it also is important to note the policies of major U.S. allies in the region and around the world, most notably Japan, were fully consistent with U.S. policy toward the referendum. During a late December 2007 summit in Beijing then–Japanese Prime Minister Yasuo Fukuda would reject the referendum and any other unilateral attempt to change the status quo in cross-Strait relations. Although Fukuda was a bit more reserved than U.S. officials, stating that he did "not support" the referendum rather than stating that he "opposed" it, he clearly implied that he saw the referendum in a very negative light. Taipei newspapers, including the pro-DPP *Taipei Times*, were quick to note Fukuda's negative stance and placed little if any significance on the difference in wording between the American and Japanese criticisms of the referendum.[81]

The concerted international criticism of the referendum apparently did real damage both to the popularity of the referendum in Taiwan (an intended consequence of U.S. public diplomacy) and the popularity of politicians who remained associated with the referendum in the March 2008 elections (an unintended consequence). The coordinated effort also helped clarify something important: when the United States, its allies, and like-minded states insist that cross-Strait differences be settled peacefully, this is not some sort of code for their support for Taiwan independence, as some pessimists in Beijing might have it. In order to avoid the opposite problem—that Beijing might somehow mistake U.S. and allied opposition to Chen's initiatives as indifference to Taiwan's security—clear messages opposing the use of force and criticizing the mainland buildup across from Taiwan as destabilizing were included alongside the criticisms of the referendum.

The consistency of allied positions on cross-Strait relations also served to bolster the U.S. diplomatic argument that upgrades in the U.S.-Japan alliance

[79] For expert coverage of the U.S.-PRC-Taiwan triangle in this period, see Alan Romberg, "Applying to the U.N. in the Name of Taiwan," *China Leadership Monitor* no. 22 (fall 2008), at http:// media.hoover.org/documents/CLM22AR.pdf. For the text of the author's September 11 speech to the U.S. Taiwan-Business Council's Defense Industry Conference in Annapolis, Maryland, see "A Strong and Moderate Taiwan," at http://hongkong.usconsulate.gov/ustw_state_2007091101.html.

[80] Thom Shanker and Helene Cooper, "Rice Has Sharp Words for Taiwan, as Gates Does for China," *New York Times*, December 22, 2007, at http://www.nytimes.com/2007/12/22/world/ asia/22diplo.html .

[81] For Japanese opposition to the referendum expressed by Prime Minister Fukuda to Premier Wen Jiabao during the former's late December 2007 trip to Beijing, see AFP, Beijing, "Tokyo Opposes Taiwan UN Referendum," in *Taipei Times*, December 29, 2007, p. 1 at http://www.taipeitimes.com/ News/front/archives/2007/12/29/2003394696.

were not designed to contain China or provide explicit or tacit support for provocative political initiatives on the island. In fact, several weeks after President Ma Ying-jeou's electoral victory in Taipei, President Hu made the first trip to Japan by a Chinese leader in over ten years. In the days leading up to the trip an influential Chinese scholar at the CCP Central Party School, Gong Li, pointed out that one reason for the much improved relations between the two sides was the fact that Japan had rejected Taiwan's UN referendum and clarified its opposition to unilateral changes to the status quo by Taiwan during Prime Minister Fukuda's trip to China in 2007. Gong places Japan's assurances in an overall strategic setting, stating that

> Japan has always treated itself as Asia's bellwether. It is understandable that Japan found it hard to accept, or was worried about, China's gradual rise and even the trend of China surpassing Japan. Judging from the contact between the two countries over the past two years, Japan has basically passed the period of adaptation—it is feeling accustomed to China's rise and finds it acceptable." [82]

In the article, Gong exaggerates the change in atmosphere in Tokyo by claiming that Japan under Fukuda somehow rejected the February 2005 "2+2" U.S.-Japan alliance statement. But despite such hyperbole, it is quite clear that Japan's diplomatic efforts had successfully reassured influential elites in Beijing that the purpose of the strengthening of the U.S.-Japan alliance was not to promote Taiwan independence or to contain the growth of China's influence on the international stage. Japan's signaling of assurances was certainly enabled by the fact that Koizumi's immediate successors, Prime Ministers Abe, Fukuda, and Aso all avoided visits to the Yasukuni Shrine and generally avoided making provocative statements about Japan's imperial history in the region.

A similar story could be told about U.S. arms sales to Taiwan. Having notified Congress regarding $3 billion (USD) in weapons sales in fall 2007, the Bush administration a year later notified Congress again about its intention to sell an additional $6.5 billion in arms to Taiwan. As always, this sparked great criticism in Beijing and led to a counterproductive cancelation of military-to-military dialogue between the two countries until early 2009, after President Obama's inauguration. But despite the large size of the packages, Beijing's response was rather limited and did not derail in any fundamental way relations across the Pacific. Nor did the weapons sales notification spoil quickly improving relations across the Taiwan Strait following the election and inauguration of President Ma Ying-jeou in March 2008 and May 2008, respectively. The two sides concluded

[82] Ma Hao-liang, "China-Japan Relations Will Usher in a Period of Relative Stability," *Ta Kung Pao*, Internet Version, May 3, 2008 in FBIS T08:50:47Z.

economic agreements during shuttle diplomacy between unofficial representatives of the two governments and party leaders of the KMT and CCP.[83]

Despite a rocky start in U.S.-China relations for the Bush administration, it seems that it found a way fairly early in its tenure to bolster regional security in a very challenging time in cross-Strait relations. It was clear on the one hand that Washington and its partners fully insisted upon peaceful resolution of cross-Strait differences and have maintained the military strength and political will to deter efforts by the mainland to coerce Taiwan toward unification against its will, while at the same time it was clear that the purpose of these efforts was not to harm Beijing's fundamental interests by promoting Taiwan independence. Such a mix is the key to solving the alliance security dilemma.

Of course, the policies of the United States and Japan toward Taiwan are only part, albeit an important part, of the overall story of why cross-Strait relations and Sino-Japanese relations have stabilized. Adjustments in China's own approach to both Taiwan and Japan under Hu Jintao since 2004 also played an important, stabilizing role. On the negative side of the equation was the persistent and arguably destabilizing buildup across the Taiwan Strait across the Jiang and Hu years. But in his first several years in office, President Hu has shown more patience than his predecessor in addressing cross-Strait relations. Since Hu consolidated his leadership position, rumors in knowledgeable circles in Beijing of a timetable for unification under Jiang Zemin have been replaced with rumors of acceptance of the status quo long into the future, as long as Taiwan does not take significant steps in the direction of de jure independence.[84]

THE U.S.-JAPAN-CHINA SECURITY DILEMMA IN THE OBAMA ADMINISTRATION

The Obama administration in its first two years in office inherited very stable cross-Strait relations and quite sound U.S.-PRC relations. For a new administration that came into office claiming to want to refocus U.S. energies toward Asia, it was unfortunate, however, that the U.S.-Japan alliance would quickly come under some real strain with the electoral defeat of the LDP in August 2009 elections and the selection of Yukio Hatoyama of the Democratic

---

[83] For comprehensive coverage of trends in cross-Strait relations during this period, see Alan D. Romberg, "Cross-Strait Relations: Ascend the Heights and Take a Long-term Perspective," at http://media.hoover.org/documents/CLM27AR.pdf ; and "Cross-Strait Relations: First the Easy, Now the Hard," at http://media.hoover.org/documents/CLM28AR.pdf, in *China Leadership Monitor*, no. 28 (spring 2009).

[84] Alan Romberg, "Cross-Strait Relations: In Search of Peace."

Party of Japan as prime minister in August. Hatoyama had promised to scrap and then renegotiate a deal struck by his LDP predecessors in 2006 to adjust the U.S. military presence on Okinawa. That deal, which included relocation on the island of the U.S. Marine airbase at Futenma and the relocation of thousands of marines from Okinawa to Guam, was some thirteen years in the making. So, when Hatoyama carried through with his party's campaign promise after assuming office and called for renegotiation of the deal, he placed an enormous strain on U.S.-Japan security relations. The United States initially adopted a tough posture on the issue. Secretary of Defense Robert Gates claimed in Tokyo that the United States would not renegotiate the deal nor would it implement aspects of the deal that were popular in Japan, such as the relocation of marines to Guam, if the unpopular aspects, including the relocation of Futenma, were scrapped by Tokyo.[85] Eventually, the United States softened its stance slightly and Hatoyama accepted the basics of the 2006 deal with minor adjustments in late May 2010; but his reversal itself was very controversial and the highly unpopular prime minister resigned from office almost immediately thereafter.

In early June Hatoyama was replaced by fellow DPJ member and party chief, Naoto Kan, who, thanks to Hatoyama's last-minute compromise, had the luxury of agreeing to honor the basic agreement without being blamed directly for having negotiated it or being the first DPJ leader to approve it. Because of national and local (Okinawan) resistance to the U.S.-Japan agreement in Japan, at the time of this publication it is not at all clear that Kan will, in the end, be able to implement the basics of the 2006 agreement without revisions. What is notable for this study is that Kan is reported to have justified his own stance toward Hatoyama's last-minute compromise by pointing to the need to shore up the U.S.-Japan alliance and maintain the U.S. basing in Okinawa as a way to enhance deterrence. He then expressed concern about the fast-paced growth of Chinese military power.[86] In a related fashion, at the Hiroshima memorial in

[85] For the Joint Press Conference release which quotes Gates's comments, see http://www.mod. go.jp/e/pressrele/2009/091021.html. Also see "Gates Prods Tokyo on Futenma Relocation," *The Daily Yomiuri*, October 22, 2009; "U.S. Raises Pressure Over Plan to Move Japan Base," *The Financial Times*, October 22, 2009; "No Futenma Relocation, No Transfer of Marines to Guam, Gates Says," *Japan Today*, October 21, 2009, at http://www.japantoday.com/category/politics/view/japan-stays-elusive-on-us-base-issue-ahead-of-gates-visit.

[86] A transcript of PM Kan's statements on the alliance in response to a reporter at a multi-party press conference on on June 22, 2010, is not available, but the comments were reported in both the Chinese and U.S. press. For Chinese coverage, see Hao Yalin and Tan Jingjing, "China Says Japanese-US Alliance Should Not Aim at Third Countries," *Xinhua News Service*, June 24, 2010, OSC Translated text, World News Connection document no. 201006241477.1_3a3e002e002e04b0b68e. For press coverage in the United States, see Eric Talmadge, "U.S.-Japan Security Pact Turns 50, Faces New Strains," *Washington Times*, June 22, 2010, at http://www.washingtontimes.com/news/2010/jun/22/us-japan-security-pact-turns-50-faces-new-strains/?page=1.

August 2010, Kan reasserted the utility of nuclear deterrence and the U.S. nuclear umbrella for Japan's security in the twenty-first century.[87]

Chinese reactions to the weakening of the alliance under Hatoyama were very muted. This is almost certainly not merely a function of increased Chinese power, although that factor probably has contributed somewhat to the result. Also likely quite important is that, while Chinese analysts recognized the real strains in the alliance over the Futenma air base issue, they did not expect this to lead to a permanent fissure in the alliance. So, absent were the concerns that Japan might go it alone in the foreseeable future that we saw in the early 1970s and the early 1990s.[88] In fact, the U.S.-Japan alliance for most of Hatoyama's short tenure in office might be seen as perfect from the perspective of many Chinese security analysts: the United States would surely stay in Japan as a "cork in the bottle," but the alliance would remain sufficiently off balance that it might have added difficulty effectively carrying out roles and missions in areas like Taiwan or the East China Sea, in which it might be taking on Chinese forces. The focus of the disputes between Tokyo and Washington—the bases in Okinawa—are, after all, perhaps the most relevant for potential future engagements with Chinese forces in these areas. So, it is not surprising that analyses in Chinese official news sources were generally positive about Hatoyama in his early months in office, and that commentary in official news service publications were quite critical of his last-minute decision to accept the basics of the 2006 deal on Okinawa force relocation.[89]

The aforementioned media reports of Kan's public effort to maintain and bolster the U.S.-Japan alliance's "deterrence" capacity, with special reference to concerns about China's quickly growing military power, was unusually forward for a Japanese leader and, of course, led to a negative official and unofficial reaction in China. The Foreign Ministry spokesman, Qin Gang, rejected the notion

[87] See "Kan Says Nuclear Deterrence Necessary for Japan," *Japan Today*, August 6, 2010, at http://www.japantoday.com/category/politics/view/kan-says-nuclear-deterrence-necessary-for-japan.

[88] For a compilation of Chinese reactions to the strains in the U.S.-Japan alliance during Prime Minister Hatoyama's time in office, see "Asia-Pacific Reaction to Strains in U.S.-Japan Ties Limited: Updated Version, Adding Topi Countries," OSC Report, Thursday April 29, 2010, World News Connection, document no. 201004291477.1_bc280215065bfa6d; for an analysis of the alliance in this period in the online version of the authoritative Party organ, the *People's Daily*, see Yu Qing, "Important Stage in Japan-US Relations," *Renmin Ribao Online*, March 31, 2010, OSC Translated Text, document no. 201003311477.1_37db03a4eflc9b93 .

[89] For an example of a laudatory Chinese analysis of Hatoyama after he came into office and scrapped the 2006 deal, see Shi Ren, "The Development of China-Japan Relations Will Usher in New Opportunities," *Zhongguo Tongxun She*, January 9, 2010, OSC Translated Text, World News Connection, document no. 201001091477.1_69d4009eda8a5b38; for a very critical analysis of Hatoyama after he agreed to the deal, in the official Xinhua News Service, see Jonathan Day, "New Base Deal Drives Wedge Through Japanese Government, Alienated Islanders," *Xinhua*, May 29, 2010, OSC Transcribed Text, World News Connection, document no. 201005291477.1_10270208ce 8607d4.

that China posed a threat to Japan and said that the United States-Japan alliance should not target third countries. He said that Beijing had expressed its concern over Kan's statement directly to Tokyo.[90] But we have not seen anywhere near the severity or persistence of the reaction that we saw in the second half of the 1990s to the Guidelines review, the joint development of TMD, and the U.S. encouragement of a more active Japanese role as part of the alliance. Relative Chinese calm in this case can likely be attributed to several factors, including the rise of Chinese power and self-confidence and the perceived decline of U.S. power because of the financial crisis and the wars in Afghanistan and Iraq. But our study of the history suggests that there might be an even more important factor behind Beijing's relatively calm reaction to Kan's statement: the positive trends in relations across the Taiwan Strait.

Since 2008 Taipei has adopted more moderate and even accommodating policies toward the mainland than at any time since 1992. Under these conditions, the details of adjustments in the U.S.-Japan alliance and the justifications for them are not nearly as sensitive as they would be if cross-Strait relations were as tense as they were from the early 1990s through 2008. It is still important to understand these dynamics in part because there is no guarantee that pro-independence leaders might not return to office in Taipei or, more generally, that cross-Strait relations will remain stable. Moreover, even under conditions of quite positive cross-Strait relations, there are still some traditional hints of neuralgia in Beijing about the purpose and power of U.S. alliances in Asia. After an international review board determined that a North Korean submarine had torpedoed an ROK naval ship in March 2010, killing 46 South Korean sailors, Washington and Seoul planned and executed a naval exercise off the Korean coast. To the great frustration of South Korea, China maintained an agnostic position about the sinking and refused to study the evidence in the weeks that followed. The Chinese government then reacted very negatively to media articles about the planning for the exercises, directly warning Washington and Seoul not to threaten Chinese security during the exercises. In addition to this official governmental warning, a series of rather threatening statements by Chinese scholars and military officers appeared in the Chinese media reports regarding the U.S.-ROK exercises.[91]

---

[90] For Qin Gang's statement see Hao Yalin and Tan Jingjing, "China Says Japanese-US Alliance Should Not Aim at Third Countries," *Xinhua News Service*, June 24, 2010, OSC Translated text, World News Connection document no. 201006241477.1_3a3e002e002e04b0b68e.

[91] For the unusually strident official response from the Chinese Foreign Ministry about public reports regarding the upcoming exercises, see "Foreign Ministry Spokesperson Qin Gang's Remarks on the ROK-U.S. Joint Military Exercises," July 21, 2010 at http://www.fmprc.gov.cn/eng/xwfw/s2510/t718571.htm. For a scholar's view that any exercises by the United States and the ROK in the Yellow Sea would be comparable to Soviet transfer of nuclear weapons to Cuba prior to the Cuban Missile Crisis, see Shen Dingli, "US-ROK Yellow Sea Military Exercises and

One can only imagine how strident the Chinese reaction to the U.S.-ROK exercises and the statements of Prime Minister Kan about the purpose of the U.S.-Japan alliance might have been if relations across the Taiwan Strait had been very tense at the same time. Under such circumstances it might prove much more difficult for the United States to balance the need to maintain strong alliances in the region and reassure Beijing that the purpose of those alliances is not fundamentally contradictory to China's own security interests. So, although the alliance-related security dilemmas are not very sharp at the time of this publication, they could become so again, with implications for the management of coercive diplomacy between the United States and China in the Taiwan Strait or in future Korea scenarios.

Understanding the Cold War history of U.S.-PRC relations from an analytic perspective can help us understand these more contemporary phenomena, but we must remember to be careful in drawing these analogies too easily. The United States and China are very far from being in a Cold War in this century. China is both much less ideological in its foreign policy and much less aggressive with its military in the reform era than it was under Mao. Moreover, the United States and China enjoy extremely deep economic and social ties in contrast to the hostile mutual isolation of the two countries in the 1950s, 1960s, and even most of the 1970s. Finally, through the hard work of diplomats and top leaders on both sides, the United States and China have managed to cooperate on a broad range of projects from North Korea to the Gulf of Aden in this century in ways that would have been simply unimaginable in the Mao years.[92] That being said, there are still classic and complex elements of mutual coercive diplomacy between the United States and the PRC over issues such as relations across the Taiwan Strait. As long as one fully recognizes the differences in context, some of the theoretical and historical lessons drawn from the earlier chapters of this book can still help us understand the challenges to U.S. coercive diplomacy and alliance management in the region that will likely persist into the early decades of this century.

---

the Cuban Missile Crisis," *Dongfang Zaobao Online*, July 12, 2010, OSC Translated Text, document no. CPP201000718138004. For a critical analysis by a high-ranking PLA officer, who accuses the United States and the Republic of Korea of using the sinking of the *Cheonan* as a pretext to practice containment against China, see "Major General Luo Yuan Discusses U.S.-ROK Military Exercise in Yellow Sea" CPP20100713787008 Beijing *Renmin Wang*, July 13, 2010, OSC (Open Source Center) Translated Text, CPP20100713787008.

[92] For a review of the Bush administration years along these lines, see Christensen, "Shaping the Choices of a Rising China."

Chapter 8

CONCLUSION

Disorganization and discord in alliance politics has made the maintenance of peace through coercive diplomacy in Asia very difficult. In fact, the Cold War likely would have been less hot if the U.S.-led alliance and the Soviet-led alliance had consistently been more closely coordinated and better organized and, therefore, had sent clearer and more coherent signals in their security policies. In the case of the communist alliance in particular, Cold War stability in Asia and beyond would likely have been greater had the alliance been more tightly controlled by the more moderate and globally oriented revisionist state, the Soviet Union, than it was when the Soviets were in competition for leadership of the alliance with the more radical Chinese communists. That competition often catalyzed the alliance to be more aggressive toward its enemies than the Soviets would have preferred.

After the collapse of the Soviet Union in 1991, as in the early 1950s and the period 1969–72, Beijing was uncertain about the strength, resolve, and purpose of U.S. alliances and security partnerships that were left as legacies of the Cold War. Even though the United States and China were far from being enemies after the Cold War, they did practice coercive diplomacy toward each other over the issue of the mainland's relations with Taiwan. In that context, some of the same problems of coordination and burden-sharing within the U.S.-led system complicated that coercive diplomacy in potentially destabilizing ways. Both sides successfully managed those problems and at the time of this writing, both the cross-Strait relationship and the U.S.-PRC relationship are basically stable, despite some ongoing tensions related to Washington's relations with Tokyo, Seoul, and Taipei.

Since the onset of the Cold War, two separate sets of dynamics made alliances in Asia "worse than a monolith" from the perspective of maintaining peace and limiting the scope and duration of conflict once it occurred. Both sets of dynamics carry theoretically important lessons for the study of international relations. The first set of dynamics relate to alliance coordination, problems of burden-sharing within an alliance, and unclear signaling in an alliance's coercive diplomacy. A lack of strong alliance coordination can easily lead to unintended and exaggerated signals of current weakness, indifference, or lack of resolve; it can also signal threatening levels of ambition and belligerence, to be unleashed either now or in the future. Either way, the absence of cohesion and clear alliance leadership can undercut one or both of the key elements of successful coercive diplomacy: credible threats that the target state(s) will be punished if proscribed actions are taken; and credible assurances to the target state(s) that, if the target forgoes proscribed action, the target's core interests will not suffer in the present or future from a decision to comply with the demands of the coercing state(s). In their formative years both alliances in the Cold War unintentionally sent signals that they were either too weak in the present to prevent aggressive actions by their adversaries or that they would be stronger and much more threatening to the other camps' core interests in the future if they were not somehow knocked off their current trajectory through near-term military pressure and belligerence. In the worst case, when signals of current weakness are combined with the prospect of future strength and aggression, real or imagined, then a cold war will quite likely become a hot one, and existing conflicts or crises will be more likely to escalate and will prove much harder to draw to a close.

The first set of dynamics is straightforward and is consistent with much of the literature on deterrence theory and rational actor theories about the causes of war.[1] In a nutshell, when alliances are poorly coordinated or have not established clear burden-sharing roles, they can send unclear signals about physical strength, resolve, and aggressiveness both for the present and future. These problems are typical when alliances are first forming or when they are adjusting to periods of significant international change, as were the U.S. alliances when the United States prepared to withdraw from Vietnam in the late 1960s and early 1970s and after the precipitous collapse of the Soviet Union. Particularly in politically adversarial relationships such as those between hostile alliances, unclear signals are likely to be read in infelicitous ways by actors on the oppos-

---

[1] For the original theoretical work that specifies the need for both credibility of threat and credibility of reassurance, see Schelling, *Arms and Influence*. Also see James Davis, *Threats and Promises: The Pursuit of International Influence* (Baltimore: Johns Hopkins University Press, 2000). For more recent related work in the rational choice tradition on the importance of two factors—transparency (complete information) and enforceable commitments—in preventing conflict, see Fearon, "Rationalitst Explanations for War," 379–414; Gartzke, "War Is in the Error Term," 567–87; and Powell, "Bargaining Theory and International Conflict," 1–30. In a sense, the alliance politics discussed here make alliances both less transparent and less capable of making clear, credible, and enforceable commitments to enemies. Jonathan Kirshner offers an approach that is critical of the recent rational choice arguments emphasizing transparency and credibility of commitment, and that instead emphasizes the danger of miscalculation in the highly complex world of international security politics, even if actors were to somehow enjoy perfect information and enforceable commitments. In Kirshner's thesis, the alliance politics discussed here would simply add greatly to the complexity of international relations and would, therefore, increase the likelihood of dangerous miscalculations by members of an alliance or their enemies, regardless of the robustness of available information. See Kirshner, "Rational Explanation for War?" 143–50.

ing side.[2] Better coordinated alliances with clearer leadership are not only apt to send clearer signals than less coordinated and more diffuse ones, thereby reducing the chances for unintended escalation from cold war to conflict, but they are also better at reaching enforceable settlements, so that existing conflicts can be brought to an end more easily.

Coordination and burden-sharing problems apply to all alliances, whether they are basically defensive alliances designed to preserve the status quo or revisionist ones designed to overturn the status quo. The alliance security dilemmas discussed in the introduction feed into these dynamics. Alliance members, who must worry about the opposite dangers of abandonment and entrapment, face tough choices about issues such as the conditionality and tightness of their alliance commitments, and degrees of burden sharing. Real or perceived change in the details of alliance politics or even the prospect of future change can create unintended instability in the mutual coercive diplomacy of the two opposing sides.

A second, and arguably theoretically more interesting set of dynamics, applies only to revisionist alliances. These dynamics involve potentially differential levels of devotion to specific revisionist conflicts, a premium on aggressive behavior for new candidate allies for the purpose of proving their worthiness as accepted members of the international movement, and the catalytic effects of competition for leadership and legitimacy in a hierarchical movement by means of activism in assaulting the perceived illegitimate status quo.

By definition, members of revisionist alliances share a common purpose in overturning the status quo; but not all actors necessarily have the same equities in specific regional conflicts and, therefore, not all actors will exhibit the same willingness to compromise to keep the peace, a prerequisite for stable, successful coercive diplomacy. An overall strategy that maximizes clarity, transparency, and moderation for the purposes of successful coercive diplomacy short of war with the opposing camp requires discipline and coordination that, all things being equal, is most likely to come from the capital of the strongest and most globally oriented actor, even when that actor shares the overall revisionist goals of its alliance partners (as stated in the introduction, there is no guarantee that the most powerful member of the alliance will be the most far-sighted and moderate and, when it is not, coercive diplomacy and containment short of war is extremely unlikely in any case).

Revisionism and civil wars pose related problems. Regardless of their level of devotion to the global movement, local revisionist actors are often less compro-

---

[2] For an excellent review of the psychological literature on this score and its implications for deterrence theory, see Jonathan Mercer, *Reputation and International Politics* (Ithaca: Cornell University Press, 1996), chs. 1–2.

mising than their foreign allies because they are involved in civil wars over their survival and their core interests and, therefore, are either more highly motivated to revise the status quo and/or more highly concerned about the security implications of allowing their domestic enemies to survive, or both. This makes them generally less willing than their foreign partners to accept negotiated peace, territorial compromises, and other conciliatory measures. In the same way, revisionist states that are located near a civil war zone may be more dedicated to actively supporting the revisionists in that civil war than more geographically distant states in the same revisionist alliance.

In the process of forming or expanding a revisionist alliance, new candidate members need to prove that they are dedicated to the revisionist international cause, not just their own parochial national interests. This process of certification can also increase the likelihood of local belligerence and generally obstreperous behavior by new alliance members toward members of the opposing camp in ways that can cause conflicts that did not previously exist and intensify and prolong conflicts that are already underway.

Finally, and most interesting, in hierarchically ordered alliances with revisionist ideologies, the process of competition for prestige with both domestic and foreign audiences among candidate leaders of the movement can have a catalytic effect on the revisionism of the alliance overall. Alliance members will make decisions to support aggression for reasons that go beyond the corporate interests of the alliance as a whole. They will sometimes want to improve their own image as keepers of the revisionist flame with relevant audiences within member states and in the broader group of revisionist or revolutionary parties around the world. Ambitious allies in such competitions for fervor adopt radical postures to improve their positions in the revisionist hierarchy. In the process, they can catalyze more moderate and patient parties, including the most powerful members of the alliance, to adopt more aggressive policies than they might otherwise prefer. While such a catalytic competition for ideological purity might eventually weaken the overall alliance, particularly if and when intramural rivalry escalates to direct conflicts among the members, in the near term such a competition will make it much more difficult for the revisionist alliance's enemies to contain the movement peacefully through coercive diplomacy.

STILL BETTER THAN A MONOLITH? DISTINGUISHING BETWEEN ENEMY STRENGTH AND THE EFFECTIVENESS OF COERCIVE DIPLOMACY

Although it is the focus of this study, coercive diplomacy is only one aspect of security politics. There are other important aspects that must be taken into consideration as well. To ensure their own security, states need not only to successfully deter aggression by other actors, they must also prepare to fight and win if deterrence fails. For this and other reasons, the overall physical strength of the opposing alliance matters greatly for one's own security. Internally divided alliances—whether they are fraught with poor coordination and organization, battles over burden-sharing, or intramural rivalries for leadership—should, all things being equal, be physically weaker than their more coordinated and hierarchically organized counterparts. More poorly organized alliances may be less capable of winning over additional adherents because they fail to create the strong "bandwagon" effect that Randall Schweller has described among revisionist alliances.[3] The overall strength or weakness of an alliance itself is an element of deterrence, not only of war-fighting, as it is a key element of the credibility of coercive threats.[4] Moreover, divisions can eventually lead to all-out splits and open hostilities as they did in the Sino-Soviet case, making what might otherwise have been a very long struggle winnable in the near term for the other camp.

Another limit to the approach here is that in some cases of adversarial relationships coercive diplomacy simply does not apply. In those instances all that matters is the relative power of the two enemy camps. In cases like opposition to Hitler and Osama Bin Laden, for example, the minimally acceptable goals of the revisionists may be anathema to their enemies, making the compromises and assurances that are essential to coercive diplomacy impossible, as there is no mutually acceptable "win set" between the revisionists and status quo actors— and so there is no coercive bargain to be struck. In such cases, generally speaking, the more divisions, distrust, and rivalry in the enemy camp, the better for one's own camp. The only exception would be cases in which a more unified and organized enemy might be easier to locate, disrupt, and attack than a more diffuse enemy. But this is still a question of brute force, not coercion.[5]

Fortunately, however, such situations of entirely incompatible interests are relatively rare. During the Cold War, for example, Moscow and Washington had very little overlap in their common interests, but they had enough to allow both superpowers to be deterred from direct adventurism against the other, at least in a world of mutual second-strike capabilities (or Mutually Assured Destruction).[6] Given such superpower caution at the top, mutual coercive diplomacy and mutual containment between the two camps likely would have been much less costly if there had been better coordination within each allied camp, clearer and more assertive superpower leadership of each camp, and, in the case of the communists, less intramural rivalry.

---

[3] Schweller, "Bandwagoning for Profit," 72–107.

[4] Walt, *The Origins of Alliances*, 35–37.

[5] For a novel analysis of the Al Qaeda network along these lines, see Jacob N. Shapiro, "The Terrorist's Challenge: Security, Efficiency, Control," Ph.D. diss., Stanford University, 2007.

[6] For a classic work, see Jervis, *The Meaning of the Nuclear Revolution*.

The benefits of enemy cohesion can exist even if, for all the reasons listed here, policy advisors would hardly ever intentionally prescribe actions that would foster coherence in the enemy alliance. Even if contemplating how to assist enemies to unify might prove a bridge too far for policy advisors, they should think twice about sowing distrust and discord in the enemy camp, particularly when they are trying to deter aggression by a revisionist alliance or reach some sort of enforceable peace accord in an existing conflict. If preventing or ending conflict rather than simply fighting and winning are priorities in one's security policy, then discord and rivalry can make the enemy camp more dangerous, not less so. Unless the discord is so severe that the revisionist actors turn their guns on each other, divisions in the revisionist camp can serve to catalyze the aggressive use of violence against status quo actors and, once conflict has begun, make negotiated peace settlements harder for status quo actors to reach with the revisionist movement.

THE CASES IN THE BOOK AS ILLUSTRATIONS OF THEORETICAL APPROACHES

The formation and adjustment of U.S. alliances and security partnerships in Asia complicated coercive diplomacy between the two Cold War camps in Asia in ways that increased the likelihood of crises and conflicts. In the 1950s in Korea, the Taiwan Strait, and Indochina, aspects of the U.S. policy toward Japan, Taiwan, and Southeast Asian allies rendered less effective the coercive diplomacy of the United States and its anticommunist allies. Concerns in Beijing about long-term potential trends in U.S. alliances in the region also delayed the process of rapprochement and normalization between the United States and China in the 1970s. In comparable but much less dramatic ways, similar dynamics in the U.S.-Japan and U.S.-Taiwan security relations negatively affected U.S.-China relations in the immediate post–Cold War decades.

It is clear that rather than making the movement less aggressive and more prone to compromise, internal communist alliance disorganization and division made the movement harder for the United States and its allies to contain through coercive diplomacy. In the 1950s the alliance-formation process caused a higher degree of revolutionary activism and less clarity in crisis-signaling than we likely would have seen had the movement been more directly under Soviet control. The lack of clear coordination and control helped not only to cause the Korean War, as Kim Il-sung successfully manipulated Stalin and Mao into supporting his invasion of the South, but also contributed to its escalation in the fall of 1950. The alliance formation process also arguably guaranteed more foreign assistance to Ho Chi Minh in the early 1950s than he would have been likely to receive in a more clearly coordinated and Soviet-dominated movement.

In the late 1950s and throughout the 1960s intramural rivalries in the communist camp created a stronger, more determined Vietnamese communist movement with fewer international restraints than Moscow likely would have placed on Hanoi if the former could have had its druthers. Not only did intramural competition produce increased foreign aid for Ho's movement, it made Ho much less likely to accept suggestions of early peace talks in 1965–67 made by the Soviet Union and its Eastern European satellites. The turn from intramural competition for leadership to direct Sino-Soviet confrontation in 1968–69 eventually greatly benefited the United States. The Sino-Soviet confrontation broke the catalytic process of ideological competition between the Soviets and the Chinese within the communist camp, and provided the United States with a new alignment partner in the Cold War, thereby, sapping the overall strength of Washington's most powerful enemy, the Soviet Union. But the decade-long rivalry that preceded the Sino-Soviet confrontation in 1969 did nothing but catalyze the Chinese and the Soviets to become more deeply involved in the Vietnam War and less supportive of a peace settlement there than they otherwise might have been. That competition also catalyzed Soviet support for revolutionary movements in far-flung portions of the developing world in ways that did not serve U.S. interests.

ALLIANCE FORMATION AND CONFLICT: FURTHER ILLUSTRATION OF THE FIRST SET OF THEORETICAL ISSUES

The United States, NATO, Georgia, and Russian Belligerence in August 2008

In 2008 the United States and its NATO allies found themselves in a position in the Caucasus that was quite reminiscent of the U.S. position in Asia in the summer of 1954. At that time Washington attempted to create a new alliance system in Southeast Asia and was actively considering an alliance with Chiang Kai-shek's Republic of China on Taiwan. This triggered PRC belligerence and a major Cold War crisis from 1954 to 1955. Similarly, the appearance of a trend in 2008 that might eventually lead to Georgia's accession to NATO helped spark a Russian attack on the former Soviet republic.

At the 20th NATO Summit in Bucharest in early April 2008, President George W. Bush made it clear that the United States wanted to admit Georgia to NATO sometime in the future. After several rounds of NATO expansion, including former Soviet republics in the Baltics, Russians could not dismiss Bush's words as empty. The president's position reflected a bipartisan view in the United States, including on Capitol Hill, that Washington and NATO should protect the young Georgian democracy against threats from an increasingly authoritarian post-Soviet Russia.[7] Despite considerable sympathy for Georgia in Europe, many NATO allies were not yet ready to accept the addition of Ukraine and Georgia to the Membership Action Plan (MAP). However, nervous analysts in Moscow apparently believed that, unless halted or reversed, trend-lines were moving in the direction of eventual Georgian and Ukrainian membership in NATO, with the prospect that these former Soviet republics might even provide basing for great power rivals. At the same time that long-term assurances to Russia were being undermined, the ambivalence of many NATO allies and the very weak U.S. and NATO military posture in the region in comparison to that of neighboring Russia all served to undercut the credibility of deterrent threats against Russian belligerence in the near term. This fateful combination simultaneously created an opening window of vulnerability and a closing window of opportunity for Russia. Given this incentive structure, it is not very surprising that Moscow decided to use military force to settle differences with Georgia over separatists in the Georgian region of South Ossetia. Following the fighting, Moscow would recognize the independence of South Ossetia and another breakaway province, Abkhazia, from the rule of the Georgian government.

The Georgian government's decision to use force to put down separatists in those regions in August 2008 provided the occasion for Russia to invade Georgia, but it is fairly clear the Russians were preparing for an attack on Georgia for months and that these deeper alliance-based factors were the main drivers of Russian activity.[8] Moreover, even Georgian President Mikheil Saakashvili's ill-conceived decision to take military action in South Ossetia appears to be deeply rooted in the uncoordinated proto-alliance politics between Washington and Tbilisi. Saakashvili apparently took false hope of international support from the signals of strong backing he had recently received from Washington. These signals included verbal support for Georgia's membership in NATO; congressional approval of hundreds of millions of dollars of assistance to the Georgian military, which had expanded upon U.S. urging; the dispatch of Israeli military advisors to help train Georgian forces, a program adopted by Israel at Washington's request; and general gratitude expressed in the United States for Georgia's

---

[7] For a March 2008 review of the policy in place leading up to the Bucharest Summit, see Jim Nichol, "Georgia and NATO Enlargement: Issues and Implications," *Congressional Research Service Reports for Congress*, March 2008, at http://fpc.state.gov/documents/organization/103692.pdf .

[8] For a convincing analysis along these lines, see Pavel Felgenhauer "NATO's Ambiguous Signaling Helped Provoke Invasion," *Eurasia Daily Monitor* 5, No. 156, August 14, 2008, at http://www.cdi.org/russia/johnson/2008-156-2.cfm.

rather large contingent in the coalition in Iraq. That false hope apparently led Saakashvili to believe that the much more powerful Russia could be deterred from responding militarily to Georgia's crackdown on pro-independence forces in South Ossetia and Abkhazia.[9]

The Russian invasion of Georgia seemed designed to punish not only Georgia but NATO itself, which had taken actions in recent months and years that aggravated Russian nationalism, posed what appeared to some in Moscow to be an increased security threat, and seemed to run against U.S. assurances at the end of the Cold War about the future of NATO expansion. From the perspective of many in Moscow, the entire expansion of NATO into Eastern Europe was a violation of agreements that President George Herbert Walker Bush had made to the final Soviet leader, Mikhail Gorbachev, at the end of the Cold War. Moreover, the inclusion in NATO of republics that were part of the former Soviet Union, including liberated Baltic States, was a further affront to Russian national pride.

The public discussions of eventually adding Georgia and Ukraine to the NATO alliance came after two recent events that touched on Russian security and prestige in the region. The first was the beginning of formal negotiations in May 2007 between the United States and Poland—a former Soviet satellite that is now a NATO member—to establish U.S. missile defense systems on Polish soil. The second and, arguably, more important, was the declaration of independence by Kosovo in February 2008 and the recognition of the new state by the United States and the European Union. As was the case for many critics in China, the entire 1999 Kosovo War was seen by many in Russia as a geostrategic move by the United States and NATO taken under humanitarian pretenses. Assurances had been made by the NATO allies during the war that the purpose of the operation was not to break up a sovereign state, Yugoslavia, but simply to prevent ethnic cleansing against the oppressed Albanians in Kosovo. Russia, a major supporter of Belgrade during the war in 1999, saw the 2008 recognition of an independent Kosovo both as a reason to punish Georgia and NATO, with whom Georgia was flirting, and as a rather cynically deployed pretext for attacking Georgia and declaring the independence of South Ossetia and Abkhazia on "humanitarian grounds"—ostensibly to protect the populations of South Ossetia and Abkhazia from crimes against humanity by Tbilisi.[10]

---

[9] Helene Cooper and Thom Shankhar, "After Mixed Messages, a War Erupted in Georgia," *New York Times*, August 13, 2008, p. 1.

[10] Ronald D. Asmus, "How the West Botched Georgia," The German Marshall Fund of the United States, August 12, 2008, at http://www.gmfus.org/publications/article.cfm?id=462, accessed July 25, 2009. In justifying the Russian invasion, Russian Foreign Minister Sergei Lavrov said that the Georgian government, in putting down separatism by force, "gave an order which led to an act of genocide, which resulted in war crimes, ethnic cleansing. And this, of course, cannot go

Of course, we lack the kind of documentary evidence of Russian decision-making in 2008 that we have for Moscow, Beijing, and other capitals in the first half of the Cold War. But a quick review of informed opinion on the August 2008 war between Russia and Georgia shows that nascent alliance politics among the United States, NATO, and Georgia are widely seen as the main factor behind the Russian attack. In an article entitled "NATO's Ambiguous Signaling Helped Provoke Invasion," Russia expert Pavel Felgenhauer argued that "it seems the main driver of the Russian invasion was Georgia's aspiration to join NATO, while the separatist problem was only a pretext." He goes on to cite the Bucharest summit as the key precipitating event.[11] Although not self-consciously theoretical in his analysis, Ronald Asmus, a renowned expert on Central European security, argues that both credibility of threats and credibility of assurances were simultaneously undercut by the still ambiguous relationship of NATO to Georgia. Asmus writes that there is "a direct line" between the outbreak of conflict and

> NATO's failure to send a unified signal on Georgia and Ukraine at the Bucharest summit. ... After a spectacular row over whether to grant Georgia and Ukraine a Membership Action Plan (MAP), the alliance agreed to drop the MAP option but offered "intensive engagement" along with a vague promise of membership sometime in the future. At the time, diplomats [trying] to put a positive spin on this outcome, claimed it was creative ambiguity. Well, that ambiguity turned out to be more destructive than creative. As opposed to reassuring Georgia and deterring Russia, NATO's decision may have provoked the Russians and probably accelerated the path to war.[12]

As Vladimir Socor summarizes, "NATO's firm promise of eventual membership without a time-frame or a mechanism was enough to provoke Russia and not enough to restrain it."[13] In other words, poor alliance coordination and signaling undercut both legs of coercive diplomacy. These analysts' views are supported by many other knowledgeable observers.[14]

---

unanswered." Quoted in "Georgia and Russia Agree on a Truce," BBC, August 13, 2008, 10:24 a.m. GMT. I am grateful to Gary Bass for calling this article to my attention.

[11] Pavel Felgenhauer "NATO's Ambiguous Signaling Helped Provoke Invasion," *Eurasia Daily Monitor* 5, no. 156, August 14, 2008, at http://www.cdi.org/russia/johnson/2008-156-2.cfm.

[12] Ronald D. Asmus, "How the West Botched Georgia," *The German Marshall Fund of the United States.* Aug. 12, 2008, at http://www.gmfus.org/publications/article.cfm?id=462 accessed July 25, 2009.

[13] Vladimir Socor, "Ukrainian and Georgian ANPs Are Also Testing NATO," *Eurasia Daily Monitor* 5, no. 234, December 9, 2008, at http://www.jamestown.org/single/?no_cache=1&tx_ttnews%5Btt_news%5D=34244.

[14] See, for example, Michael Hirsh, "Pushing Russia's Buttons," *Newsweek,* August 12, 2008, at http://www.newsweek.com/id/152087/page/1, accessed July 25, 2009; Helene Cooper and Thom Shankhar, "After Mixed Messages, a War Erupted in Georgia," p. 1.

POOR COORDINATION, RADICALISM, AND COMPETITION IN REVISIONIST MOVEMENTS: FURTHER ILLUSTRATIONS OF THE SECOND SET OF THEORETICAL INSIGHTS

Pan-Arabism and the Six Day War, 1967

The Cold War communist alliance in East Asia is not the only historical example of an alliance in which lack of coordination and internal competition led to higher levels of external aggression than we might have expected from a more monolithic movement. Malcolm Kerr's work on the Arab states' policies leading up to the 1967 war with Israel strongly suggests that the rivalry among the members of the pan-Arab movement made key members more belligerent and adventurist toward the Israelis than a more unified and hierarchically ordered pan-Arab movement would likely have been, thus complicating greatly Israeli coercive diplomacy.[15] More recent work by Michael Barnett fully supports Kerr's conclusions about the war. Both agree that Syrian and Jordanian competitors for leadership in the pan-Arab movement dragged a reluctant Egyptian President Gamal Abdel Nasser, the ostensible leader of the movement, into a more belligerent stand against Israel.[16]

As Kirk Beattie argues, Nasser had helped create pan-Arabism as a revisionist movement in the 1950s and had placed his personal prestige at home and abroad on Egypt's leadership in maintaining the unity of the movement and achieving its goals against its enemies.[17] In an important history of the Six Day War, Michael Oren shows how pan-Arab disunity in the period 1964–67, under the unsteady leadership of Nasser's Egypt, allowed weaker Arab states, such as Syria, Jordan, and Saudi Arabia to drive the Arab allies into war with Israel by competing with the Egyptians to be the most resolutely pro-Palestinian and anti-Israeli.[18] In Oren's account, Syria in particular played the biggest catalytic role in the alliance. He writes, "[I]nsecure at home and in fierce competition with Egypt and Jordan, Syria's rulers had tried to earn prestige by picking fights with Israel."[19] In his review of Oren's book, political scientist Gary Bass sums up well Cairo's response to this Syrian intramural challenge to "the most powerful and self-confident state" in the Arab world, Egypt:

[15] See Malcolm Kerr, *The Arab Cold War, 1958–67: A Study of Ideology in Politics* (New York: Oxford, 1967); and *The Middle East Conflict*, Foreign Policy Association, Headline Series, No. 191, October 1968.

[16] Michael N. Barnett, *Dialogues in Arab Politics: Negotiations in Regional Order* (New York: Columbia University Press, 1998), ch. 5.

[17] Kirk Beattie, *Egypt during the Nasser Years* (Boulder, Colo.: Westview Press, 1994), 117–18.

[18] Michael B. Oren, *Six Days of War: June 1967 and the Making of the Modern Middle East* (New York: Oxford University Press, 2002), ch. 1.

[19] Ibid., 27.

Before the crisis, Nasser was mostly preoccupied with inter-Arab politics, trying to maintain his position as the leader of the Arab revolutionaries. With 50,000 Egyptian soldiers bogged down fighting Saudi-backed royalists in Yemen, Nasser was in no hurry to tangle with the Israeli Army. Yet it was all but impossible for Nasser to claim leadership of the Arab world without addressing the Palestinian issue. So, pushed forward by Syrian-sponsored Palestinian raids and fierce Israeli reprisals, egged on by Jordan and Syria, and duped by false Soviet reports of an imminent Israeli invasion of Syria, Nasser began his ill-starred escalation.[20]

Political biographer Joel Gordon similarly writes that "Egypt remained center stage and Nasser was still the impresario but he found himself compelled to take increasing risks to retain his dominant role.. .... A close examination of the run up to the June 1967 war reinforces the conclusion, amid a plethora of mis-steps and mis-cues, Nasser had been drawn into a showdown that he had hoped to avoid."[21]

It is fairly clear from these accounts that, if Nasser had enjoyed more control over the movement he led and was able to pursue his preferred strategy, war with Israel would have been less likely. But poor coordination and internal competition in the pan-Arab movement, combined with Nasser's domestic political legitimacy concerns within Egypt itself, allowed a relatively weak member of the movement, Syria, to be a catalytic player. To protect his image at home and abroad, Nasser adopted what otherwise would have been an irrational set of policies toward Israel. Historian Laura James writes of the strained intra-alliance rivalry between Syria and Egypt and the predictably more moderate Egyptian posture toward Israel:

> Nasser appears to have hoped that this treaty [the Egypt-Syria Defense Agreement of November 1966] would enhance his control over the spiraling militancy of his competitor. In fact, it increased the risk that Egypt would be faced with a stark choice between war with Israel and severe Arab censure.[22]

Oren agrees with James's assessment of Nasser's motives: the mutual defense treaty, proposed by Nasser in 1966, would enhance "Syria's ability to lure Egypt into a conflict, [but] would also enable Egypt to limit Syria's maneuverability."[23] As discussed in the introductory chapter, alliances can play such a restraining role on their most war-prone members, particularly in status quo alliances. But in a

---

[20] Gary Bass, "Days that Shook the World: Michael Oren's New Look at the 1967 War," *New York Times Book Review*, June 16, 2002, p. 30.

[21] Joel Gordon, *Nasser: Hero of the Arab Nation* (Oxford: Oneworld Publications, 2006), 70–71.

[22] Laura M. James, *Nasser at War: Arab Images of the Enemy* (New York: Palgrave MacMillan, 2006), 103.

[23] Oren, *Six Days of War*, 31.

competitive, revisionist alliance such as this, in which prestige was measured by one's anti-Israeli vigor, the "chain gang" effect was much stronger than the "drunk tank" effect, and rather than Egypt playing the role of "veto player," Syria was able to play the role of "catalytic player." In James's account, Nasser neither wanted nor expected war when the crisis began but felt obliged to take actions that eventually set Egypt on the course for war because a combination of Israeli bluster against Syria and sharp criticism of Egypt from inside the Arab world, particularly in Jordan and Saudi Arabia, had "attacked Nasser's prestige."[24]

In this sense, from a rational choice perspective, war, even a losing war with Israel, may have had positive utility for Nasser in the end even if it was harmful to the goals of the overall pan-Arab movement, which he claimed to lead. In retrospect, and from an Israeli perspective, one might say that Israel's national security benefited from this process because Israel's fantastically successful campaign severely weakened the pan-Arab movement; but from the more narrow perspective of coercive diplomacy and war avoidance explored in this book, the pan-Arab movement was "worse than a monolith" for those in Israel who hoped to deter, rather than simply defeat, Arab aggression. And at the time, war was feared quite widely in Israel, and no one was sure that the outcome would be nearly as triumphant as it was. For example, the battle-hardened veteran Yitzhak Rabin, then Chief of Staff of the Israeli Defense Forces, famously suffered a nervous breakdown on the eve of war.[25]

ISRAEL, TWO STATES, AND A DIVIDED PALESTINE: 1997–2009

The violent deadlock across the Israel-Palestine divide has persisted with tremendous human costs, especially for the Palestinian population in Gaza. But even though their enemies are divided and therefore, arguably, "weaker" by the measures of brute force struggles for survival, a strategic net assessment of Israel's national situation would quickly reveal that national security for Israel in the broadest sense of the word, beyond mere national survival, is harmed tremendously by the intramural rivalries among the Palestinians. In the 1990s and the first decade of the twenty-first century, Israel's pursuit of a negotiated peace and, eventually, a two-state solution with Palestinian leaders suffered greatly from the internal divisions and rivalries of Israel's revisionist opponents.

Efforts by the United States and Europe to support the most moderate and secular elements of one Palestinian faction, Fatah (as opposed to the more radical Islamist movement, Hamas) as a negotiating partner for Israel, would

---

[24] James, *Nasser at War*, 107.
[25] Oren, *Six Days of War*, 91.

backfire. Hamas and other radical elements of the Palestinian polity were cata-lytic players who undercut the moderates' legitimacy and authority every time they sought accommodation with Israel on the road to a permanent peace. In this case, the third-party audience was mostly domestic, not foreign. But as in international revisionist movements like international communism and pan-Arabism, intramural fighting only serves to radicalize the Palestinian move-ment and scuttle any hopes for a lasting and meaningful negotiated peace. So, as Jonathan Schanzer argues about the split in the late 1990s, "Palestinian unity unquestionably suffered during this period. On balance, so did Israeli security."[26] Schanzer concludes his analysis of three decades of internal struggle between Fatah and Hamas writing, "So long as their internal struggle persists, a negoti-ated settlement to the Palestinian-Israeli conflict will be elusive."[27]

As was the case with the Soviets in Southeast Asia in the 1960s and Nasser in 1967, in the late 1990s the accusations of selling out by fellow revisionists of a more radical ilk catalyzed Fatah to adopt a very hostile position toward Israel on the ground, even as Fatah continued to pursue peace negotiations with the Israelis. In addition, the ideologically charged environment would render Fa-tah incapable of making the kinds of compromises at the negotiating table that would be required to hammer out a lasting peace between Israel and a neighbor-ing Palestinian state. Like Nasser, Arafat and his successors in Fatah might have actually benefited from the promotion of ongoing violence with Israel, even if the Palestinian people as a whole suffer from it. Support for such violence, or at a minimum, the eschewing of politically unpopular forms of compromise with Israel, may have made sense not only for Hamas, but for a beleaguered Fatah.[28]

Very much in the spirit of this book, Dennis Ross writes of the destructive impact of intra-Palestinian rivalries during the Intifada in 2000, "Arafat sud-denly had less control than previously over forces building up from the Palestin-ian street. This was so because the different Palestinian forces—the Fatah activists, elements of the security forces, and Hamas—seemed to regard the new Intifada as an opportunity to gain greater power for themselves. Arafat, the master of maneuvering, would want to control the violence lest any one group become too powerful. At the same time he would try to ride the emotional wave that the violence reflected in Palestinian society."[29]

---

[26] Jonathan Schanzer, *Fatah vs. Hamas: The Struggle for Palestine* (New York: Palgrave MacMillan, 2008), 46.

[27] Schanzer, *Fatah vs. Hamas*, 196.

[28] As Daniel Pipes writes, "[T]he competition raging between Fatah and Hamas since 1987 for the backing of the Palestinian street. ... impelled Fatah not to be seen as going easy on Israel but as aggressively anti-Zionist as Hamas." Daniel Pipes, in foreword for Schanzer, *Fatah vs. Hamas*, xiv.

[29] Dennis Ross, *The Missing Peace: The Inside Story of the Fight for Middle East Peace* (New York, Farrar, Straus, and Giroux, 2004), 733.

Eventually, Hamas's ability to sell itself as promoters of clean government, social services, and a more resolute "anti-Zionist" posture toward Israel would lead to electoral victories that gave Hamas a majority in the Palestinian Legislature in January 2006 and clear control over Gaza in 2007. With Fatah controlling the territory of the West Bank and the executive branch under the government of President Mahmoud Abbas, these results left the Palestinian polity divided politically and territorially, thus precluding meaningful negotiations with Israel. Not only could Fatah not afford to be forthcoming, the total lack of coordination in the Palestinian political movement and the very strong political divisions between Gaza and the West Bank meant that no agreement made by Fatah could be implemented and enforced effectively in any case. In fact, given how radical the political environment in Gaza has become in this catalytic process, Almut Moller wonders whether Hamas itself would be able to contain other radical groups in the territory if it were ever to decide to seek accommodation with Fatah and join in serious negotiations with Israel.[30]

British Foreign Minister David Miliband recently recognized how internal Palestinian divisions often preclude serious peace negotiations. He states, "The reunification of the Palestinian people with a single voice to speak with them, to speak for the West Bank and Gaza is absolutely essential."[31] The German observer Michael Bronig agrees, stating that "the internal schism between the more moderate and secular Fatah and the Islamist Hamas has not only left the Palestinian leadership in disarray, but has also made meaningful peace negotiations with Israel next to impossible."[32] In accord with this logic, in late 2008 scholars Shibley Telhami and Steven Cook argued that the incoming Obama administration's best strategy to achieve peace between Israel and Palestine would be to foster unity, not divisions, in the Palestinian polity both as a way of establishing a credible negotiating partner for Israel and to break the cycle of competitive anti-Israeli bidding in Palestinian politics.[33] In other words, the catalytic anti-Israeli competition in Palestinian politics makes the movement worse than a mono-

[30] Almut Moller, "After Gaza: A New Approach to Hamas." *Austrian Institute for European and Security Policy FOKUS.* February 2009, at http://www.aicgs.org/documents/advisor/moeller0309.pdf.

[31] David Milliband quoted in Almut Moller, "After Gaza: A New Approach to Hamas," *Austrian Institute for European and Security Policy FOKUS,* February 2009, at http://www.aicgs.org/documents/advisor/moeller0309.pdf.

[32] Michael Bronig, "Meaningful Peace Won't Work if Hamas and Fatah are at Odds." *The Daily Star,* April 24, 2009. Available online at http://www.zawya.com/Story.cfm/sidDS240409_dsart66/Meaningful%20peace%20won't%20work%20if%20Hamas%20and%20Fatah%20are%20at%20odds/.

[33] Shibley Telhami and Steven A. Cook, "Addressing the Arab-Israeli Conflict." In ch. 5 of *Restoring the Balance: A Middle East Strategy for the Next President.* CFR-Saban Center at Brookings Institution, December 2008, at http://www.cfr.org/content/publications/attachments/CFR-Saban_Executive_Summaries.pdf.

lith from the perspective of Israeli coercive diplomacy, and, therefore, from the perspective of the U.S. national interests. Outside efforts to sow discord in the Palestinian polity would likely only undercut the prospect of successful coercive diplomacy and a lasting negotiated peace.[34]

The Global War on Terror: Is Violent Islamic Fundamentalism Worse Than a Monolith?

Globally, one can imagine such a catalytic dynamic taking hold in militant fundamentalist religious movements. Each member of the movement might seek recognition for sanctity by encouraging the movement's aggression toward "infidel" regimes. We should be cautious, however, with such analogies. The argument laid out in this book applies only to coercive diplomacy before and during wars, not brute force efforts to fully destroy an enemy. Coercive diplomacy implies the ability to bargain with one's enemy and reach a modus vivendi short of war. In terms of bargaining theory, to establish a relationship of coercive diplomacy, there must be some overlap in the win sets of the two actors involved. Such a situation almost certainly does not exist between Al Qaeda and the vast majority of states around the world. Since the relationship between actors like Al Qaeda and the United States is one of brute force, Washington should actively sow discord within Al Qaeda and between Al Qaeda and its allies. This can weaken the overall movement and make it less physically dangerous to the United States and others, and less politically capable of destabilizing important states around the world, such as Pakistan. Since unity in any coalition involving Al Qaeda will be unlikely to breed moderation or the ability to negotiate an acceptable settlement, there are really no significant opportunity costs to such a strategy of sowing internal distrust and discord in that movement.

This is not true of all enemies, however, most of whom can be coerced and contained through the proper mix of credible threats and assurances. Understanding how coercive diplomacy is affected by the dynamics of alliance politics among adversaries is a useful tool in the study of diplomacy, and potentially in the practice of diplomacy. I hope that this book contributes to both of these endeavors. I would note, however, that we need to be sober about the practical

---

[34] Broad swathes of the Palestinian public had never fully accepted the notion of recognizing Israel as a legitimate state, let alone compromising on many other detailed elements of any workable peace deal. Hamas was able to get many former supporters of Fatah and many neutral observers to side with Hamas in the internal struggle by adopting a more radical and violent posture toward Israel and by accusing Fatah of both corruption and selling-out to Israel. The two types of accusations against Fatah were easily linked via rumors that Fatah's leaders had accepted personal bribes from Israel in exchange for accommodation on core issues of Palestinian sovereignty. For discussion of these rumors, see Yossef Bodansky, "Arafat: Between Jihad and Survival," *The Maccabean*, Freemen Center for Strategic Studies, May 1997, at http://www.freeman.org/m_online/may97/bodansky.htm.

benefits to policymakers of knowing about the nature of divisions within the camp of one's enemy. In some cases, it may help one understand the disastrous circumstances that one faces without providing ready solutions. For example, in the early and mid-1960s some leading academic analysts and key government advisors in the United States fully recognized how Sino-Soviet rivalry could have a catalytic effect on Vietnamese revolution, yet the United States was still unable to avoid the disasters of Vietnam.[35] Although understanding of communist alliance dynamics in U.S. elite circles was much more sophisticated at the time than scholars would later allow, that understanding did not prevent the escalation of the Vietnam War, the inability to find a quick negotiated settlement once the war had escalated, nor the ultimate U.S. defeat in that war. Comprehending one's enemy is always a good thing compared to the alternatives, but it is not always a panacea.

[35] See chapter 6 for internal discussion by USG officials who, in 1965–66, understood the basic alliance dynamics in the communist camp that prevented a negotiated end to the fighting in Vietnam. For extremely sagacious scholarly writings from the 1960s on related themes, see Zbigniew Brzezinski, "The Challenge of Change in the Soviet Bloc," *Foreign Affairs* (April 1961), and Donald Zagoria, *Vietnam Triangle: Moscow, Peking, Hanoi* (New York: Pegasus 1967), ch. 4.

Bibliography

Chinese Documents

Dang de Wenxian [Party Documents] no. 4. Beijing: Central Documents Publishing House, 1989.

Foreign Ministry Archives of the People's Republic of China, Beijing, Accessed August 2004.

Jianguo Yilai Liu Shaoqi Wengao [The Manuscripts of Liu Shaoqi since the Founding of the Nation]. Vol. 1. Beijing: Central Documents Publishing House, 1998.

Jianguo Yilai Mao Zedong Wengao [The Manuscripts of Mao Zedong since the Founding of the Nation]. Vols. 1–12. Beijing: Central Documents Publishing Company, 1987–90. Internally circulated.

Mao Zedong, *"Jianguo Hou Wo Dui Wai Junshi Yuanzhu Qingkuang"* [Our Foreign Military Aid Situation After the Founding of the Nation"]. In *Junshi Ziliao* [Military Materials] no. 4 (1989): chs. 2–3. Internally circulated.

Mao Zedong Junshi Wenxuan [Selections of Mao's Writings on Military Affairs]. Beijing: Liberation Army Soldiers Press, 1981. Internally circulated.

Mao Zedong Waijiao Wenxuan [Selected Diplomatic Documents of Mao Zedong]. Beijing: Central Literature Publishing House, 1994.

Peng Dehuai Junshi Wenxuan [Selected Military Writings of Peng Dehuai]. Beijing: Central Documents Publishing House, 1988.

"Riben Fandongpai QinMei LianSu FanHua, Zhengba Yazhou de Waijiao Luxian" (Neibu Jiaoxue Cankao Ziliao) [The Pro-America, Align with the Soviets, Oppose China, Struggle for Asian Hegemony Diplomatic Line of the Japanese Reactionary Faction]. Beijing University, International Politics Department, December 1971. Internally circulated educational reference materials.

"Riben Fandui Chaoji Daguo Kongzhi, Weixie he Qifu de Douzheng" [The Japanese Struggle to Oppose Superpower Control, Threats, and Bullying]. Beijing University, International Politics Department, 1975. Internally circulated educational reference materials.

Shen Zhihua, ed., *Chaoxian Zhanzheng: Eguo Danganguan de Jiemi Wenjian* [The Korean War: Declassified Documents from the Russian Archives]. Vols. 1–3. Taipei, Academia Sincia, Institute of Modern History, Historical Material Collection No. 48, 2003.

Zhanhou ZhongRi Guanxi Wenxian Ji: 1945–70 [Collected Documents on Postwar Sino-Japanese Relations]. Beijing: Zhongguo Shehui Kexue Chubanshe, 1996.

Zhou Enlai Junshi Wenxuan [A Selection of Zhou Enlai's Writings on Military Affairs]. Vol. 4. Bejing: Renmin Chubanshe, 1997.

Zhou Enlai Nianpu [Zhou Enlai's Chronicle].Vols. 1 and 2. Beijing: Zhongyang Wenxian Chubanshe, May 1997.

Zhou Enlai Waijiao Huodong Da Shiji [A Chronicle of Major Events in Zhou Enlai's Diplomatic Activities], *1949–75*. Beijing: Shijie Zhishi Publishers, 1993.

Zhang Wentian Nianpu [Chronicle of Zhang Wentian]. Volumes 1 and 2. Beijing: Chinese Communist Party History Publishers, 2000.

Chinese Communist Party Memoirs

Hong Xuezhi. *KangMei YuanChao Zhanzheng Huiyi* [Memories from the Korean War].
Beijing: Liberation Army Press, 1991.

Li Yueran. *Waijiao Wutai Shang de Xin Zhongguo Lingxiu* [New China's Leaders on the
Diplomatic Stage]. Beijing: Liberation Army Press, 1989.

Liu Xiao. *Chushi Sulian Banian* [Eight-Year Mission to the Soviet Union]. Beijing: Central Archives and Historical Records Press, 1986.

Shi Zhe. *Zai Lishi Juren Shenbian: Shi Zhe Huiyilu* [At the Side of History's Giants: Shi
Zhe's Memoirs]. Beijing: Chinese Central Documents Press, 1991.

Wu Lengxi, *Shinian Lunzhan: 1956-66, ZhongSu Guanxi Huiyilu* [Ten-Year Theory Battle:
1956–66, A Memoir of Sino-Soviet Relations]. Vols. 1 and 2. Beijing: Central Documents Publishing House, May 1999.

Xu, Marshall Xiangqian. "The Purchase of Arms from Moscow." In Xiaobing Li, Allan R.
Millett, and Bin Yu, *Mao's Generals Remember Korea*. Lawrence: University of Kansas
Press, 2001.

Ye Fei. *Ye Fei Huiyilu* [Memoirs of Ye Fei]. Beijing: Liberation Army Press, 1988.

Chinese-Language Books, Articles, and Interviews

Cai Zuming, ed. *Meiguo Junshi Zhanlüe Yanjiu* [Studies of American Military Strategy].
Beijing: Academy of Military Sciences Press, 1993. Internally circulated.

Chai Chengwen and Zhao Yongtian. *Banmendian Tanpan* [Negotiations at Panmunjom].
Beijing: Liberation Army Press, 1989.

Feng Xianzhi and Jin Chongji. *Mao Zedong Zhuan: 1893–1949* [The Biography of Mao
Zedong]. Vols. 1 and 2. Beijing: China Central Documents Publishing House, 2003.

Gao Heng, "Shijie Junshi Xingshi" [The World Military Scene]. *Shijie Jingji yu Zhengzhi*
[World Economy and Politics] no. 2 (February 1995): 14–18.

Gao Qiufu, ed. *Xiao Yan Weigan: Kesuowo Zhanzheng yu Shijie Geju* [The Kosovo War
and the World Structure]. Beijing: Xinhua Publishers, July 1999. Internally circulated.

Guo Ming, *ZhongYue Guanxi Yanbian Sishi Nian* [40-Year Evolution of Sino-Vietnamese
Relations]. Guangxi: Guangxi People's Publishers, 1992. Internally circulated.

Halisen Houpu [Hope Harrison]. "Zhongguo yu 1958–1961 Nian de Bolin Weiji" [China
and the 1958–1961 Berlin Crisis]. A Chinese-language version of a talk and a paper
presented at a conference in Beijing. Also found online at http://www.coldwarchina.
com/wjyj/blwj/001801.html.

Han Nianlong and Xue Mouhong, eds. *Dangdai Zhongguo Waijiao* [Contemporary Chinese Diplomatic Relations]. Beijing: Social Science Press, 1987. Internally circulated.

He Fang, "Lengzhan Hou de Riben Duiwai Zhanlüe" [Japan's Post–Cold War International Strategy]. *Waiguo Wenti Yanjiu* [Research on Foreign Problems] no. 2 (1993).

Jin Chongji. *Zhou Enlai Zhuan 1949–76, Xia* [Biography of Zhou Enlai 1949–76, vol 2].
Beijing: Chinese Central Documents Publishing House, 1998.

Jin Guangyao. "Gu Weijun yu Mei Tai Guanyu Yanhai Daoyu Jiaoshe (1954.12–1955.2)"
[Wellington Koo and the U.S.-Taiwan Negotiations Regarding the Offshore Islands].

Paper presented to the Scholarly Conference on China in the 1950s, Fudan University, Shanghai, August 2004.

Li Danhui. "ZhongSu Guanxi yu Zhonggguo de YuanYue KangMei" [Sino-Soviet Relations and the Chinese Aid Vietnam, Resist America (Effort)]. *Dangshi Yanjiu Ziliao* no. 251 (June 1998): 1–18. Internally circulated.

———. "ZhongSu Zai YuanYue KangMei Wenti Shang de Maodun yu Chongtu (1965–72)" [Sino-Soviet Contradictions and Conflicts over the Question of Aiding Vietnam and Resisting America]. In Zhang Baijia and Niu Jun, eds., *Lengzhan yu Zhongguo de Zhoubian Guanxi* [The Cold War and China's Relations on Its Periphery]. Beijing: World Knowledge Publishers, 2002.

Li Jie. *ZhongSu Lunzhan de Qiyin: Guocheng Ji Qi Yingxiang* [Origins of the Sino-Soviet Ideological Conflict: Process and Its Influence]. Paper presented at the October 1997 Conference on Sino-Soviet Relations and the Cold War, Chinese Central Committee Documents Research Office, Beijing.

Li Shusheng, "Sulian de Jieti yu MeiRi zai Yatai Diqu de Zhengduo" [The Disintegration of the Soviet Union and U.S.-Japan Rivalry in the Asia Pacific]. *Shijie Jingji yu Zhengzhi* [World Economy and Politics] no. 7 (1992).

Liang Zhensan. "Chaoxian Zhanzheng Qijian ZhongChao Gao Ceng de Maodun, Fenqi ji qi Jiejue: Lengzhan Zhong Shehuizhuyi Zhenying Nei Guojia Guanxi Yanjiu Anli zhi Yi" [Contradictions, Conflicts and Their Resolution among Chinese and Korean Elites during the Korean War Period: A Case of International Relations within the Socialist Camp during the Cold War]. The Central Research Center of the Modern History Center, Taipei, Paper No. 40 (June 2003): 60–61.

Liu Liping, "Jilie Zhendanzhong de Meiguo Duiwai Zhengce Sichao" [The Storm over Contending Positions on U.S. Foreign Policy]. *Xiandai Guoji Guanxi* [Contemporary International Relations] no. 6 (1992): 15–18.

Liu Shilong, "Dangqian Rimei Anbao Tizhi de San Ge Tedian" [Three Special Characteristics of the Current U.S.-Japan Security Structure]. *Riben Yanjiu* [Japan Studies] no. 4 (1996): 18–30.

Liu Zhihui. *Chashang Chibang de Long* [Dragon with Wings]. Beijing: Liberation Army Arts Press, 1992.

Lu Zhongwei. "Yazhou Anquanzhong de ZhongRi Guanxi" [Sino-Japanese Relations in the Asian Security Environment]. *Shijie Jingji yu Zhengzhi* [World Economy and Politics] no. 3 (March 1993).

Niu Jun. "San Ci Taiwan Haixia Junshi Douzheng Juece Yanjiu" [A Study of Decision-making during the Three Military Conflicts across the Taiwan Strait]. Unpublished manuscript presented at the Scholarly Conference on China in the 1950s. Fudan University, Shanghai, August 2004.

———. "Zhanzheng Jubuhua yu Tingzhan Tanpan Juece: Kangmei YuanChao Zhanzheng Juece Yanjiu Zhi San" [The Localization of War and the Decision for Ceasefire Talks: Three Studies of Decision-Making in the Korean War]. A paper presented at the Scholarly Conference on China in the 1950s, Fudan University, Shanghai, August 2004.

Pan Sifeng, ed. *Riben Junshi Sixiang Yanjiu* [Research on Japanese Military Thought]. Beijing: Academy of Military Sciences Press, 1992. Internally circulated.

Qu Aiguo, Bao Mingrong, and Xiao Zuyue. *Zhongguo Zhiyuan Budui YuanYue KangMei Junshi Xingdong Gaishu* [A Narrative of the Military Activities of the Chinese

Volunteer Units in the Assist Vietnam Oppose America War]. Beijing: Junshi Kexue Chubanshe, 1995.

Qu Xing. "ZhongYue Zai Yinzhi Zhan Wenti Shang de Zhanlüe Yi Zhi yu Celüe Chayi" [Sino-Vietnamese Strategic Coordination and Tactical Differences on the Question of the Indochina War]. In Zhang and Niu, eds., *Lengzhan yu Zhong Guo de Zhoubian Guanxi*. Beijing: World Knowledge Publishers, 2002.

Quan Yanchi. *Gongheguo Mi Shi* [Secret History of the Republic]. Hohhot, PRC: Inner Mongolia Publishers, 1998.

Shen Zhihua. "Bo Xiong Shijian yu Zhongguo: Zhongguo Zai 1956 Nian 10 Yue Weiji Chuli Zhong de Juese yi Yingxiang" [China and the Polish-Hungarian Incident: China's Role and Influence in Handling the October 1956 Crisis]. A paper presented to the Scholarly Conference on China in the 1950s, Fudan University, Shanghai, August 2004.

————. *Mao Zedong, Si Dalin yu Chao Zhan: Zhong Su Zui Gao Jimi Dangan* [Mao Zedong, Stalin and the Korean War: The Top Secret Sino-Soviet Archives]. Hong Kong: Cosmos Books, 1998.

————. "Sulian dui 'Da Yue Jin' He Renmin Gongshe de Taidu ji Jieguo" [The Soviet Attitudes about the "Great Leap Forward" and the People's Communes and Their Results]. *Zhonggong Dangshi Ziliao* [Chinese Communist Party History Research Materials] 1 (2003): 118–39.

————. "Su Lian Kongjun Chudong: Chaozhan Chuqi Zhong Su Chao Tongmeng de Nei Zai Guanxi" [The Soviet Air Force Goes into Action: Relations within the Chinese-Soviet-Korean Alliance in the Early Stages of the Korean War]. Paper presented at the International Symposium on "The Cold War in East Asia," Hokkaido University, Japan, June 24–27. 2008.

————. "Yi Jiu Wu Ba Nian Paoji Jinmen Qian, Zhongguo Shi Fou Gaozhi Sulian" [Did China Notify the Soviet Union in Advance Before Shelling Quemoy in 1958?]. *Zhonggong Dangshi Yanjiu* [Chinese Communist Party History Studies] no. 3 (2004): 35–40.

————. "Yuanzhu yu Xianzhi: Sulian yu Zhongguo de Hewuqi de Yanzhi" [Assistance and Limitations: Sino-Soviet Nuclear Weapons Research and Development]. *Lishi Yanjiu* [Historical Studies] (March 2004): 110–31.

————. *Zhonggong Dangshi Yanjiu* [Chinese Communist Party History Studies]. 2004.

Song Liansheng, *KangMei YuanChao Zai Huishou* [Looking Back on the Korean War]. Yunnan People's Press, January 2002.

Wang Shu, Xiao Xiangqian, et al. *Bu Xunchang de Tanpan* [Unusual Negotiations]. Jiangsu People's Press, 1996.

Wang Suhong and Wang Yubin. *Kongzhan Zai Chaoxian* [Aerial Combat in Korea]. Beijing: Liberation Army Press, 1992.

Wang Yanyu, ed. *Riben Junshi Zhanlüe Yanjiu* [Research on Japanese Military Strategy]. Beijing: Academy of Military Sciences Press, 1992. Internally circulated.

Wang Xiangen. *ZhongGuo Mimi Da Fabing: Yuan Yue Kang Mei Shilu* [China's Secret Large Dispatch of Troops: The Real Record of the War to Assist Vietnam and Resist America]. Jinan: Jinan Publishers, November 1992.

Wang Yulin. *DaGuo de Duikang: Ershi Shiji Zhongmei Guanxi* [Great Power Confrontation]. Beijing: Shishi Chubanshe, 1997.

Wu Peng. "Riben Weihe Jianchi Xiang Haiwai Paibing" [Why Japan Insisted on Sending Forces Abroad]. *Shijie Jingji yu Zhengzhi* [World Economy and Politics] no. 12 (December 1992): 46–50.

Wu Xuewen, Lin Liande, and Xu Zhixian. *Dangdai ZhongRi Guanxi* [Contemporary Sino-Japanese Relations]. Beijing: Shishi Chubanshe, 1995.

Xu Yan. *Di Yi Ci Jiaoliang: KangMei YuanChao Zhanzheng Lishi Huigu yu Fansi* [The First Trial of Strength: Reflections on the History of the Korean War]. Beijing: Chinese Broadcast Television Press, 1990.

———. *Jinmen Zhizhan* [The Battle over Quemoy]. Beijing: Chinese Broadcast Television Press, 1992.

———. *Mao Zedong yu KangMei YuanChao Zhanzheng: Zhengque er Huihuang Yunchou Weiwo* [Mao Zedong and the Korean War: A Correct and Glorious Mapping Out of Strategy]. Beijing: PLA Publishing House, December 2003.

Yang Gongsu. *Zhonghua Renmin Gongheguo Waijiao Lilun yu Shixian* [The Theory and Practice of PRC Diplomacy]. A limited edition Beijing University textbook, 1996 version. Internally circulated.

Yang Xuejun and Li Hanmei. "Yingxiang Weilai Riben Dui Wai Zhanlüe he Xingwei de Zhongyao Yinsu" [Important Factors Influencing Future Japanese Foreign Strategy and Conduct]. *Zhanlüe yu Guanli* [Strategy and Management] no. 1 (1998): 17–22.

Yang Yunzhong. "Meiguo Zhengfu Jinyibu Tiaozheng dui Ri Zhengce" [Further Adjustments in America's Japan policy]. *Shijie Jingji yu Zhengzhi* [World Economy and Politics] no. 7 (July 1995): 61–65.

Ye Yumeng. *Chubing Chaoxian* [The Korean War]. Beijing: Beijing Shiyue Wenyi Chubanshe, 1990.

Ye Yonglie, *Chen Boda Zhuan* [Biography of Chen Boda]. Beijing: Zuojia Chubanshe, 1996.

Zhang Baijia. "Mao Zedong yu ZhongSu Tongmeng He Zhong Su Fenlie" [Mao Zedong and the Sino-Soviet Alliance and the Sino-Soviet Split]. Unpublished manuscript presented to the Chinese Communist Party Central Party History Research Office's International Scholars Research Forum, Beijing, October 1997.

Chinese Newspapers and Magazines

Beijing Review
People's Daily Online
Xinhua News
Jiefangjun Bao
Hong Kong Ta Kung Pao
Beijing Renmin Wang
Beijing Liaowang
Hong Kong Tongxun She
Nanfang Dushi Bao
Canwang Xinwen Zhoukan
Huanqiu Shibao (Global Times)

English-Language Primary Source Documents

Clinton-Hashimoto Joint Communiqué. April 17, 1996. Available at http://www.ioc
.u-tokyo.ac.jp/~worldjpn/documents/texts/docs/19960417.D1E.html.
Cold War International History Project Bulletins: Vols. 6–7, 16, and 17.
Foreign Relations of the United States, 1950, Vols. 3 and 6. Washington, D.C.: U.S. Depart-
ment of State.
Foreign Relations of the United States, 1952–54. Vol. 14, China and Japan. Washington,
D.C.: U.S. Department of State.
Foreign Relations of the United States, 1964–68. Washington, D.C.: U.S. Department of
State.
"The Guidelines for U.S.-Japan Defense Cooperation," pp. 55–72 in Green and Mochizuki,
The U.S.-Japan Security Alliance in the Twenty-first Century. Washington, D.C.: Coun-
cil on Foreign Relations, 1998.
Harry S. Truman Library
        Papers of Dean Acheson
        Meetings of the National Security Council: Memoranda for the President
National Archives
        Documents of the National Security Council
        United States Department of State
            Record Group 59 (RG59)
                Confidential U.S. State Department Central Files, Library and
                Contract Microfilm (LM)
                    China: Foreign Affairs
                    China: Internal Affairs
                Decimal Files
                Records of the Executive Secretariat
                Records of the Office of Chinese Affairs, 1945–55 (microfilm C0012)
                Records of the Policy Planning Staff
            Record Group 84 (RG84)
                China Post Files
National Intelligence Council. *Estimative Products on Vietnam, 1948–75*. Washington,
D.C.: National Intelligence Council, April 2005.
National Intelligence Council. *Tracking the Dragon: National Intelligence Estimates
on China During the Era of Mao, 1948–78*. Washington, D.C.: National Intelligence
Council, 2004.
National Security Archives
        Burr, William, ed, "The Beijing-Washington Back Channel and Henry Kissinger's
        Secret Trip to China." *National Security Archive Electronic Briefing Book* No. 66.
        February 27, 2002.
        ———. "Secret U.S. Chinese Diplomacy behind Nixon's Trip," *National Security
        Archive Electronic Briefing Book* No. 145. December 21, 2004.
        Burr, William, Sharon Chamberlain, Gao Bei, and Zhao Han, eds. "Negotiating
        U.S.-Chinese Rapprochement." *National Security Archive Electronic Briefing
        Book* No. 70. May 22, 2002.
Papers of John Foster Dulles, Firestone Library, Princeton University, Princeton, N.J.

The Pentagon Papers: The Gravel Edition 1 (1971): 108–46. Available at http://25thaviation .org/id966.htm.

United States Security Strategy for the East Asia-Pacific Region. Washington, D.C.: U.S. Department of Defense, 1998.

Westad, Odd Arne, Chen Jian, Stein Tonneson, Nguyen Vu Tung, and James Hershberg, eds. "77 Conversations between Chinese Foreign Leaders of the Wars in Indochina." Cold War International History Project (CWIHP) Working Paper No. 22 (May 1998).

English-Language Books and Articles

Accinelli, Robert. "'A Thorn in the Side of Peace': The Eisenhower Administration and the 1958 Offshore Island Crisis." In Robert S. Ross and Jiang Changbin, eds. *Re-examining the Cold War: U.S.-China Diplomacy, 1954–73.* Cambridge: Harvard University Asia Center, 2001.

Acheson, Dean. *Present at the Creation: My Years at the State Department.* New York: Norton, 1969.

Akira Iriye, "Chinese-Japanese Relations," in Christopher Howe, ed., *China and Japan: History, Trends, Prospects* (Oxford: Clarendon Press, 1996).

Ang Cheng Guan. *Vietnamese Communists' Relations with China.* Jefferson, N.C.: McFarland, 1997.

Asher, David L. "U.S.-Japan Alliance for the Next Century." *Orbis* 41, no. 3 (summer 1997): 343–75.

Asmus, Ronald D. "How the West Botched Georgia," *The German Marshall Fund of the United States.* Aug. 12, 2008, at http://www.gmfus.org/publications/article. cfm?id=462. Accessed July 25, 2009.

Bachman, David. *Bureaucracy, Economy and Leadership in China: The Institutional Origins of the Great Leap Forward.* Cambridge: Cambridge University Press, 1991.

Barnett, Doak. *China and the Major Powers in East Asia.* Washington, D.C.: Brookings Institution, 1977.

Barnett, Michael N. *Dialogues in Arab Politics.* New York: Columbia University Press, 1998.

Barnhart, Michael. *Japan Prepares for Total War.* Ithaca: Cornell University Press, 1987.

Bass, Gary. "Days that Shook the World: Michael Oren's New Look at the 1967 War." *New York Times Book Review,* June 16, 2002, p. 30.

Beattie, Kirk. *Egypt during the Nasser Years.* Boulder, Colo.: Westview Press, 1994).

Betts, Richard K. "Wealth, Power, and Instability." *International Security* 18, no. 3 (winter 1993/94), 34–77.

Blight, James G., Robert S. McNamara, and Robert K. Brigham. *Argument Without End: In Search of Answers to the Vietnam Tragedy.* New York: Public Affairs, 1999.

Blum, Robert. *Drawing the Line: The Origins of American Containment Policy in East Asia.* New York: Norton, 1982.

Bodansky, Yossef. "Arafat: Between Jihad and Survival." *The Maccabean,* Freemen Center for Strategic Studies, May 1997, at http://www.freeman.org/m_online/may97/ bodansky.htm.

Brigham, Robert K. *Guerrilla Diplomacy: The NLF's Foreign Relations and the Vietnam War.* Ithaca: Cornell University Press, 1998.

Brigham, Robert K. "Vietnamese-American Peace Negotiations: The Failed 1965 Initiatives." *Journal of East Asian Studies* 4 (2005): 377–97.

Bronig, Michael. "Meaningful Peace Won't Work if Hamas and Fatah Are at Odds." *The Daily Star.* April 24, 2009. Available online at http://www.zawya.com/Story.cfm/sidDS240409_dsart66/Meaningful%20peace%20won't%20work%20if%20Hamas%20and%20Fatah%20are%20at%20odds/.

Brzezinski, Zbigniew. "The Challenge of Change in the Soviet Bloc." *Foreign Affairs* (April 1961).

Buckley, Roger. *US-Japan Alliance Diplomacy, 1945–90.* Cambridge: Cambridge Studies in International Relations, 1992.

Bush, Richard. *Untying the Knot: Making Peace in the Taiwan Strait.* Washington, D.C.: Brookings Institution, 2005.

Campbell, Kurt M. "The Official U.S. View." In Michael J. Green and Mike M. Mochizuki, eds., *The U.S.-Japan Security Alliance in the Twenty-first Century.* New York: Council on Foreign Relations Study Group Papers, 1998.

Cha, Victor D. *Alignment despite Antagonism: The U.S.-Korea-Japan Security Triangle.* Stanford: Stanford University Press, 1999.

———."Powerplay: Origins of the U.S. Alliance System in Asia." *International Security* 34, no. 3 (winter 2009/2010): 158–96.

Chang, Gordon. *Friends and Enemies: The United States, China, and the Soviet Union, 1948–72.* Stanford: Stanford University Press, 1990.

Chen Jian. "China's Changing Aims during the Korean War, 1950–1951." *Journal of American-East Asian Relations,* 1, no. 1 (spring 1992).

———. "China and the First Indo-China War, 1950–54." *China Quarterly* (1993): 85–110.

———. "China's Involvement in the Vietnam War, 1964–69." *China Quarterly* 142 (June 1995): 356–87.

———. *China's Road to the Korean War.* New York: Columbia University Press, 1996.

———. *Mao's China and the Cold War.* Chapel Hill: University of North Carolina Press, 2000.

Christensen, Thomas J. "Chinese Realpolitik." *Foreign Affairs* 75, no. 5 (September/October 1996): 37–52.

———. *Useful Adversaries: Grand Strategy, Domestic Mobilization, and Sino-American Conflict, 1947–58.* Princeton: Princeton University Press, 1996.

———. "Perceptions and Alliances in Europe, 1865–1940." *International Organization* 51, no. 1 (winter 1997): 65–97.

———. "China, the U.S.-Japan Alliance and the Security Dilemma in East Asia." *International Security* 23, no. 4 (spring 1999): 49–80.

———. "Theater Missile Defense and Taiwan's Security." *Orbis* 44, no.1 (winter 2000): 79–90.

———. "Posing Problems Without Catching Up: China's Rise and Challenges for U.S. Security Policy," *International Security* 25, no. 4 (2001): 5–40.

———. "China," pp. 51–94 in Richard Elllings and Aaron Friedberg with Michael Wills, eds., *Asian Aftershocks: Strategic Asia 2002–2003.* Seattle, Washington: National Bureau of Asian Research, 2002.

———. "The Contemporary Security Dilemma: Deterring a Taiwan Conflict." *Washington Quarterly* 25, no. 4 (2002): 7–22.

———. "Taiwan's Legislative Yuan Elections and Cross-Strait Relations: Reduced Tensions and Remaining Challenges." *China Leadership Monitor* no. 13 (winter 2005) at http://media.hoover.org/documents/clm13_tc.pdf .

———. "Windows and War: Trend Analysis and Beijing's Use of Force," pp. 50–86 in Robert Ross and Alastair Iain Johnston, eds., *New Directions in the Study of Chinese Foreign Policy*. Stanford: Stanford University Press, 2006.

———. "Shaping the Choices of a Rising China: Some Recent Lessons for the Obama Administration." *Washington Quarterly* 32, no. 3 (July 2009): 89–104.

Christensen, Thomas J., and Peter Gries. "Power and Resolve in U.S. China Policy." *International Security* 26, no. 2 (fall 2001): 155–65.

Christensen, Thomas J., and Jack Snyder. "Chain Gangs and Passed Bucks: Predicting Alliance Patterns in Multipolarity." *International Organization* 44 (1990): 137–68.

Cohen, Warren I., et al. "Symposium: Rethinking the Lost Chance in China." *Diplomatic History* 21 (winter 1997): 71–116.

Crawford, Timothy W. *Pivotal Deterrence: Third-Party Statecraft and the Pursuit of Peace*. Ithaca: Cornell University Press, 2003.

Cumings, Bruce. *Origins of the Korean War*. Vol. 2. Ithaca: Cornell University Press, 2004.

Cunningham, Kathleen Gallagher. "Divided and Conquered: Why States and Self-Determination Groups Fail in Bargaining over Autonomy." Ph.D. dissertation, University of California, San Diego, 2007.

Davis, James. *Threats and Promises: The Pursuit of International Influence*. Baltimore: Johns Hopkins University Press, 2000.

Dower, John. *Embracing Defeat: Japan in the Wake of World War II*. New York: Norton, 1999.

Fearon, James. "Rationalitst Explanations for War." *International Organization* 49, no. 3 (summer 1995): 379–414.

Felgenhauer, Pavel. "NATO's Ambiguous Signaling Helped Provoke Invasion." *Eurasia Daily Monitor* 5, no. 156, August 14, 2008, at http://www.cdi.org/russia/johnson/2008-156-2.cfm.

Finkelstein, David. *Washington's Taiwan Dilemma: From Abandonment to Salvation*. Fairfax, Va.: George Mason University Press, 1993.

Finnegan, Michael. *Managing Unmet Expectations in the US-Japan Alliance*. National Bureau of Asian Research Special Report, November 2009.

Foot, Rosemary. *The Wrong War: American Policy and the Dimensions of the Korean Conflict, 1950–1953*. Ithaca: Cornell Univesrity Press, 1985.

Ford, Harold P. *CIA and the Vietnam Policymakers: Three Episodes*. Langley, Va.: History Staff, Center for the Study of Intelligence, Central Intelligence Agency, 1998.

Fravel, M. Taylor. *Strong Border, Secure Nation: Cooperation and Conflict in China's Territorial Disputes*. Princeton: Princeton University Press, 2008.

Friedberg, Aaron. *In the Shadow of the Garrison State*. Princeton: Princeton University Press, 2000.

Friedberg, Aaron L. "Ripe for Rivalry: Prospects for Peace in a Multipolar Asia." *International Security* 18, no. 3 (winter 1993/94), 5–33.

Fursenko, Alexsei and Timothy Naftali. "*One Hell of A Gamble*": *Khrushchev, Castro, and Kennedy, 1958–1964, The Secret History of the Cuban Missile Crisis*. New York: Norton, 1997.

Gaddis, John Lewis. *Strategies of Containment: A Critical Appraisal of Postwar American National Security Policy*. New York: Oxford University Press, 1982.

———. *The Long Peace: Inquiries into the History of the Cold War.* New York: Oxford University Press, 1987.

Gaiduk, Ilya V. *The Soviet Union and the Vietnam War*. Chicago: Ivan R. Dee, 1996.

———. *Confronting Vietnam: Soviet Policy toward the Indochina Conflict*. Washington, D.C., and Stanford: Woodrow Wilson Center and Stanford University, 2003.

Garrett, Banning, and Bonnie Glase. "Chinese Apprehensions about Revitalization of the U.S.-Japan Alliance." *Asian Survey* 37, no. 4 (April 1997): 383–402.

Gartzke, Erik. "War is In the Error Term." *International Organization* 53, no. 3 (summer 1999): 567–87.

Garver, John. "Paradigms, Polemics, Responsibility, and the Origins of the U.S.-PRC Confrontation in the 1950s." *Journal of American-East Asian Relations* (1994): 1–34.

———. "The Tet Offensive and Sino-Vietnamese Relations," pp. 45-61 in Marc Jason Gilbert and William Head, eds., *The Tet Offensive*. Westport, Conn.: Praeger, 1996,.

Gelpi, Christopher. "Alliances as Instruments of Intra-Allied Control." In Helga Haftendorn, Robert O. Keohane, and Celeste A. Wallender, eds., *Imperfect Unions: Security Institutions over Time and Space*. Oxford: Oxford University Press, 1999: 107–39.

Glaser, Charles L., and Chaim Kaufmann. "What Is the Offense-Defense Balance and Can We Measure It?" *International Security* 22, no. 4 (spring 1998): 44–82.

Gobarev, Viktor. "Soviet policy Toward China: Developing Nuclear Weapons, 1949–69." *Journal of Slavic Military Studies* 12, no. 4 (December 1999): 1–53.

Goldgeier, James, and Michael McFaul. "A Tale of Two Worlds." *International Organization* 46, no. 2 (Spring 1992): 467–92.

Goldstein, Lyle. "Research Report: Return to Zhenbao Island: Who Started Shooting and Why Does It Matter." *China Quarterly*, no. 168 (2001): 985–97.

Goldstein, Steven M. "Sino-American relations, 1948-50: Lost Chance or No Chance." In Yuan Ming and Harry Harding, eds., *Sino-American Relations 1945–55: A Joint Assessment of a Critical Decade*. Wilmington, Del.: Scholarly Resources, 1989.

Goncharov, Sergei N. "Stalin's Dialogue with Mao Zedong." *Journal of Northeast Asian Studies* (winter 1991–92): 45–76.

Goncharov, Sergei N., John W. Lewis, and Xue Litai. *Uncertain Partners: Stalin, Mao and the Korean War*. Stanford: Stanford University Press, 1993.

Gong Li. "Tensions in the Taiwan Straits: Chinese Strategies and Tactics," in Robert S. Ross and Jiang Changbin, eds. *Re-examining the Cold War: U.S.-China Diplomacy, 1954–73*. Cambridge: Harvard University Asia Center, 2001.

Gordon, Joel. *Nasser: Hero of the Arab Nation*. Oxford: Oneworld Publications, 2006.

Green, Marshall. "The State Department Perspective." In Marshall Green, John H. Holdridge, and William N. Stokes. *War and Peace with China: First-Hand Experiences in the Foreign Service of the United States*. Washington, D.C.: Dacor Press, 1994.

Hao Yufan and Zhai Zhihai. "China's Decision to Enter the Korean War: History Revisited." *China Quarterly*, no. 121 (March, 1990): 94–115.

Harding, Henry. *A Fragile Relationship: The United States and China since 1972*. Washington, D.C.: Brookings Institution, 1992.

Harrison, Hope M. "Driving the Soviets Up the Wall: A Super-Ally, a Superpower, and the Building of the Berlin War, 1958–1961." *Cold War History* 1, no. 1 (autumn 2000): 53–74.

———. *Driving the Soviets Up the Wall, Soviet-East German Relations, 1953–61*. Princeton: Princeton University Press, 2003.

He, Di, "The Evolution of the People's Republic of China's Policy toward the Offshore Islands," pp. 222–45 in Warren Cohen and Akira Iriye, eds., *The Great Powers in East Asia, 1953–1960*. New York: Columbia University Press, 1990.

He, Yinan. *The Search for Reconciliation: Sino-Japanese and German-Polish Relations since WWII*. Cambridge: Cambridge University Press, 2009.

Heginbotham, Eric, and Richard J. Samuels. "Mercantile Realism and Japanese Foreign Policy." *International Security* 22, no. 4 (spring 1998): 171–203.

Herring, George C. "The Indochina Crisis: 1954–55." In Richard H. Immerman, ed., *John Foster Dulles and the Diplomacy of the Cold War*. Princeton: Princeton University Press, 1990.

Hershberg, James G. "Who Murdered Marigold: New Evidence on the Mysterious Failure of Poland's Secret Initiative to Start U.S.-North Vietnamese Peace Talks, 1966," Cold War International History Project (CWIHP) Working Paper No. 27 (April 2000).

Hirsh, Michael. "Pushing Russia's Buttons," *Newsweek*, August 12, 2008. At http://www.newsweek.com/id/152087/page/1, accessed July 25, 2009.

Holdridge, John. *Crossing the Divide: An Insider's Account of the Normalization of U.S.-China Relations*. Lanham, Md.: Rowman & Littlefield, 1997.

Holsti, Ole P., Terrence Hopmann, and John D. Sullivan. *Unity and Disintegration in International Alliances: Comparative Studies*. New York: John Wiley and Sons, 1973.

Hundred Year Statistics of the Japanese Economy. Tokyo: Bank of Japan, Statistical Department, July 1966, pp. 292–93.

Immerman, Richard H. *John Foster Dulles: Piety, Pragmatism, and Power in U.S. Foreign Policy*. Wilmington, Del.: Scholarly Resources, 1999.

Iriye, Akira. "Chinese-Japanese Relations." In Christopher Howe, ed., *China and Japan: History, Trends, Prospects*. Oxford: Clarendon Press, 1996.

Jain, Rajendra Kumar. *China and Japan, 1949–76*. Atlantic Highlands, N.J.: Humanities Press, 1977.

James, Laura M. *Nasser at War: Arab Images of the Enemy*. New York: Palgrave MacMillan, 2006.

Jansen, Marius B. *Japan and China: From War to Peace, 1894–1972*. Chicago: Rand McNally College Publishers, 1975.

"Japan-U.S. Joint Declaration on Security, Alliance for the 21st Century." *"The World and Japan" Database Project*. Database of Japanese Politics and International Relations. Institute of Oriental Culture of the University of Tokyo, 1996. At http://www.ioc.u-tokyo.ac.jp/~worldjpn/documents/texts/docs/19960417.DIC.html..

Jervis, Robert. *Perception and Misperception in International Politics*. Princeton: Princeton University Press, 1967.

Jervis, Robert. *The Illogic of American Nuclear Strategy*. Ithaca: Cornell University Press, 1984.

———. *The Meaning of the Nuclear Revolution: Statecraft and the Prospect of Armageddon*. Ithaca: Cornell University Press, 1989.

Johnston, Alastair Iain. "Cultural Realism and Strategy in Maoist China," pp. 216–68 in Peter J. Katzenstein, ed., *The Culture of National Security*. New York: Columbia University Press, 1996.

Kastner, Scott. *Commerce in the Shadow of Conflict: Political Conflict and Economic Interdependence across the Taiwan Strait and Beyond*. Stanford: Stanford University Press, 2009.

Kaufman, Burton I. *The Korean War: Challenges in Crisis, Credibility and Command*. New York: Alfred A. Knopf, 1986.

Keefe, John. *Anatomy of the EP-3 Incident, April 2001*. Alexander, Va.: Center for Naval Analysis, 2001.

Kennedy, Andrew. "The Origins of Audacity: National Efficacy Beliefs, International Structures, and the Curious Rises of China and India." Paper presented at the annual meeting of the ISA's 49th Annual Convention, Bridging Multiple Divides, San Francisco, Calif..

Kerr, Malcolm. *The Arab Cold War, 1958–67: A Study of Ideology in Politics*. New York: Oxford, 1967.

Kirshner, Jonathan D. "A Rational Explanation for War?" *Security Studies* 10, no. 1 (autumn 2000): 143–50.

Kissinger, Henry. *White House Years*. Boston: Little, Brown, 1979.

Kydd, Andrew, and Barbra F. Walter. "Sabotaging the Peace: The Politics of Extremist Violence," *International Organization*, 56, no. 2 (spring 2002): 263–96.

LaFeber, Walter. *The Clash: U.S.-Japan Relations Throughout History*. New York: Norton, 1997.

Lam, Willy Wo-Lap. "Dual Edge to 'Liberation' Timetable." *South China Morning Post*, March 1, 2000.

Lampton, David M. *Same Bed, Different Dreams: Managing U.S.-China Relations, 1989–2000*. Berkeley: University of California Press, 2001.

Levesque, Jacques. *The USSR and the Cuban Revolution: Soviet Ideological and Strategical [sic] Perspectives*. Translated from the French by Deanna Drendel Lebeouf. New York: Praeger Publishers, 1978.

Lewis, John W., and Xue Litai. *China's Strategic Seapower: The Politics of Force Modernization in the Nuclear Age*. Stanford: Stanford University Press, 1994.

Lewis, John Wilson, and Xue Litai. *China Builds the Bomb*. Stanford: Stanford University Press, 1988.

Li Danhui. "Vietnam and Chinese Policy toward the United States." In William C. Kirby, Robert S. Ross, and Gong Li, eds. *Normalization of U.S.-China Relations: An International History*. Cambridge: Harvard University Asia Center, 2005.

Li Yaqiang, "What Is Japan Doing Southward?" *Beijing Jianchuan Zhishi* [Naval and Merchant Ships] no. 6 (June 6, 1997): 7–8, in Foreign Broadcast Information Service Daily Report China, September 4, 1997.

Lilley, James, and Jeffrey Lilley. *China Hands: Nine Decades of Adventure, Espionage, and Diplomacy in Asia*. New York: Public Affairs, 2004.

Lind, Jennifer. *Sorry States: Apologies in International Politics*. Ithaca: Cornell University Press, 2008.

Liska, George. *Nations in Alliance: The Limits of Interdependence*. Baltimore: Johns Hopkins University Press, 1968.

Liu, Jiangyong, "New Trends in Sino-U.S.-Japan Relations." *Contemporary International Relations* 8, no. 7 (July 1998): 1–13.

Logevall, Frederick. *Choosing War: The Lost Chance for Peace and the Escalation of War in Vietnam*. Berkeley: University of California Press, 1999.

Luthi, Lorenz. *The Sino-Soviet Split*. Princeton: Princeton University Press, 2008.

MacFarquhar, Roderick. *The Origins of the Cultural Revolution, Vol. 2: The Great Leap Forward 1958–60*. New York: Columbia University Press, 1983.

———. *The Origins of the Cultural Revolution, Volume 3: The Coming of the Cataclysm, 1961–66*. New York: Columbia University Press, 1997.

Mann, James. *About Face: A History of America's Curious Relationship with China, From Nixon To Clinton*. New York: Knopf, 1998.

Mansfield, Edward, Helen Milner, and Jon C. Pevehouse. "Vetoing Cooperation: The Impact of Veto Players on International Trade Agreements." A paper prepared for the 2005 International Studies Association Conference, Honolulu, Hawaii, January 26, 2005.

Mansfield, Edward, and Jack Snyder. "Democratization and the Danger of War." *International Security* 20, no. 1 (summer 1995): 5–38.

Mansourov, Alexandre. "Stalin, Mao, Kim, and China's Decision to Enter the Korean War, September 16-October 15, 1950." *Cold War International History Project Bulletin* (*CWIHPB*) nos. 6–7 (winter 1995–96).

May, Ernest. *"Lessons" of the Past: The Use and Misuse of History in American Foreign Policy*. New York: Oxford University Press, 1973.

Mayers, David. *George Kennan and the Dilemmas of U.S. Foreign Policy*. New York: Oxford University Press, 1988.

Mercer, Jonathan. *Reputation and International Politics*. Ithaca: Cornell University Press, 1996.

Mochizuki, Mike M. "American and Japanese Strategic Debates," pp. 43–82 in Mochizuki, ed., *Toward a True Alliance: Restructuring U.S.-Japan Security Relations*. Washington, D.C.: Brookings Institution Press, 1997.

———. "New Bargain for a New Alliance," pp. 5–40 in Mochizuki, ed., *Toward a True Alliance: Restructuring U.S.-Japan Security Relations*. Washington, D.C.: Brookings Institution Press, 1997.

Moller, Almut. "After Gaza: A New Approach to Hamas." *Austrian Institute for European and Security Policy FOKUS* . February 2009. At http://www.aicgs.org/documents/advisor /moeller0309.pdf.

Momoi, Makato. "Basic Trends in Japanese Security Policies." In Robert Scalapino, ed., *The Foreign Policy of Modern Japan*. Berkeley: University of California Press, 1977.

Morgenthau, Hans J. "Alliances in Theory and Practice." In Arnold Wolfers, ed., *Alliance Policy in the Cold War*. Baltimore: Johns Hopkins University Press, 1959.

Morrow, James D. "Arms versus Allies: Trade-offs in the Search for Security." *International Organization* 47, no. 2 (spring 1993): 207–33.

Naughton, Barry. "The Third Front: Defense Industrialization in China's Interior." *China Quarterly* no. 115 (September 1988): 351–86.

Nichol, Jim. "Georgia and NATO Enlargement: Issues and Implications." *Congressional Research Service Reports for Congress*, March 2008. At http://fpc.state.gov/documents/organization/103692.pdf .

Nitze, Paul. "Coalition Policy and the Concept of World Order." In Arnold Wolfers, ed., *Alliance Policy and the Cold War*. Baltimore: Johns Hopkins University Press, 1969.

Nitze, Paul. *From Hiroshima to Glasnost: At the Center of Decision*. New York: Grove Weidenfeld, 1989.

Nixon, Richard M. *RN: The Memoirs of Richard Nixon*. New York: Grosset & Dunlap, 1978.

Ogata, Sadako. "The Business Community and Japanese Foreign Policy: Normalization of Relations with the People's Republic of China." In Robert Scalapino, ed., *The Foreign Policy of Modern Japan*. Berkeley: University of California Press, 1977.

O'Hanlon, Michael. "Restructuring U.S. Forces and Bases in Japan," pp. 149–78 in Mike M. Mochizuki, ed., *Toward a True Alliance: Restructuring U.S.-Japan Security Relations*. Washington, D.C.: Brookings Institution, 1997.

Oren, Michael B. *Six Days of War: June 1967 and the Making of the Modern Middle East*. New York: Oxford University Press, 2002.

Paige, Glenn. *The Korea Decision June 24–30, 1950*. New York: Free Press, 1968.

Peng Dehuai. *Memoirs of a Chinese Marshal: The Autobiographical Notes of Peng Dehuai (1898–1974)*. Beijing: Foreign Languages Press, 1984.

Person, James F. "We Need Help From the Outside: The North Korean Opposition Movement of 1956." Cold War International History Project (CWIHP) Working Paper No. 52 (August 2006).

Pipes, Daniel. Foreword to Jonathan Schanzer, *Fatah vs. Hamas: The Struggle for Palestine*. New York: Palgrave MacMillan, 2008.

Pleshakov, Constantine, and Vladislav Zubok. *Inside the Kremlin's Cold War: From Stalin to Khrushchev*. Cambridge: Harvard University Press, 1996.

Pollard, Robert A. *Economic Security and the Origins of the Cold War*. New York: Columbia University Press, 1985.

Porter, Bruce D. *The USSR in Third World Conflicts: Soviet Arms and Diplomacy in Local Wars, 1945–1980*. Cambridge: Cambridge University Press, 1984.

Posen, Barry. *The Sources of Military Doctrine: France, Britain, and Germany between the Wars*. Ithaca: Cornell University Press, 1984.

Powell, Robert. "Bargaining Theory and International Conflict." *Annual Review of Political Science* 5 (June 2002): 1–30.

Pressman, Jeremy *Warring Friends: Alliance Restraint in International Politics*. Ithaca: Cornell University Press, 2008.

Radchenko, Sergey, and David Wolff. "To the Summit by Proxy Summits: New Evidence from Soviet and Chinese Archives on Mao's Long March to Moscow, 1949." *Cold War International History Project Bulletin* (*CWIHPB*) no. 16 (fall 2007/winter 2008): 105–12.

Radtke, Kurt Werner. *China's Relations with Japan, 1945–83: the Role of Liao Chengzhi*. Manchester, UK: Manchester University Press, 1990.

Restall, Hugo. "'Opposing the Sun': Japan Alienates Asia." *Far Eastern Economic Review* (April 2005).

Richter, James G. *Khrushchev's Double Bind: International Pressures and Domestic Coalitions.* Baltimore: Johns Hopkins University Press, 1994.

Romberg, Alan. *Rein In at the Brink of the Precipice.* Washington, D.C.: Stimson Center, 2003.

———. "Cross-Strait Relations: Ascend the Heights and Take a Long-term Perspective." *China Leadership Monitor* no. 27 (winter 2008), at http://media.hoover.org/documents /CLM27AR.pdf.

———. "Cross-Strait Relations: In Search of Peace." *China Leadership Monitor* no. 23 (winter 2008), at http://media.hoover.org/documents/CLM23AR.pdf.

———. "Applying to the U.N. in the Name of Taiwan." *China Leadership Monitor* no. 22 (fall 2008), at http://media.hoover.org/documents/CLM22AR.pdf.

———. "Cross-Strait Relations: First the Easy, Now the Hard." *China Leadership Monitor* no. 28 (spring 2009), at http://media.hoover.org/documents/CLM28AR.pdf.

Ross, Dennis. *The Missing Peace: The Inside Story of the Fight for Middle East Peace.* New York: Farrar, Straus, and Giroux, 2004.

Ross, Robert S. *Negotiating Cooperation: The United States and China, 1969–1989.* Stanford, CA: Stanford University Press, 1995.

Ross, Robert S., and Jiang Changbin. *Re-examining the Cold War: U.S.-China Diplomacy, 1954–73.* Cambridge: Harvard University Asia Center, 2001.

Rotter, Andrew. *The Path to Vietnam: Origins of the American Commitment to Southeast Asia,* Ithaca: Cornell University Press, 1987.

Samuels, Richard J. *Rich Nation, Strong Army: National Security and the Technological Transformation of Japan.* Ithaca: Cornell University Press, 1994.

———. *Securing Japan: Tokyo's Grand Strategy and the Future of East Asia.* Ithaca: Cornell University Press, 2007.

Sanger, David E. "New Missile Defense in Japan under Discussion with U.S." *New York Times,* September 18, 1993.

Saunders, Phillip C., and Erica Strecker Downs. "Legitimacy and the Limits of Nationalism: China and the Diaoyu Islands." *International Security* 23, no. 3 (winter 1998/1999): 114–46.

Schaller, Michael. *Altered States: The United States and Japan since the Occupation.* New York: Oxford University Press, 1997.

Schanzer, Jonathan. *Fatah vs. Hamas: The Struggle for Palestine.* New York: Palgrave MacMillan, 2008.

Schelling, Thomas C. *Arms and Influence.* New Haven: Yale University Press, 1966.

Schnabel, James, and Robert J. Watson. *History of the Joint Chiefs of Staff. Vol. 3: The Korean War,* Part 1. Wilmington, Del.: Glazier, 1979.

Schroeder, Paul W. "Alliances, 1814–1945: Weapons of Power and Tools of Management," pp. 227–62 in Klaus Knorr, ed., *Historical Dimensions of National Security Problems.* Lawrence: University of Kansas Press, 1976.

Schweller, Randall L. "Bandwagoning for Profit: Bringing the Revisionist State Back In." *International Security* 19, no. 1 (1994): 72–107.

Segal, Gerald. *Defending China.* Oxford: Oxford University Press, 1984.

Selverstone, Marc. J. *Constructing the Monolith: The United States, Great Britain, and International Communism, 1945–50.* Cambridge: Harvard University Press, 2009.

Shapiro, Jacob N. "The Terrorist's Challenge: Security, Efficiency, Control." Ph.D. diss., Stanford University, 2007.

Shen Zhihua. "Khrushchev, Mao, and the Unrealized Sino-Soviet Military Cooperation." Unpublished manuscript written for the Parallel History Project on Nation and the Warsaw Pact, China and the Warsaw Pact, October 2002. At http://www.isn.ethz.ch/php/documents/collection_11/texts/Zhiua_engl.pdf.

———. "Sino-North Korean Conflict and Its Resolution During the Korean War." *Cold War International History Project Bulletin (CWIHPB)* nos. 14/15 (fall 2003/winter 2004): 9–24.

———. "China and the Soviet Union Dispatch Troops to Aid Korea: The Establishment of the Chinese-Soviet-Korean Alliance in the Early Stages of the Korean War." Translated by Yang Jingxia and Douglas A. Stiffler. Unpublished working paper, 2008.

Shen Zhihua, ed. *The Korean War: Declassified Documents from the Russian Archives,* Vols. 1–3. Taipei: Institute of Modern History, Academia Sinica

Sheng, Michael. *Battling Western Imperialism: Mao, Stalin, and the United States.* Princeton: Princeton University Press, 1997.

Shimizu Sayuri. "Perennial Anxiety: Japan-US Controversy over Recognition of the PRC, 1952–1958." *Journal of American-East Asian Relations* 4, no. 3 (fall 1995): 223–48.

Snyder, Glenn H. "The Balance of Power and the Balance of Terror," pp. 184–201 in Paul Seabury, ed., *The Balance of Power.* San Francisco: Chandler, 1965.

———. "The Security Dilemma in Alliance Politics." *World Politics* 36 (July 1984): 461–95.

———. *Alliance Politics.* Ithaca: Cornell University Press, 1997.

Snyder, Jack. *From Voting to Violence: Democratization and Nationalist Conflict.* New York: Norton, 2000.

Socor, Vladimir. "Ukrainian and Georgian ANPs Are Also Testing NATO." *Eurasia Daily Monitor* 5, no. 234, December 9, 2008, at http://www.jamestown.org/single/?no_cache=1&tx_ttnews%5Btt_news%5D=34244.

Soeya, Yoshihide. "U.S.-Japan-China Relations and the Opening to China: The 1970s," Working Paper No. 5, U.S.-Japan Project, National Security Archive.

Stanley, Elizabeth A. *Paths to Peace: Domestic Coalition Shifts, War Termination and the Korean War.* Stanford: Stanford University Press, 2009.

Stokes, Bruce, and James Shinn. *The Tests of War and the Strains of Peace: The U.S.-Japan Security Relationship.* New York: Council on Foreign Relations Study Group Report, January 1998.

Stokes, William N. "War in Korea." In Marshall Green, John H. Holdridge, and William N. Stokes, eds., *War and Peace in China.* Bethesda, Md.: DACOR Press, 1994.

Stolper, Thomas. *China, Taiwan, and the Offshore Islands.* Armonk, N.Y.: ME Sharpe, 1985.

Stueck, William Whitney. *The Road to Confrontation: American Policy toward Korea and China, 1947–1950.* Chapel Hill: University of North Carolina Press, 1981.

———. *The Korean War: An International History.* Princeton: Princeton University Press, 1997.

Suettinger, Robert L. *Beyond Tiananmen.* Washington, D.C.: Brookings Institution Press, 2003.

Tatsumi, Yuki. "U.S.-Japan Security Consultative Committee: An Assessment." *Pacific Forum PacNet Newsletter* #10, Center for Strategic and International Studies, March 2005.

Telhami, Shibley, and Steven A. Cook. "Addressing the Arab-Israeli Conflict." Ch. 5 in *Restoring the Balance: A Middle East Strategy for the Next President.* CFR-Saban Center at Brookings Institution. December 2008. At http://www.cfr.org/content/publications/attachments/CFR-Saban_Executive_Summaries.pdf.

Tian, John Qunjian. *Government, Business, and the Politics of Interdependence and Conflict across the Taiwan Strait.* New York: Palgrave, 2006.

Truman, Harry S. *Memoirs. Vol. 2, Years of Trial and Hope.* Garden City, N.Y.: Doubleday, 1956.

Tsebelis, George. *Veto Players: How Political Institutions Work.* Princeton: Princeton University Press, 2002.

Tucker, Nancy Bernkopf. *Patterns in the Dust: Chinese-American Relations and the Recognition Controversy, 1949–50.* New York: Columbia University Press, 1983.

———. "John Foster Dulles and the Taiwan Roots of the 'Two Chinas' Policy.'" In Immerman, ed., *John Foster Dulles and the Diplomacy of the Cold War.* Princeton: Princeton University Press, 1990.

———. *Taiwan, Hong Kong, and the United States, 1945–1992.* New York: Twayne Publishers, 1994.

———. *Strait Talk: United States-Taiwan Relations and the Crisis with China.* Cambridge: Harvard University Press, 2009.

Twomey, Christopher P. "The Military Lens." Ph.D. diss. Department of Political Science, Massachusetts Institute of Technology, October 2004.

Tyler, Patrick. *A Great Wall: Six Presidents and China, An Investigative History.* New York: Century Foundation Books, 1999.

Valentino, Benjamin. "Small Nuclear Powers and Opponents of Ballistic Missile Defenses in the Post–Cold War Era." *Security Studies* 7, no. 2 (winter 1997/98): 229–32.

Van Evera, Stephen. "Primed for Peace: Europe after the Cold War." *International Security* 15, no. 3 (winter 1990/91): 7–57.

———. *Causes of War: Power and the Roots of Conflict.* Ithaca: Cornell University Press, 1999.

Vanderpool, Guy R. "COMINT and the PRC Intervention in the Korean War." DOCID: 340650, in National Security Archive, George Washington University. At http://www.gwu.edu/~nsarchiv/NSAEBB/NSAEBB278/index.htm, document no. 21, p. 15.

Vasquez, John. *The War Puzzle.* Cambridge: Cambridge University Press, 1993.

Walt, Stephen M. *The Origins of Alliances.* Ithaca: Cornell University Press, 1987.

Waltz, Kenneth N. *Theory of International Politics.* Reading, Mass.: Addison-Wesley, 1979.

Wang Zhongchun, "The Soviet Factor in Sino-American Normalization," in William C. Kirby, Robert S. Ross, and Gong Li, eds., *Normalization of U.S.-China Relations: An International History.* Cambridge: Harvard University Asia Center, 2005.

Weathersby, Kathryn. "New Findings on the Korean War." *Cold War International History Project Bulletin (CWIHPB)* 3 (fall 1993).

———. "Soviet Aims in Korea and the Origin of the Korean War, 1945–50: New Evidence From Russian Archives." Cold War International History Project (CWIHP) Working Paper No. 8 (1993).

Weathersby, Kathryn. "New Russian Documents on the Korean War: Introduction and Translations." *Cold War International History Project Bulletin (CWIHPB)* nos. 6/7 (1996).

———. "Stalin, Mao and the End of the Korean War." In Odd Arne Westad, ed., *Brothers in Arms; The Rise and Fall of the Sino-Soviet Alliance*. Washington, D.C.: Woodrow Wilson Center Press, 1998.

———. "Should We Fear This? Stalin and the Danger of War with America," Cold War International History Project (CWIHP) Working Paper No. 39 (July 2002).

Westad, Odd Arne, ed. *Brothers in Arms: The Rise and Fall of the Sino-Soviet Alliance*. Washington, D.C.: Woodrow Wilson Center Press, 1998.

———. *Decisive Encounters: The Chinese Civil War: 1946-1950*. Stanford: Stanford University Press, 2003.

———. "The Sino-Soviet Alliance and the United States." In Odd Arne Westad, ed., *Brothers in Arms: The Rise and Fall of the Sino-Soviet Alliance*. Washington, D.C.: Woodrow Wilson Center, 1998.

Weiss, Jessica Chen. "Powerful Patriots: Nationalism, Diplomacy, and the Strategic Logic of Anti-Foreign Protest." Ph.D. diss., University of California, San Diego, 2008.

Weitsman, Patrica A. *Dangerous Alliances: Proponents of Peace, Weapons of War*. Stanford: Stanford University Press, 2004.

Whiting, Allen S. *China Crosses the Yalu: The Decision to Enter the Korean War*. Stanford: Stanford University Press, 1960.

———. *The Chinese Calculus of Deterrence*. Ann Arbor: University of Michigan, 1975.

———. *China Eyes Japan*. Berkeley: University of California Press, 1989.

Wu Chunsi. "Tactical Missile Defense, Sino-U.S.-Japanese Relationship, and East Asian Security." *Inesap Information Bulletin* no. 16 (November 1998): 20–23.

Wu Xinbo. "The End of the Silver Lining: A Chinese View of the U.S.-Japan Alliance." *Washington Quarterly* 29, no. 1 (September 2005): 119–30.

Xia, Yafeng. *Negotiating with the Enemy: U.S.-China Talks during the Cold War, 1949–1972*. Bloomington: Indiana University Press, 2006.

Yang Bojiang, "Why (a) U.S.-Japan Joint Declaration on (the) Security Alliance?" *Contemporary International Relations* 6, no. 5 (May 1996): 1–12.

Yang Kuisong. "Changes in Mao Zedong's Attitude toward the Indochina War, 1949–1973." Cold War International History Project (CWIHP) Working Paper No 34 (February 2002).

Yoshitsu, Michael M. *Japan and the San Francisco Peace Settlement*. New York: Columbia University Press, 1983.

Young, Kenneth T. *Negotiating with the Chinese Communists: The United States Experience, 1953-1967*. New York: McGraw Hill, 1968.

Zagoria, Donald. *Vietnam Triangle: Moscow, Peking, Hanoi*. New York: Pegasus 1967.

Zhai Qiang. "China and the Geneva Conference of 1954." *China Quarterly* no. 129 (March 1992): 103–22.

———. "China and Johnson's Escalation of the Vietnam War, 1964–65." Paper presented at the conference "New Evidence on the Cold War in Asia," Hong Kong, January 1996.

————. "Beijing and the Vietnam Peace Talks, 1965–68: New Evidence From Chinese Sources." Cold War International History Project (CWIHP) Working Paper No. 18 (June 1997).

————. *China and the Vietnam Wars, 1950–1975*. Chapel Hill: University of North Carolina Press, 2000.

Zhang Baijia. "The Changing International Scene and Chinese Policy Toward the United States," in Robert S. Ross and Jiang Changbin, eds. *Re-examining the Cold War: U.S.-China Diplomacy, 1954–73*. Cambridge: Harvard University Asia Center, 2001.

Zhang Shu Guang. *Deterrence and Strategic Culture*: *Chinese-American Confrontations 1949–58*. Ithaca: Cornell University Press, 1992.

————. *Mao's Military Romanticism : China and the Korean War, 1950–1953*. Lawrence: University Press of Kansas, 1995.

Index